371
Topics in Current Chemistry

Aims and Scope

The series Topics in Current Chemistry presents critical reviews of the present and future trends in modern chemical research. The scope of coverage includes all areas of chemical science including the interfaces with related disciplines such as biology, medicine and materials science.

The goal of each thematic volume is to give the non-specialist reader, whether at the university or in industry, a comprehensive overview of an area where new insights are emerging that are of interest to larger scientific audience.

Thus each review within the volume critically surveys one aspect of that topic and places it within the context of the volume as a whole. The most significant developments of the last 5 to 10 years should be presented. A description of the laboratory procedures involved is often useful to the reader. The coverage should not be exhaustive in data, but should rather be conceptual, concentrating on the methodological thinking that will allow the non-specialist reader to understand the information presented.

Discussion of possible future research directions in the area is welcome.

Review articles for the individual volumes are invited by the volume editors.

Readership: research chemists at universities or in industry, graduate students.

More information about this series at http://www.springer.com/series/128

Harun Tüysüz • Candace K. Chan
Editors

Solar Energy for Fuels

With contributions by

A.S. Batchellor · D.K. Bediako · S.W. Boettcher · A. Braun · C.K. Chan · P.A. Crozier · G. Dodekatos · F. Gans · J.M. Gregoire · J.A. Haber · A. Kleidon · V. Krewald · P. Kurz · F. La Mantia · R. Marschall · B.K. Miller · L. Miller · D.G. Nocera · F.E. Osterloh · D.A. Pantazis · H.S. Park · C. Ranjan · M. Retegan · S. Schünemann · K. Takanabe · L. Trotochaud · H. Tüysüz · A.M. Ullman · L. Zhang

Springer

Editors
Harun Tüysüz
Max-Planck-Institut für
Kohlenforschung
Mülheim an der Ruhr, Germany

Candace K. Chan
Materials Science and Engineering
School for Engineering of Matter,
Transport and Energy
Arizona State University
Tempe, Arizona
USA

ISSN 0340-1022 ISSN 1436-5049 (electronic)
Topics in Current Chemistry
ISBN 978-3-319-23098-6 ISBN 978-3-319-23099-3 (eBook)
DOI 10.1007/978-3-319-23099-3

Library of Congress Control Number: 2015954168

Springer Cham Heidelberg New York Dordrecht London

Printed on acid-free paper

Springer International Publishing AG Switzerland is part of Springer Science+Business Media (www.springer.com)

Preface: Solar Energy for Fuels

Few scientific researchers would deny that, as a result of various factors such as funding cycles, overall political climate, and stagnation in progress, the intensity of activity in any particular research field is cyclical. This is particularly true for research areas related to renewable energy conversion and storage, with the return of interest in solar energy related fields in the last decade corresponding with increasing concerns about climate change and depleting fossil fuel energy sources.

Solar energy can be used to create an extremely energy-dense fuel, namely hydrogen gas, from water. The hydrogen can be used in hydrogen fuel cell vehicles or to create syngas with carbon monoxide to make synthetic liquid fuels in the Fischer–Tropsch process. Additionally, solar energy can be used to photoelectrochemically or electrolytically drive the reduction of carbon dioxide to carbon-based fuels such as methane, carbon monoxide, or alcohols.

The use of solar energy to create fuels is attractive, as it makes use of an abundant resource. In Chap. 1 by Kleidon et al., a detailed analysis regarding the potential of solar energy conversion using principles of thermodynamics and conversion limits is presented. The conversion of solar energy to make fuels occurs in nature via the process of photosynthesis. Substantial efforts have been made to better understand the natural photosynthetic mechanisms in order to discover how artificial photosynthetic systems can be engineered, especially for overcoming the kinetic bottlenecks in the water oxidation reaction. In Chap. 2, Pantazis et al. provide an overview of the key features of Photosystem II, the natural enzyme that catalyzes water oxidation. The adoption of these concepts to form biomimetic water oxidation catalysts is described in Chap. 3 by Kurz.

Several different approaches and device configurations are possible for realizing solar fuel production. Slurry photocatalysts are a low-cost extension of photoelectrochemical cells, whereby a semiconductor particle decorated with co-catalysts for the hydrogen and oxygen evolution reactions can be used to split water and co-generate the products simultaneously. Takanabe describes the fundamental processes involved during overall water splitting on slurry photocatalysts in Chap. 4.

Since the first demonstrations of solar fuel production in the 1970s, the global scientific community has become equipped with several advantages to better tackle the energy problem. For instance, the development of nanoscience as a field and its accompanying synthetic and characterization tools have enabled a fresh look at old materials, as well as completely novel approaches. Nanostructured materials have unique and different physical and chemical properties in comparison to their bulk counterparts. In Chap. 5, Osterloh describes the key parameters in nanomaterials that can be exploited to improve photocatalytic performance.

Aside from using nanostructuring strategies, researchers have also developed new ways to boost the overall performance of the photocatalysts. In Chap. 6, Marschall describes how advanced heterojunctions can be engineered in powdered materials to produce composite photocatalysts with better efficiency. Another strategy is to load the surface of a light-harvesting semiconductor with a co-catalyst that provides catalytic sites for the oxidation or reduction reactions and promotes separation of the photogenerated charges. In Chap. 7, Nocera et al. gives a detailed overview of the properties of water oxidation co-catalysts based on cobalt oxido thin films. Then, in Chap. 8, Tüysüz et al. describes how surface plasmons, an optical property found in nanostructured metals, can be used to improve the visible light harvesting efficiency and enhance the photocatalytic activity of semiconducting particles.

Finally, in operando and in situ techniques have also been applied to photocatalysis and photoelectrochemistry to gain a better understanding of catalytic processes and monitor materials under reaction conditions. Increasingly advanced setups have allowed for unprecedented experiments involving light excitation, electrical biasing, and introduction of water for conducting microscopy and spectroscopy on electrocatalyst and photocatalyst materials during solar-to-fuel reactions. An overview of several in situ characterization tools that have recently been developed is given in Chap. 9, as well as different strategies for implementing high throughput screening of solar fuel materials properties.

Harun Tüysüz
Candace K. Chan

Contents

Top Curr Chem (2016) 371: 1–22
DOI: 10.1007/128_2015_637

Published online: 20 March 2015

Physical Limits of Solar Energy Conversion in the Earth System

Axel Kleidon, Lee Miller, and Fabian Gans

Abstract Solar energy provides by far the greatest potential for energy generation among all forms of renewable energy. Yet, just as for any form of energy conversion, it is subject to physical limits. Here we review the physical limits that determine how much energy can potentially be generated out of sunlight using a combination of thermodynamics and observed climatic variables. We first explain how the first and second law of thermodynamics constrain energy conversions and thereby the generation of renewable energy, and how this applies to the conversions of solar radiation within the Earth system. These limits are applied to the conversion of direct and diffuse solar radiation – which relates to concentrated solar power (CSP) and photovoltaic (PV) technologies as well as biomass production or any other photochemical conversion – as well as solar radiative heating, which generates atmospheric motion and thus relates to wind power technologies. When these conversion limits are applied to observed data sets of solar radiation at the land surface, it is estimated that direct concentrated solar power has a potential on land of up to 11.6 PW (1 PW $= 10^{15}$ W), whereas photovoltaic power has a potential of up to 16.3 PW. Both biomass and wind power operate at much lower efficiencies, so their potentials of about 0.3 and 0.1 PW are much lower. These estimates are considerably lower than the incoming flux of solar radiation of 175 PW. When compared to a 2012 primary energy demand of 17 TW, the most direct uses of solar radiation, e.g., by CSP or PV, have thus by far the greatest potential to yield renewable energy requiring the least space to satisfy the human energy demand. Further conversions into solar-based fuels would be reduced by further losses which would lower these potentials. The substantially greater potential of solar-based renewable energy compared to other forms of renewable energy simply reflects much fewer and lower unavoidable conversion losses when solar radiation is directly converted into renewable energy.

A. Kleidon (✉), L. Miller, and F. Gans
Max-Planck-Institute for Biogeochemistry, Hans-Knoell-Str. 10, 07745 Jena, Germany
e-mail: akleidon@bgc-jena.mpg.de

Keywords Carnot limit · Global estimates · Photosynthesis · Solar energy · Theoretical potentials · Thermodynamic limits · Thermodynamics

Contents

1 Introduction

Renewable energy provides a sustainable form of energy generation which is not associated with the emission of greenhouse gases. It thereby forms a critical component to avoid greenhouse-induced global climate change and to avoid the consequences of depleting fossil fuel resources. Among the different forms of renewable energy, solar energy is seen as the most abundant source of renewable energy [1]. Its potential is often described as being so huge [2] that it is not a limiting factor in meeting human energy demands within this century. The deployment of solar energy technology, mostly in form of photovoltaic electricity generation, is rapidly growing, so that solar energy already contributes a sizable percentage to electricity generation in some industrialized countries.

Yet, as with any other resource on the planet, the generation of renewable energy by the Earth system is limited, and so is the potential of using solar radiation as an energy resource. Quantitative estimates of the theoretical potential of solar energy are based on physical limitations rather than what is currently technologically possible. These estimates play an important role in allowing us to assess which form of renewable energy has the greatest potential to meet human energy demands. The difference between the theoretical potential and what is currently technologically possible also informs us about the extent to which technology can be developed further. Hence, theoretical potentials provide an important foundation for establishing the options for future sustainable energy developments.

Theoretical potentials of renewable energy are derived from the combination of thermodynamics, which provides a general, physical formulation of the limits associated with energy conversion, and Earth system science, which provides the context in which energy conversion takes place. Thermodynamics describes general rules for energy conversions, and yields well-established, fundamental conversion limits, such as the Carnot limit of a heat engine. This limit is set by the heat flux through the engine, but also by the entropy exchange, which limits how much of the heat can be converted into physical work (which is also referred to as exergy). Analogous to the heat engine example, it is possible to calculate a thermodynamic conversion limit for a radiation energy converter [3–5] which can be used to quantify how much of the solar radiative flux can be converted into renewable energy. In addition to the magnitude of the solar radiative flux at the surface, a critical component that is needed is the entropy associated with the radiative flux. This property of radiation, however, is not commonly available and needs to be estimated from available data. Nevertheless, the combination of this theoretical conversion limit with the conditions at the Earth's surface allows us to quantify a theoretical potential for solar renewable energy which is solely based on the limits of physical conversion processes within the Earth's environment and which is independent of the available technology. This conversion limit is not constrained solely to physical conversions either, so that it also applies to any form of photochemical conversion, including photosynthesis. It thus sets an upper limit to the potential by which solar radiation can supply renewable energy for human energy use.

What has just been described sets the limit for the conversion of solar radiation into a general, non-radiative form of energy. In the case of photovoltaics, for instance, the resulting form of energy is in electric form. The further conversion of this energy into some chemical form to yield solar-based fuels would likely be associated with further conversion losses, thus lowering the potential to convert solar radiation into chemical fuels. Here, however, we focus on the limits imposed on the first step from solar radiation to energy in non-radiative form, and discuss the implications for conversions to solar-based fuels at the end of this chapter.

In this chapter, we describe the theoretical background of these conversion limits for solar radiation and combine these with climate data sets based on observed solar radiation fluxes at the Earth's surface to obtain estimates of solar renewable energy potential. We also describe the closely related thermodynamic conversion limit which applies to the radiative heating after solar radiation is being absorbed by the Earth's surface, which drives atmospheric convection. The subtle difference of the latter limit is that it does not convert solar radiation, but rather solar radiative heating, so that this thermodynamic limit represents the well-established Carnot limit of a heat engine. This limit constrains the generation of atmospheric motion and thus provides an estimate for the theoretical potential of wind energy. We also briefly describe the theoretical and observed limits associated with photosynthesis, which also act to convert solar radiation into another energy form which can be used as a renewable form of energy in the form of biofuels. We then estimate and compare these solar-based renewable energy potentials using observed solar radiative fluxes at the surface. This comparison shows that the direct use of solar radiation represents by far the most efficient way to generate renewable energy as it involves the fewest conversions of solar radiation to renewable energy. We close with a brief summary and conclusion, where we also outline the possible application of such thermodynamic considerations to further conversions into solar-based fuels.

2 Thermodynamic Background

Thermodynamics is a physical theory which describes rules for energy conversions at a very fundamental level. At the center are the first and second laws of thermodynamics. The first law essentially states the conservation of energy during the conversion of one form of energy into another. The second law requires that the entropy during an energy conversion process does not decrease. Although entropy seems to be an abstract thermodynamic property, it can loosely be seen as a measure for the dispersal of the energy at the microscopic scale of atoms and molecules [6]. The more a given amount of energy is dispersed within a system, the higher its entropy is going to be. This requirement stated by the second law is so profound in physics that it has been labeled the arrow of time [7].

When the first and second laws are combined, they provide a constraint on the magnitude of energy conversion. This is schematically illustrated in Fig. 1, in which a fraction of an incoming flux of energy, J_{in}, is converted into another form of energy at a rate G which represents some form of free energy which can be used to perform work. This conversion rate, G, represents the power of the engine, although we use G here as it also refers to the generation rate of free energy. For instance, a classical heat engine converts a fraction of the influx of heat (J_{in}) into mechanical work (G). For solar energy conversion, the influx is represented by the flux of solar radiation (J_{in}), and the conversion rate G is associated with the rate by which, for instance, electric energy is generated.

The first law of thermodynamics then states that energy is conserved during the conversion process. In a steady state of the conversion process in which the energy content inside the converter is unchanged, this translates into the requirement that

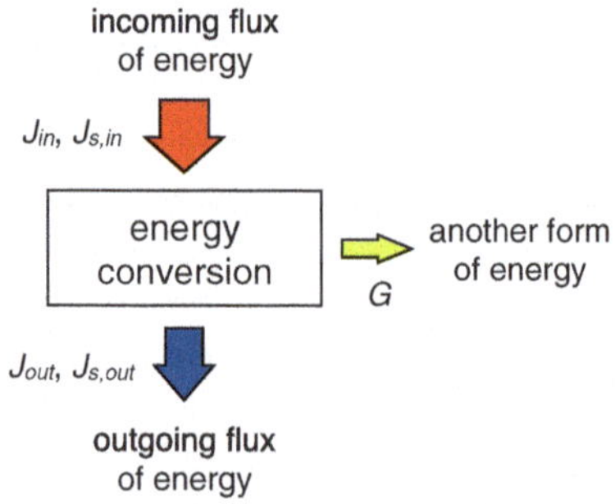

Fig. 1 Illustration of a generic energy converter which generates power at a rate G. The magnitude of energy conversion is constrained by the combination of the two laws of thermodynamics: the first law states energy conservation associated with the energy fluxes J_{in}, J_{out}, and G (i.e., $J_{in} = J_{out} + G$), and the second law requires that the entropy export, $J_{s,out}$, is greater or equal to the entropy import, $J_{s,in}$ (i.e., $J_{s,out} \geq J_{s,in}$). The ideal case of $J_{s,out} = J_{s,in}$ sets the upper limit on the conversion rate

the incoming flux of energy, J_{in}, is balanced by the generation rate and a waste flux of energy, J_{out} which leaves the converter:

$$0 = J_{in} - J_{out} - G. \tag{1}$$

The second law of thermodynamics adds a further constraint on the conversion process. It requires that the entropy does not decrease during the conversion process. In the steady-state setting considered here, the second law imposes a constraint on the entropy balance of the conversion process. This entropy balance consists of the import of entropy, $J_{s,in}$, associated with the incoming flux, J_{in}, with the entropy export, $J_{s,out}$, associated with the outgoing flux, J_{out}, and, potentially, of the entropy production, σ, within the system. We thus obtain for the entropy balance in steady state

$$0 = J_{s,in} - J_{s,out} + \sigma. \tag{2}$$

In the context of this entropy balance, the second law requires that $\sigma \geq 0$. In the ideal case, where $\sigma = 0$, there are no inefficiencies during the conversion process which would constitute irreversible losses, such as frictional or diffusive losses. This condition then sets the upper limit on an energy conversion process.

To illustrate how the combination of the two laws yields a conversion limit, we use the common example of a heat engine. For heat fluxes, the entropy fluxes are simply given by $J_{s,in} = J_{in}/T_{in}$ and $J_{s,out} = J_{out}/T_{out}$, where T_{in} and T_{out} are the temperatures at which the heat fluxes J_{in} and J_{out} are added or removed from the conversion process. When the expressions for the entropy fluxes are used in the entropy balance in the ideal case of $\sigma = 0$, this then yields the condition

$$\frac{J_{in}}{T_{in}} - \frac{J_{out}}{T_{out}} = 0. \tag{3}$$

When we now use the first law (1) to replace J_{out} in (3) and solve it for G, we obtain the upper limit of G that is permitted by the second law:

$$G = J_{in} \times \frac{T_{in} - T_{out}}{T_{in}} = J_{in} \times \eta_{Carnot}, \tag{4}$$

which is the well-known Carnot limit of a heat engine. In this equation, the expression $\eta_{Carnot} = (T_{in} - T_{out})/T_{in}$ is the so-called Carnot efficiency and informs us about how much of the heat flux J_{in} can be converted into free energy.

The conversion limit expressed by (4) only used the first and second law in its derivation for a generic energy conversion process in steady state, without any details on how this conversion is being performed. It thus represents a fundamental limit on energy conversion which is set by the laws of thermodynamics. However, the expression given by (4) is only applicable to the conversion of heat into work, because it specified the entropy fluxes as being thermal entropy fluxes. When thermodynamics is applied to derive limits on the conversion process of solar radiation, we need to use expressions of radiative entropy fluxes, which are somewhat different than for heat.

3 Limits to Converting Solar Radiation

To obtain theoretical limits of the conversion of solar radiation, we need expressions of the entropy associated with radiative fluxes. These expressions go back to the original work by Max Planck [8], who introduced statistical mechanics to the description of radiation to derive radiation laws. Without going into the details of this formulation, one can show that the entropy of a radiative flux is of similar form to a heat flux, so that it is proportional to the magnitude of the radiative flux, and inversely related to the temperature at which the radiation was emitted. Because solar radiation is emitted at a much higher temperature of about $T_{sun} = 5{,}760$ K compared to when radiation is re-emitted by the Earth system of about $T_r = 255$ K, solar radiation has a much lower entropy than the terrestrial radiation emitted by Earth. This low entropy of solar radiation is reflected in the much shorter wavelength of solar radiation around 550 nm in the visible range, whereas the emitted radiation from Earth is centered around 11 μm in the infrared range. When one uses the particle view of radiation, then a given flux of solar radiation represents a flux of fewer and more energetic photons, whereas terrestrial radiation represents a flux of many more, but less energetic photons. This difference in entropy in the radiative exchange between Earth and space ultimately drives the dynamics of the Earth's climate system [9, 10]. However, in contrast to the entropy associated with heat fluxes, there is an additional contribution by radiation pressure to the entropy flux, and the spatial concentration over the solid angle also plays a role. This latter aspect in particular plays an important role in deriving theoretical limits, as it yields different limits for the potential of photovoltaic and direct concentrated solar power. For details on the formulation of radiative entropy within the Earth system, see, e.g., the recent review by Wu and Liu [11]. The following derivations of the thermodynamic limit of solar energy conversion follow the works by Petela [3], Press [4], and Landsberg and Tonge [5].

The incoming solar radiative flux at the Earth's surface can be expressed as

$$R_{s,in} = B\varepsilon\sigma_b T_{sun}^4, \tag{5}$$

where B is a geometric factor, ε is a so-called dilution factor, $\sigma_b = 5.67 \times 10^{-8}$ W m^{-2} K^{-4} is the Stefan–Boltzmann constant, and $T_{sun} = 5{,}760$ K is the emission temperature of the Sun. The two factors B and ε describe the concentration of solar radiation over a small region of the sky. At the top of the atmosphere, where solar radiation has not been altered since it was emitted, the dilution factor is $\varepsilon = 1$, that is, it is nearly undiluted blackbody radiation. The geometric factor is $B = \Omega_{sun}/\pi$, where $\Omega_{sun} = 6.8 \times 10^{-5}$ sr (steradian, the SI unit for solid angles) is the solid angle

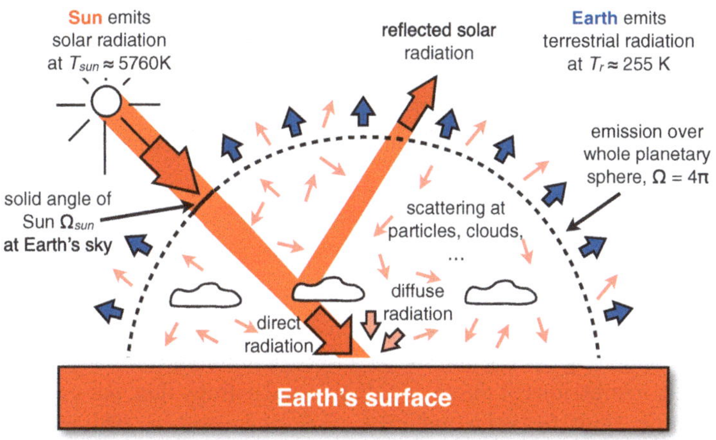

Fig. 2 Illustration of the solar radiative flux at the Earth's surface and the properties that affect its magnitude and entropy. The solar radiative flux at the surface consists of the direct component (*red arrow*) which is non-scattered solar radiation which still maintains its narrow solid angle, $\Omega_{\text{sun}} \approx 6.8 \times 10^{-5}$ sr, and a diffuse component (*pale red arrow*), which represents scattered solar radiation with approximately the same spectral composition, but reflecting a much broader solid angle of $\approx 2\pi$. Emitted radiation of the Earth (terrestrial radiation, *blue arrows*) is emitted over the whole area of the Earth and represents a solid angle of 4π

of the Sun (see Fig. 2). This solid angle expresses the size of the Sun at the orbital distance of the Earth, so that the solar radiative flux inside this solid angle is the same as the solar radiative flux when it was emitted from the solar surface. This solar radiative flux has a magnitude of $\sigma_b T_{\text{sun}}^4 = 62.4 \times 10^6\,\text{W}\,\text{m}^{-2}$ (i.e., the Sun is really bright), but its brightness is constrained into a very narrow section of the sky. When combined, these factors yield a mean incoming flux of solar radiation at the top of the atmosphere of about $R_{\text{s,in}} = 1,370\,\text{W}\,\text{m}^{-2}$, a value known as the "solar constant".

When solar radiation is scattered in the atmosphere, the magnitude of the flux does not change, but the relative values of B and ε change. In the extreme case, solar radiation is scattered over the whole hemisphere, which corresponds to a solid angle of $\Omega = 2\pi$. In this case the value of B would substantially increase and the dilution factor would become much less. The spectral composition of this diluted radiation would still be relatively unaffected (although some scattering processes alter the wavelength). This scattered radiation is then referred to as "diluted" blackbody radiation. In the following, scattered, diluted solar radiation is referred to as diffuse solar radiation, whereas the non-scattered solar radiation is referred to as direct radiation.

The entropy of the solar radiative flux, J_s, is then described as

$$J_s = \frac{4}{3} B X(\varepsilon) \sigma_b T_{sun}^3, \tag{6}$$

where the function $X(\varepsilon)$ expresses the effect of dilution on the entropy flux. This function can be derived analytically, and a numerical approximation of this function is given by [5] as

$$X(\varepsilon) = \varepsilon \times (0.9652 - 0.2777 \ln \varepsilon + 0.0348\varepsilon). \tag{7}$$

Note that the expression of the radiative entropy flux is similar to the entropy of a heat flux, except for the factor 4/3, which originates from the contribution of radiation pressure (e.g., [11]), and for the effect of dilution, $BX(\varepsilon)$.

With these expressions of the solar radiative and entropy flux, one can derive the thermodynamic limit for solar radiative energy conversions. This derivation is analogous to the derivation of the Carnot limit in the previous section. The first law is represented in steady state by

$$R_{in} = R_{out} + J + G, \tag{8}$$

where R_{in} is the incoming flux of solar radiation, R_{out} is the emission of longwave radiation, J is a heat exchange flux with the surroundings, and G is the potential to generate free energy from the conversion process. The entropy balance of the conversion process is given by the import of entropy by R_{in}, the export of entropy by radiative emission, R_{out}, and the entropy exchange produced by J:

$$J_{s,in} = J_{s,out} + \frac{J}{T}, \tag{9}$$

where T is the environmental temperature. To obtain the upper limit, it is again assumed that there is no irreversible entropy production associated with the conversion process. This then yields a limit on the generation rate, G, given by

$$G = R_{in} - R_{out} - T(J_{s,in} - J_{s,out}). \tag{10}$$

A specific expression for the generation limit is obtained when the expressions for the radiative energy and entropy fluxes from the above are used. With (5) and (6) we obtain

$$\begin{aligned} G &= R_{in}\left(1 + \frac{1}{3}\left(\frac{T}{T_{sun}}\right)^4 - \frac{4}{3}\frac{X(\varepsilon)}{\varepsilon}\frac{T}{T_{sun}}\right) \approx R_{in}\left(1 + \frac{4}{3}\frac{X(\varepsilon)}{\varepsilon}\frac{T}{T_{sun}}\right) \\ &= R_{in} \times \eta_{rad}. \end{aligned} \tag{11}$$

The second term in the first expression on the right hand side can be neglected for conditions on Earth because $T \ll T_{sun}$. The last expression in (11) again formulates the generation rate in terms of the influx of solar radiation, R_{in}, and a conversion efficiency, η_{rad}:

$$\eta_{rad} = 1 - \frac{4}{3}\frac{X(\varepsilon)}{\varepsilon}\frac{T}{T_{sun}}. \tag{12}$$

This expression describes the theoretical maximum efficiency of converting solar radiation into other forms of energy.

To obtain estimates for the maximum efficiency from (12), we not only need information on the surface temperature, T, but also on the nature of the solar radiation in terms of how much it has been diluted so that we can determine the value of ε and $X(\varepsilon)$. For this we consider two cases. The first efficiency we derive is for the conversion of direct, i.e., non-scattered solar radiation at the surface. Direct radiation is used, for instance, by solar technologies in which solar radiation is focused by mirrors onto a small area to reach a high temperature ("direct concentrated solar", DCS). Direct radiation is characterized by a value of epsilon of $\varepsilon = 1$ and $X(1) = 1$. With a mean surface temperature of about $T = 288$ K, this yields a maximum efficiency of $\eta_{direct} = 93\%$. Note how the value of the efficiency differs from the Carnot limit because of the effect of the radiation pressure. The efficiency is almost 1, indicating that it can almost be completely converted into another form of energy. This high efficiency is ultimately produced by the high radiative temperature difference between the Sun and the radiative temperature of the Earth. The second efficiency we derive concerns the use of diffuse solar radiation. An example of this conversion type is photovoltaics or photochemistry, as these do not make specific use of the concentrated nature of direct radiation. We consider diffuse radiation that is completely scattered, i.e., diluted to a solid angle of $\Omega_{diffuse} = 2\pi$, but for which the radiative flux is unchanged, so that the product $B\varepsilon$ is the same for direct and diffuse radiation. This yields the condition $B_{dir}\varepsilon_{dir} = \Omega_{sun}/\pi = B_{diffuse}\varepsilon_{diffuse}$, from which the value of $\varepsilon_{diffuse}$ can be inferred from the respective solid angle of completely scattered solar radiation, $\Omega_{diffuse} = 2\pi$. This yields an efficiency of $\eta_{diffuse} = 73\%$, which is notably less than η_{direct}. This reduced efficiency is because the scattering of solar radiation is an irreversible process which produces entropy, so the diffuse solar radiation has a higher entropy.

When these thermodynamic limits of solar energy conversion are compared with solar radiative fluxes at the Earth's surface, one can infer solar energy potentials associated with the use of direct and diffuse radiation. These estimates are obtained by multiplying the efficiencies with the respective radiative flux, as in (11).

4 Limits to Converting Solar Radiative Heating

We next consider the indirect way of converting solar radiation in terms of the heating rate which results when solar radiation is being absorbed at a surface. The limit is then derived from the consideration of how much of the radiative heating rate can maximally be used to generate mechanical work from a heat engine. For this derivation we consider a setup shown in Fig. 3. The surface is heated by the absorption of solar radiation, R_{in}, which is cooled by a net radiative exchange with the atmosphere, R_l, and a convective heat flux, J_{in}, which transports heat from the surface to the atmosphere by atmospheric motion. Note that the net radiative cooling, R_l, describes the difference between the emission of terrestrial radiation from the surface (an upward flux) and the downwelling radiative flux of radiation which was emitted by the atmosphere back to the surface (a downward flux, representing the atmospheric greenhouse effect). The heat engine is then driven by the convective heat flux, J_{in}, and by the temperature difference between the surface, T_s, and the atmosphere, T_a. This setup was used in previous studies to derive the limit of maximum convective power, which was then used to infer mean climatic characteristics and sensitivities [12–15]. These applications are rather successful, suggesting that the atmosphere in fact operates near this theoretical limit.

The derivation of the limit is based on the Carnot limit as given by (4) applied to the heat flux J_{in}, with two important extensions. First, because the generated power is dissipated by friction within the system, frictional dissipation needs to be

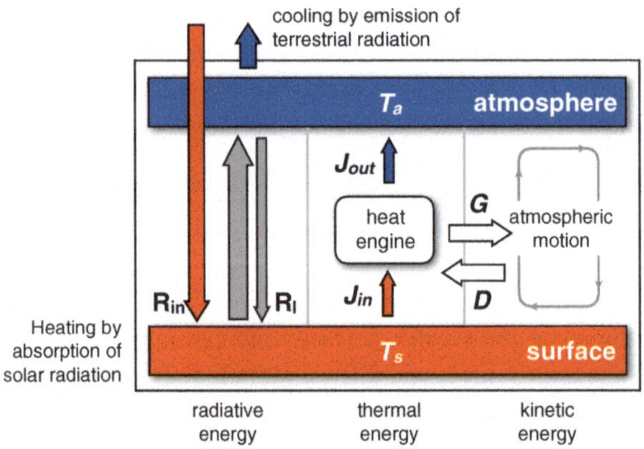

Fig. 3 Setup used to infer the conversion limit when solar radiation is used as a heating source. The illustration shows the surface-atmosphere system heated by the absorption of solar radiation, R_{in}, at the surface. This heating is compensated for in the steady state by net exchange of terrestrial radiation, R_l, between the surface and the atmosphere, and a heat flux, J_{in}, which is sustained by atmospheric motion. This heat flux drives a heat engine which generates power at a rate G and utilizes the difference in temperatures between the surface at a temperature T_s and the atmosphere at a temperature T_a. After [12] under Creative Commons license CC BY 3.0

accounted for as an additional heating term. It can be shown that this results in a slightly different form of the Carnot limit, in which the denominator in the efficiency is replaced by the temperature of the cold reservoir, T_a. Second, when a substantial fraction of R_{in} results in the heat flux J_{in}, its effect on the temperature difference $T_s - T_a$ needs to be accounted for as well. This effect can easily be derived from the surface energy balance. To do so, the net radiative cooling of the surface, R_l, is approximated by a linear relationship, $R_l = k_r(T_s - T_a)$, where k_r is an effective radiative conductance. The surface energy balance in steady state requires that $R_{in} = R_l + J_{in}$. When both expressions are combined, we obtain an expression for the temperature difference:

$$T_s - T_a = \frac{R_{in} - J_{in}}{k_r}. \tag{13}$$

Note how this expression reflects the decrease of $T_s - T_a$ with a greater flux J_{in}, with the extreme case of no temperature difference if $J_{in} = R_{in}$. This effect on the temperature difference has an important implication in that it implies that not all of the radiative heating rate can be used by a heat engine and converted into work. When this expression is used in the modified Carnot limit just described, we note that the power derived by the heat engine is a quadratic function of J_{in}:

$$G = J_{in} \times \frac{T_s - T_a}{T_a} = J_{in} \times \frac{R_{in} - J_{in}}{k_r T_a}. \tag{14}$$

The maximum power the heat engine can generate is then obtained by maximizing this expression with respect to the heat flux, which is analytically done by $\partial G / \partial J_{in} = 0$ and then solving for J_{in}, $T_s - T_a$, and for G. The value of the maximum power limit, G_{max}, is then given by

$$G_{max} = \frac{R_{in}^2}{4k_r T_a} = R_{in} \times \eta_{heat}, \tag{15}$$

with the efficiency given by

$$\eta_{heat} = \frac{R_{in}}{4k_r T_a} = \frac{1}{2} \frac{T_s - T_a}{T_a}. \tag{16}$$

Here, the value of k_r is derived from the expression of R_l at the maximum power limit, which is given by $R_l = k_r(T_s - T_a) = R_{in}/2$, so that $k_r = R_{in}/(2(T_s - T_a))$. Note that the reduction by 1/2 reflects the fact that only a fraction of the solar radiative heating can be used for the conversion. Using observed values of the surface and atmospheric temperature of $T_s = 288\,\text{K}$ and $T_a \approx T_r = 255\,\text{K}$ yields an efficiency of $\eta_{heat} = 6.5\%$. This efficiency is much lower than the efficiency obtained from the direct use of solar radiation which makes explicit use of the low radiative entropy of solar radiation. Once solar radiation is absorbed at the prevailing temperatures of the Earth, this low entropy is lost, resulting in substantial entropy production by absorption of solar radiation. Because the temperature differences on Earth are relatively low, the resulting efficiencies for converting solar radiative heating into another form are equally low.

A few aspects need to be mentioned about the power generated from solar radiative heating. First, power is also generated from horizontal gradients in solar radiative heating. The difference in solar radiative heating between the tropics and the polar regions, for instance, drives large-scale atmospheric flow, yielding additional power. This power is constrained in a similar way by thermodynamics. As the temperature difference is of similar magnitude as used here, but the heat flux associated with large-scale flow is substantially smaller, the generated power is a fraction of the power associated with the vertical difference in radiative heating [12]. For the first order estimates derived here, this contribution is neglected. Second, a sizable fraction of the maximum power derived from the vertical difference in radiative heating is used in the atmosphere to drive the hydrologic cycle. The heat flux described by J_{in} includes the flux of latent heat, which represents about 80% of J_{in} [16]. The resulting work is dissipated by several irreversible processes within the water cycle [17], such as frictional dissipation of falling raindrops, but also generates the potential energy of precipitation which drives river flow on land or the desalination of ocean water. The resulting conversion efficiency into the kinetic energy of motion is thus likely to be smaller than the 6.5% estimated above, and more likely to be 20% of this value, which would result in an efficiency of about 1.3%. Third, when kinetic energy of motion is used as wind energy and is converted further by wind turbines, only a fraction of the kinetic energy can be converted into renewable energy. Using considerations of momentum balance and maximum conversion limits set the efficiency of this conversion to about 38% of the generated kinetic energy [18, 19]. This would then imply an overall conversion efficiency of about 0.5% in deriving a wind energy potential from absorbed solar radiation. As a last point, it is mentioned here that some of the kinetic energy of atmospheric motion is not dissipated by friction, but converted further, for instance in wave power over oceans or the kinetic energy associated with the wind-driven oceanic circulation. Observation-based estimates [20] suggest that about 0.05% of the absorbed solar radiation at the surface result in the generation of ocean waves, and only 0.002% drive the wind-driven oceanic circulation. These constitute further energy conversions originating from solar radiation and providing other forms of renewable energy, but their potentials are subsequently lower.

5 Limits to Converting Solar Radiation by Photosynthesis

Photosynthesis directly converts solar radiation into chemical energy and represents another possibility to convert solar energy into some other form. In principle, the same theoretical limit as derived in Sect. 3 applies to photosynthesis and photochemical conversions in general, as these directly use diffuse solar radiation rather than heat when absorbed. The theoretical limit of photosynthesis is, however, substantially lower, mostly because chlorophyll uses a relatively narrow range of the solar spectrum. Observations of agricultural crops grown in the absence of limiting resources further suggest that the actual efficiency is even lower, although the fundamental physical reasons for this lower observed efficiency remain unclear.

As a photochemical reaction, photosynthesis converts carbon dioxide (CO_2) and water into carbohydrates and molecular oxygen using light, so the availability of CO_2 and light set in principle the primary limitations on the upper limit on how much CO_2 can be taken up [21–23]. The requirement for light is associated with the energetic minimum requirement of four photons of wavelengths of 680 and 700 nm each to reduce one molecule of CO_2, yielding a total requirement of eight photons. This corresponds to an energetic requirement of about 1,370 kJ of radiative energy per mole of fixed carbon, and yields about 479 kJ of chemical free energy which is bound in carbohydrates. This energy conversion proceeds with an efficiency of 479 kJ/1,370 kJ = 35% and was found by Hill and Rich [24] to be operating near the maximum thermodynamic efficiency at low light intensities. Considering that only about 48.5% of the solar spectrum can be used by photosynthesis and that conversion losses reduce this number further, the maximum thermodynamic efficiency of photosynthesis to convert solar radiation from theoretical considerations has been placed at around 12% [25–28]. Efficiencies derived from observations of agricultural crops place the actual photosynthetic efficiency at less than 3%. Even the most productive crops in the midwestern United States have recently been found to be display low conversion efficiencies [29]. This conversion efficiency describes the conversion from solar radiation into chemical energy. As biomass is formed out of carbohydrates there are further conversion losses, with typical rates of biomass production being half the photosynthetic CO_2 uptake. The conversion rate of carbohydrate into biofuels should thus be less than 1.5% of the absorbed solar radiation.

This conversion efficiency can, of course, only be achieved when other factors do not limit photosynthesis, particularly the availability of water on land and the availability of nutrients in the ocean. These limitations reduce the actual efficiencies of natural photosynthesis to lower values. When global estimates for photosynthetic CO_2 uptake for the marine and terrestrial biosphere are used [30], the actual efficiencies are 0.1% in the ocean and 0.7% on land.

6 Global Estimates

To obtain first-order global estimates of the different potentials of renewable energy derived from solar radiation, we use a global data set of solar radiation derived from observations. The NASA Surface meteorology and Solar Energy (NASA-SSE) data set was constructed by a combination of observations from satellites and polynomial fits to obtain a climatology of direct and diffuse solar radiation at the surface.

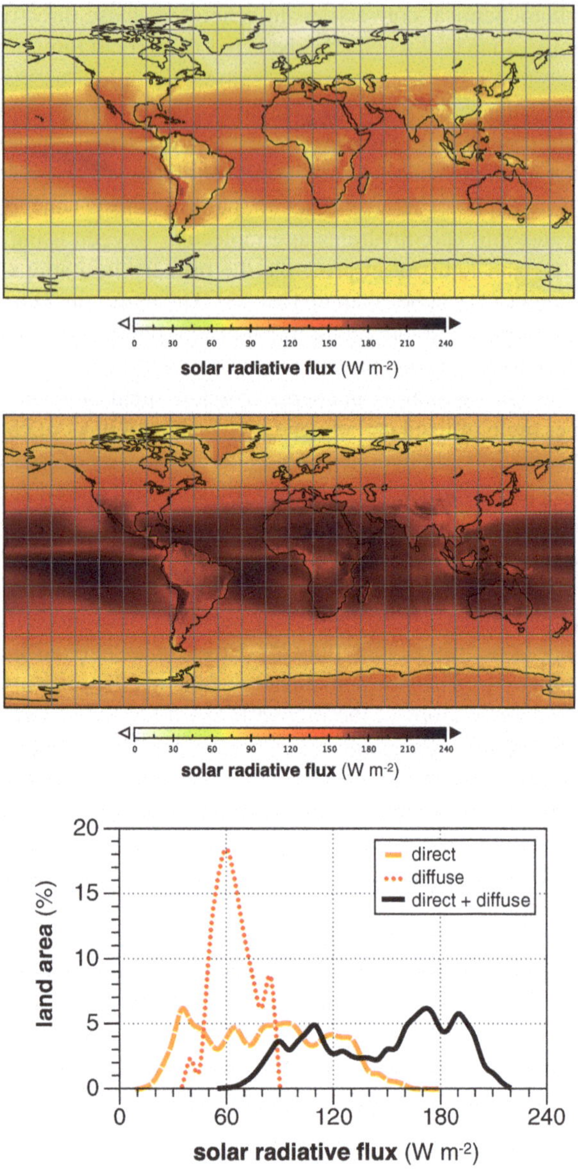

Fig. 4 Maps of the climatological annual means of downward solar radiation in terms of (**a**) its direct component and (**b**) the total downward solar radiation (direct and diffuse). (**c**) Histogram of the occurrence of the two components of solar radiation and the total over land regions, using bins of a width of 5 W m^{-2}. The diffuse downward solar radiative flux is not shown in a separate map because its variation is relatively small within about 50–90 W m^{-2}. The data is taken from the NASA Surface meteorology and Solar Energy (SSE) dataset version 6.0

The data is available for download at http://eosweb.larc.nasa.gov/sse/. The mean annual spatial distribution of the downward flux of solar radiation at the surface is shown in Fig. 4, with global means given in Table 1. The maps of solar radiation show what one would expect: the highest values of direct radiation are found in the subtropical desert regions (Fig. 4a), where the lack of cloud cover and the comparatively short path lengths through the atmosphere reduce the effect of scattering by clouds. Diffuse solar radiation is relatively uniform across regions with a value of

Table 1 Planetary averages of the direct, diffuse, and total downward solar radiative flux at the surface shown in Fig. 4 and derived estimates of the potentials of different forms of renewable energy

Description	Global	Land	Ocean	Units
Downward solar radiative fluxes at the surface				
A: Direct solar radiation	87.3	83.7	88.8	W m^{-2}
	44.5	12.5	32.1	PW
B: Diffuse solar radiation	67.7	66.5	68.2	W m^{-2}
	34.5	9.9	24.6	PW
C: Total solar radiation	155.0	150.2	157.0	W m^{-2}
(=A + B)	79.1	22.4	56.7	PW
Renewable energy potentials				
D: From direct solar radiation	81.2	77.8	82.6	W m^{-2}
(=93% of A)	41.4	11.6	29.8	PW
E: From total solar radiation	113.2	109.6	114.6	W m^{-2}
(=73% of C)	57.7	16.3	41.4	PW
F: Photovoltaics	31.0	30.0	31.4	W m^{-2}
(=20% of C)	15.8	4.5	11.3	PW
G: Biomass	2.3	2.3	2.4	W m^{-2}
(=1.5% of C)	1.2	0.3	0.9	PW
H: Wind power	0.8	0.8	0.8	W m^{-2}
(=0.5% of C)	0.4	0.1	0.3	PW
Surface area	510.0	148.9	361.1	10^{12} m^2

Estimates in lines D and E use maximum efficiencies permitted by thermodynamics, whereas F uses an efficiency from current technology. Estimate G uses the maximum observed efficiency of photosynthesis. Note that 1 PW = 10^{15} W

about 60 W m^{-2}, as shown in the histogram in Fig. 4c. The total downward flux of solar radiation is shown in Fig. 4b, which shows that the peak values in the total flux are also found in the subtropical desert regions. Not shown here are the seasonal variations in the solar radiative flux.

When these values are integrated over all land, oceans, and for the entire planet, they yield the averages shown in Table 1. These global numbers show that direct and diffuse radiation are relatively equally distributed, with a slighter greater share of about 53–55% of direct solar radiation contributing to the total downward flux. The global average downward flux of 155.0 W m^{-2} (line C in Table 1) is somewhat lower than the global average of *absorbed* solar radiation at the surface of 165 W m^{-2} from the recent estimate given by Stephens et al. [16]. This tells us that the following numbers should be seen as relatively rough, first-order estimates of the magnitudes, with uncertainties in the solar radiative flux being in the order of at least 10 W m^{-2}.

We can now use the estimated efficiencies from the previous sections to infer the theoretical potential of different solar energy use. These estimates are obtained by multiplication of the efficiency with the radiative flux, with the values provided in

Table 1. The relative geographical distribution would be as shown in Fig. 4, because the potential is simply a fraction of the shown map, with the fraction given by the efficiency used in the estimate.

The global theoretical potential for the use of direct solar radiation is obtained by combining the maximum efficiency η_{direct} with the values of the downward solar radiative flux. This yields an average potential of 81.2 W m^{-2}, or 41.4 PW when integrated over the whole surface area (Table 1, line D). When evaluated over land, the average potential is about the same (77.8 W m^{-2}), which yields a potential of 11.6 PW when integrated over all continental regions. This potential sets the upper limit to generating energy by direct concentrated solar-based technology. The currently achieved efficiency is in the range of about 15% [31] and thus much lower than what is theoretically possible.

The theoretical potential for the use of solar radiation which does not specifically utilize the direct component is shown in line E of Table 1, using the efficiency η_{diffuse} from above. The global mean potential is 113.2 W m^{-2}, or 57.7 PW, with values of 109.6 W m^{-2}, or 16.3 PW, when applied to land only. These values are notably larger than the potential for direct solar radiation because of the greater overall flux of solar radiation that could be utilized. This potential sets the upper limit to energy generation by photovoltaics. The currently most effective solar cells have an efficiency of about 44%, with a typical value at around 20% [32]. These efficiencies are also notably lower than what is theoretically possible. For comparison, the global potentials for photovoltaics using a 20% efficiency are also given in line F of Table 1.

For completeness, the potentials for biomass and wind energy are also included in Table 1. The potential for biomass should be viewed with some reservation, as it would require the absence of nutrient and water limitation in all regions, so that the number shown in the Table is more of a hypothetical potential. The potential for wind energy is based on the restrictions described at the end of Sect. 4 and yields about 114 TW, which is somewhat higher than the more detailed estimate of 16–68 TW obtained by Miller et al. [18].

To interpret these various estimates, compare them, and place them into the context of human energy demand, the estimates for land are shown in a graphical way in Fig. 5. The figure starts with the solar radiative flux at the top of the Earth's atmosphere of 1,370 W m^{-2}, that is, the "solar constant" from Sect. 3, which integrated over the illuminated cross section of the Earth yields a total flux of 174.8 PW or 174,750 TW. Because of reflection, which is about 30% of the incoming solar radiative flux on the planetary scale, and because of atmospheric absorption, this influx of solar radiation is reduced to about half the incoming flux at the top of the atmosphere. On the global scale, this provides an energy flux of 79,100 TW at the surface (line C in Table 1).

The solar radiative flux at the Earth's surface can be converted into renewable energy in three different ways. The first is shown in Fig. 5 by the red arrow on the left. Here, the solar radiative flux at the surface is converted directly into renewable energy. This is achieved, for instance, by using direct concentrated solar or photovoltaic technologies. The maximum efficiencies associated with these conversions

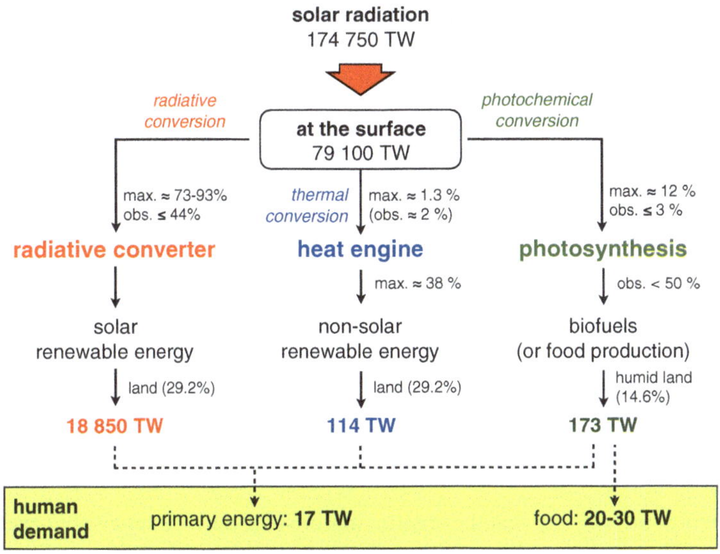

Fig. 5 Summary of the climatological estimates of different forms of solar-derived forms of renewable energy on land in comparison to the human demand in terms of primary energy and food. Note that the observed efficiency of 2% for thermal energy conversion applies to the large-scale circulation and is not directly comparable with the efficiency of atmospheric convection estimated here. The number is provided for a qualitative indication of a typical efficiency associated with atmospheric motion

are given by the efficiencies η_{direct} and η_{diffuse} from Sect. 3. In the ideal case, the efficiency η_{direct} is used for the direct part of the solar radiative flux, and η_{diffuse} is used for the diffuse part. When combined with the values of Table 1 and applied to the land area, which covers about 29.2% of the Earth's surface, these yield a total solar energy potential on land of 18,850 TW.

The second way solar radiation can be used to generate renewable energy is through its differential radiative heating which is the source for conversion into kinetic energy. Radiative heating takes place when solar radiation is absorbed at the surface and converted into heat while radiation is emitted from the atmosphere, thereby generating a differential heating and cooling source. This differential heating creates temperature differences within the Earth system which are used by the atmospheric "heat engine" to generate motion and kinetic energy, and it acts to deplete these differences. This is shown in Fig. 5 by the blue arrow in the middle of the figure. This type of conversion is associated with a very low efficiency resulting from three factors. First, not all the radiative flux can be used to drive the heat engine; second, a substantial fraction of the convective heat flux is consumed by the hydrologic cycle; and third, the temperature difference used by the heat engine is comparatively low. When the generated kinetic energy of the winds is then converted further to renewable wind energy, not all of the kinetic energy can be used. Using the reduced efficiency at the end of Sect. 4 together with the values in Table 1 yields a rough estimate of the wind energy potential on land of about 114 TW. It should be noted that the efficiency from observations of about 2% given in Fig. 5 refers to the efficiency of the large-scale atmospheric circulation, which is not the same as the efficiency associated with vertical convection derived here. This number is thus only given for a qualitative indication of a typical efficiency.

The third way to use solar radiation is through photochemical conversion, specifically photosynthesis in this case. This is shown in Fig. 5 by the green arrow on the right. As described in Sect. 5, the theoretical limit of photosynthesis imposed on the radiative conversion process is about 12%, although observations have so far reported a much lower limit of 3%. Only half of this efficiency results in the generation of biomass, which then yields either biofuel as a renewable form of energy or food. When these numbers are applied to humid regions, in which water availability does not limit natural photosynthesis and which constitutes about 14.6% of the Earth's surface, this form of renewable energy yields a potential 173 TW.

It is worth mentioning that these estimates of solar-derived renewable energy are substantially larger than those derived from the additional forcing provided by the geothermal heat flux from the Earth's interior of less than 50 TW [33], and by the tidal forcing of the Moon and the Sun, which provide about 5 TW [34].

These potentials are now placed into the context of the human demand for energy. This demand consists of two forms: the need for primary energy, which is currently about 17 TW [35], and the need for food resources provided by agricultural production, which is an indirect form of energy demand related to human metabolic energy needs. Using estimates of the human appropriation of net primary productivity of about 25–40% [36, 37] and applying it to the energy equivalent of the net primary productivity of land regions yields an estimate of about 20–30 TW [9, 10]. Although renewable energy can be derived by all three ways shown in Fig. 5, the demand for food production can only be provided by the third way through the products of photosynthesis. These two forms of human energy demand are shown at the bottom of Fig. 5.

The estimates in Fig. 5 impressively show how the potential for renewable energy by direct conversion of solar radiation into other forms of energy is much larger than by indirect means. The theoretical potential of solar energy is two orders of magnitude greater than that indirectly derived from solar radiation. Although these estimates are certainly first-order estimates and associated with some uncertainties, the orders of magnitude follow from the thermodynamic constraints discussed in the section above. Hence, the estimates shown in Fig. 5 should be rather robust.

The estimates have two important implications for the form of renewable energy technology from which the greatest impact on energy generation can be expected, and for the associated environmental impacts when used to meet human energy demands. The potential of solar renewable energy is huge even when reduced to the efficiencies in photovoltaics that are currently technologically achievable. Future improvements in the efficiencies in, e.g., photovoltaics can realize a greater fraction of the theoretical potential, which would have a substantial impact on how much more energy can be generated from solar energy. In comparison, an improvement in the efficiency of photosynthesis by bioengineering, or an expansion of agriculture into arid lands, could potentially double or quadruple the potential shown in Fig. 5. Yet, because the potential is quite small compared to that of solar energy, the resulting impact on the potential would be comparatively small. The direct use of solar energy already provides the greatest source of renewable energy and future improvements should make this source even greater.

The substantial potential of solar energy also implies that the associated environmental impacts are much smaller when used to meet the primary energy demand. This can be seen by the ratio of the present-day human energy need to the potential. This ratio of $17/18{,}850 = 0.09\%$ implies that only a minute fraction of the land surface would need to be used to meet the human energy demand. Furthermore, because most of the absorption of solar radiation at the surface results in the conversion into heat, the originally low entropy of solar radiation is swamped by the huge entropy production during absorption. This loss is reduced when solar radiation is converted into electricity before it results in radiative heating at the low prevailing temperatures of the Earth's surface. In comparison, if the human energy demand is met by wind energy, then a much greater fraction of the potential would need to be used. Given the numbers in Fig. 5, this fraction would be $17/114 = 14.9\%$. Because the dissipation of the kinetic energy by surface friction plays an important role in maintaining the turbulent exchange between the surface and the atmosphere, using a substantial fraction of the wind energy potential results in inevitable climatic impacts at the land surface [18]. Hence, the magnitude of these potentials not only informs us about how human energy demand can be met most effectively, but also about the potential environmental impacts when used.

7 Summary and Conclusions

We have provided estimates of the theoretical potentials of the different ways in which solar radiation can be used to generate renewable energy. These theoretical potentials were derived from the laws of thermodynamics in combination with global, observation-based estimates of the solar radiative flux at the Earth's surface. The estimates derived in this way show that the direct use of solar radiation provides a much greater potential for renewable energy than other forms derived in more indirect ways.

These estimates provide a basis for further energy conversions into solar-based fuels. This conversion involves chemical conversions which are likely to proceed with efficiencies of less than one, as required by the laws of thermodynamics, so that there are likely to be inevitable losses in these energy conversions. The potentials obtained here may be combined with such chemical conversion efficiencies to evaluate the overall efficiency of the energy conversion chain of a particular technology. For instance, when an efficient way to convert biomass into biofuels is compared with a comparatively inefficient conversion of electric energy generated by PV into some solar-based fuel, then we need to recognize that the theoretical potential of PV is much greater than that of biomass. What this would then imply is that the latter conversion, which may at first seem less efficient, may nevertheless provide a greater overall potential in generating fuels. What this suggests is that the comparison of the potentials of different technologies to generate solar-based fuels should be evaluated in a system's perspective which includes the whole chain of energy conversion from solar radiation to fuel.

Although we have focused here on the global scale, it is important to note that the proportion of the different potentials of renewable energy are similar when applied to the regional to national scale. This can easily be demonstrated by using the characteristics of Germany. Its area is about 360,000 km^2, its mean solar radiation about 120 W m^{-2}, and it is somewhat more windy than average, with a mean dissipation rate of kinetic energy of about 2 W m^{-2}. Using these numbers and the efficiencies derived above, the solar and wind energy potentials are 31,500 and 274 GW respectively. For comparison, in 2013 a total of 13,828 PJ of primary energy were consumed in Germany [38], yielding a mean consumption rate of 438 GW. Even though Germany is not as sunny as desert regions and windier than the mean, the theoretical potential for solar energy is still two orders of magnitude greater than for wind energy, and it can easily meet the primary energy demands at national level.

In conclusion, solar energy provides by far the greatest potential for renewable energy in the Earth system. The reason is simple and rooted deeply in the physics of energy conversion. Thermodynamics, particularly the second law, tells us that any energy conversion is associated with inevitable losses. Applied to renewable energy, this implies that the most direct use of the largest forcing of the Earth system, that is, solar radiation, is associated with the greatest magnitude and the least conversion losses, yielding the greatest potential. Solar energy is therefore sure to play a central role in future energy supply based on renewable energy.

Acknowledgements The datasets of solar radiation were obtained from the NASA Langley Research Center Atmospheric Science Data Center Surface meteorology and Solar Energy (SSE) web portal supported by the NASA LaRC POWER Project (http://eosweb.larc.nasa.gov/sse/).

References

1. Edenhofer O, Pichs-Madruga R, Sokona Y, Seyboth K, Matschoss P, Kadner S, Zwickel T, Eickemeier P, Hansen G, Schloemer S, von Stechnow C (eds) (2011) IPCC special report on renewable energy sources and climate change mitigation. Prepared by working group III of the intergovernmental panel on climate change. Cambridge University Press, Cambridge
2. Rogner HH, Barthel F, Cabrera M, Faaij A, Giroux M, Hall D, Kagramanian V, Knonov S, Lefevre T, Moreira R, Noetstaller R, Odell P, Taylor M (2000) Energy resources. In: World energy assessment. Energy and the challenge of sustainability. United Nations Development

Programme, United Nations Department of Economic and Social Affairs, and World Energy Council, New York, pp 135–171
3. Petela R (1964) Exergy of heat radiation. J Heat Trans 86:187–192
4. Press WH (1976) Theoretical maximum for energy from direct and diffuse sunlight. Nature 264:734–735
5. Landsberg PT, Tonge G (1979) Thermodynamics of the conversion of diluted radiation. J Phys A 12:551–562
6. Atkins P, de Paula J (2010) Physical chemistry, 9th edn. Oxford University Press, New York
7. Eddington AS (1928) The nature of the physical world. Macmillan, New York
8. Planck M (1906) Theorie der Wärmestrahlung. Barth, Leipzig
9. Kleidon A (2010) Life, hierarchy, and the thermodynamic machinery of planet Earth. Phys Life Rev 7:424–460
10. Kleidon A (2012) How does the Earth system generate and maintain thermodynamic disequilibrium and what does it imply for the future of the planet? Phil Trans R Soc A 370:1012–1040
11. Wu W, Liu Y (2010) Radiation entropy flux and entropy production of the Earth system. Rev Geophys 48:RG2003
12. Kleidon A, Renner M (2013) Thermodynamic limits of hydrologic cycling within the Earth system: concepts, estimates and implications. Hydrol Earth Syst Sci 17:2873–2892
13. Kleidon A, Renner M (2013) A simple explanation for the sensitivity of the hydrologic cycle to climate change. Earth Syst Dyn 4:455–465. doi:10.5194/esd-4-455-2013
14. Kleidon A, Renner M, Porada P (2014) Estimates of the climatological land surface energy and water balance derived from maximum convective power. Hydrol Earth Syst Sci 18:2201–2218
15. Kleidon A, Kravitz B, Renner M (2015) The hydrologic sensitivity to global warming and solar geoengineering derived from thermodynamic constraints. Geophys Res Lett 42:138–144
16. Stephens GL, Li J, Wild M, Clayson CA, Loeb N, Kato S, L'Ecuyer T, Stackhouse PW, Lebsock M, Andrews T (2012) An update on Earth's energy balance in light of the latest global observations. Nat Geosci 5:691–696
17. Pauluis O (2005) Water vapor and entropy production in the Earth's atmosphere. In: Kleidon A, Lorenz RD (eds) Non-equilibrium thermodynamics and the production of entropy: life, Earth, and beyond. Springer, Heidelberg, pp 173–190
18. Miller LM, Gans F, Kleidon A (2011) Estimating maximum global land surface wind power extractability and associated climatic consequences. Earth Syst Dyn 2:1–12
19. Gans F, Miller LM, Kleidon A (2012) The problem of the second wind turbine – a note on a common but flawed wind power estimation method. Earth Syst Dyn 3:79–86
20. Ferrari R, Wunsch C (2009) Ocean circulation kinetic energy: reservoirs, sources, and sinks. Annu Rev Fluid Mech 41:253–282
21. Blackman FF (1905) Optima and limiting factors. Ann Bot 19:281–295
22. Monteith JL (1972) Solar radiation and productivity in tropical ecosystems. J Appl Ecol 9 (3):747–766
23. Sharkey TD (1985) Photosynthesis in intact leaves of c3 plants: physics, physiology and rate limitations. Bot Rev 51:53–105
24. Hill R, Rich PR (1983) A physical interpretation for the natural photosynthetic process. Proc Natl Acad Sci U S A 80:978–982
25. Duysens LNM (1958) The path of light energy in photosynthesis. In: Brookhaven Symposia in Biology 1: the photochemical apparatus, its structure and function. Brookhaven Natl. Lab., Upton, pp 10–25
26. Radmer R, Kok B (1977) Photosynthesis: limited yields, unlimited dreams. Bioscience 27:599–605
27. Landsberg PT, Tonge G (1980) Thermodynamic energy conversion efficiencies. J Appl Phys 51:R1
28. Zhu XG, Long SP, Ort DR (2008) What is the maximum efficiency with which photosynthesis can convert solar energy into biomass? Curr Opin Biotechnol 19:153–159

29. Guanter L, Zhang Y, Jung M, Joiner J, Voigt M, Berry JA, Frankenberg C, Huete AR, Zarco-Tejada P, Lee JE, Moran MS, Ponce-Campos G, Beer C, Camps-Valls G, Buchmann N, Gianelle D, Klumpp K, Cescatti A, Baker JM, Griffis TJ (2014) Global and time-resolved monitoring of crop photosynthesis with chlorophyll fluorescence. Proc Natl Acad Sci U S A 111:E1327–E1333
30. Ciais P, Sabine C, Bala G, Bopp L, Brovkin V, Canadell J, Chhabra A, DeFries R, Galloway J, Heimann M, Jones C, Qúere CL, Myneni RB, Piao S, Thornton P (2013) Carbon and other biogeochemical cycles. In: Stocker TF, Qin D, Plattner GK, Tignor M, Allen SK, Boschung J, Nauels A, Xia Y, Bex V, Midgley PM (eds) Climate change 2013: the physical science basis. Contribution of Working Group I to the Fifth Assessment Report of the Intergovernmental Panel on Climate Change. Cambridge University Press, Cambridge
31. Trieb F, Schillings C, O'Sullivan M, Pregger T, Hoyer-Klick C (2009) Global potential of concentrating solar power. In: Proceedings of the SolarPACES Conference, Berlin
32. Lewis N (2007) Toward cost-effective solar energy use. Science 315(5813):798–801
33. Davies JH, Davies DR (2010) Earth's surface heat flux. Solid Earth 1:5–24. doi:10.5194/se-1-5-2010, http://www.solid-earth.net/1/5/2010/
34. Kagan BA, Sündermann J (1996) Dissipation of tidal energy, paleotides, and evolution of the Earth-Moon system. Adv Geophys 38:179–266
35. EIA (2009) International energy outlook. Tech. rep., Energy Information Administration, US Dept. of Energy, Energy Information Administration
36. Vitousek PM, Ehrlich PR, Ehrlich AH, Matson PA (1986) Human appropriation of the products of photosynthesis. Bioscience 36:368–373
37. Haberl H, Erb KH, Krausmann F, Gaube V, Bondeau A, Pluttzar C, Gingrich S, Lucht W, Fischer-Kowalski M (2007) Quantifying and mapping the human appropriation of net primary productivity in Earth's terrestrial ecosystems. Proc Natl Acad Sci U S A 104:12942–12947
38. AGEB (2014) Auswertungstabellen zur Energiebilanz Deutschland. Tech. Rep., Arbeitsgemeinschaft Energiebilanzen e.V. www.ag-energiebilanzen.de

Top Curr Chem (2016) 371: 23–48
DOI: 10.1007/128_2015_645

Published online: 23 June 2015

Principles of Natural Photosynthesis

Vera Krewald, Marius Retegan, and Dimitrios A. Pantazis

Abstract Nature relies on a unique and intricate biochemical setup to achieve sunlight-driven water splitting. Combined experimental and computational efforts have produced significant insights into the structural and functional principles governing the operation of the water-oxidizing enzyme Photosystem II in general, and of the oxygen-evolving manganese–calcium cluster at its active site in particular. Here we review the most important aspects of biological water oxidation, emphasizing current knowledge on the organization of the enzyme, the geometric and electronic structure of the catalyst, and the role of calcium and chloride cofactors. The combination of recent experimental work on the identification of possible substrate sites with computational modeling have considerably limited the possible mechanistic pathways for the critical O–O bond formation step. Taken together, the key features and principles of natural photosynthesis may serve as inspiration for the design, development, and implementation of artificial systems.

Keywords Manganese · Oxygen-evolving complex · Photosystem II · Water oxidation

Contents

V. Krewald, M. Retegan, and D.A. Pantazis (✉)
Max Planck Institute for Chemical Energy Conversion, Stiftstr. 34-36, 45470 Mülheim an der Ruhr, Germany
e-mail: dimitrios.pantazis@cec.mpg.de

1 Introduction

The biological conversion of light energy into chemical energy sustains practically all life on our planet [1]. Although not all photoautotrophs perform the same type of chemistry or have the same biochemical machinery, the fundamental functions are essentially the same: light-induced charge separation that drives electron transfer to eventually create energy-rich reduced compounds. A major differentiating factor among photosynthetic organisms is the source of electrons. In oxygenic photosynthesis performed by plants, algae and cyanobacteria, the source of electrons is water:

$$2H_2O \rightarrow O_2 + 4H^+ + 4e^-.$$

The light-driven oxidation of water takes place in the membrane-embedded pigment–protein supercomplex Photosystem II (PS-II), at an active site (the oxygen-evolving complex, or OEC) which contains an oxo-bridged tetra-manganese–calcium cluster (Mn_4CaO_5) [2–7]. Dioxygen is a waste product and is simply discarded by photosynthetic organisms. However, it is through this biochemical process that the Earth's atmosphere obtained its high oxygen content and the ozone layer, while life exploited oxygen's oxidizing power in the evolution of respiration that can sustain more complex life forms. Protons are used for the creation of a membrane gradient driving the chemiosmotic synthesis of ATP (adenosine triphosphate), an energy carrier molecule in metabolic processes. The electrons are transported along other components of the photosynthetic chain (see Fig. 1) to be utilized eventually in the synthesis of NADPH (reduced nicotinamide adenine dinucleotide phosphate). Through NADPH, a carrier of reducing equivalents, the electrons extracted from water are employed in the light-independent carbon fixation reactions that reduce CO_2 to carbohydrates:

$$CO_2 + 4H^+ + 4e^- \rightarrow (CH_2O) + H_2O.$$

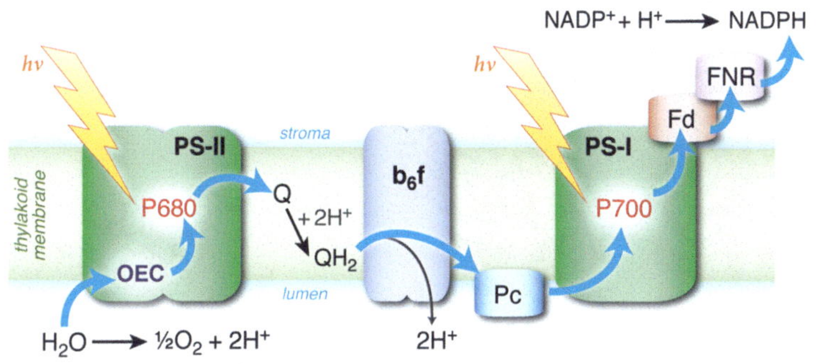

Fig. 1 Components of the light-dependent reactions and electron transfer in oxygenic photosynthesis. *Blue arrows* indicate the flow of electrons from H_2O to NADPH. *Q* plastoquinone, *Pc* plastocyanin, *Fd* ferredoxin, *FNR* ferredoxin-$NADP^+$ reductase

Among the distinctive features of biological oxygenic photosynthesis are the highly efficient light conversion, utilization of the ultimately abundant "electron source" on our planet (water), and a water oxidation catalyst composed of a readily available first-row transition metal, manganese. Faced by the current and near-future energy challenges of our civilization, and by the need to control the adverse effects of fossil fuel use, it is precisely these features of biological photosynthesis that we are trying today to reproduce in synthetic light-driven systems. The aim is to split water on a large scale to produce molecular hydrogen or other reduced compounds as energy carriers (solar fuels) [8–15]. Although the highly complicated biological system cannot be viewed as a blueprint to be faithfully reproduced in artificial systems, it remains a uniquely successful example of this type of chemistry – and a unique source of chemical insight. Several aspects of natural photosynthesis remain insufficiently understood despite decades of research, but many questions have also been answered and many fundamental principles have been clarified. Instead of attempting a broad and exhaustive overview of all facets of natural photosynthesis, in this chapter we focus on the most relevant aspects of water oxidation performed by PS-II and discuss the structural and functional principles that have emerged from recent research on biological water splitting.

2 Photosystem II Structure and Function

2.1 *Overall Structure and Electron Transfer*

The refinement of the PS-II structure has been a laborious and lengthy undertaking for X-ray crystallography. The first crystallographic model, at 3.8 Å resolution, was obtained by Zouni and coworkers in 2001 [16] and it took 10 years of effort [17–21] until a resolution of 1.9 Å was achieved [22]. Even more recently, a structure was obtained at 1.95 Å resolution using X-ray free electron laser pulses instead of

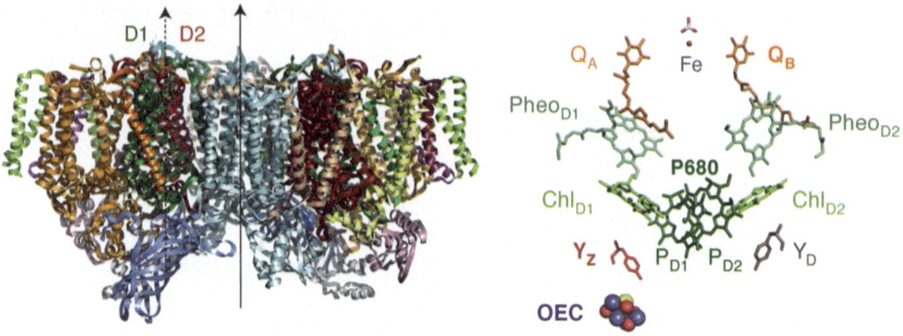

Fig. 2 *Left*: structure of the PS-II dimer. *Right*: pigments, other cofactors and important residues involved in charge separation, electron transfer, and catalytic activity within a PS-II monomer

synchrotron radiation [23], potentially avoiding the effects of extensive radiation damage which compromised earlier crystallographic models. PS-II has a dimeric structure (Fig. 2). Each monomer consists of more than 20 proteins arranged in a pseudo-dimeric fashion. A large number of cofactors are accommodated in the protein, including chlorophylls, pheophytins, carotenes, plastoquinones, lipids, bicarbonate, heme and non-heme iron centers, and an Mn_4CaO_5 complex at the OEC active site.

The core of each PS-II monomer is formed by the two homologous proteins D1 and D2, which contain most of the cofactors involved in catalysis and electron transfer (Fig. 2). Both proteins are highly conserved and very few variations exist among photosynthetic organisms. This is particularly true for D1, which harbors the oxygen-evolving manganese cofactor. Two proteins that contain several chlorophyll molecules, CP43 and CP47, surround the D1 and D2 proteins and are involved in light harvesting and transfer of excitation energy to the charge-separation site (P680), a set of four chlorophylls and two pheophytins located symmetrically around an intersection of the D1 and D2 polypeptide chains. Several other proteins surround the D1–D2 and CP43–CP47 pairs. These include "extrinsic" proteins on the lumenal side of PS-II, i.e., the side facing the interior of the thylakoid. These proteins may be necessary for catalytic function or they may bind only transiently to stabilize specific states or facilitate large-scale transformations such as active site reconstitution [24–26].

Photosystem II converts light energy into electrochemical potential which drives the oxidation of water and the reduction of plastoquinone [1, 27] according to the overall reaction

$$2H_2O + 2Q + 4H^+{}_{stroma} \rightarrow O_2 + 2QH_2 + 4H^+{}_{lumen}.$$

The dioxygen thus generated is released as a byproduct into the atmosphere, whereas plastoquinol (QH_2) diffuses into the membrane to cytochrome b_6f. Through the series of electron transport steps depicted in Fig. 1, the chain terminates with the synthesis of NADPH, the biological equivalent of reduced hydrogen. It is conceptually convenient to view the function of PS-II as the sum of four processes: (1) light harvesting and excitation energy transfer to the reaction center, (2) electronic excitation of the reaction center and charge separation, (3) transfer of the electron from the reaction center to plastoquinone, and (4) reduction of the reaction center by the Mn_4CaO_5 cluster, which in turn oxidizes water.

Following the absorption of visible light by chlorophyll molecules and their subsequent excitation, the excitation energy is transferred to the reaction center, a directed process facilitated by the spatial arrangement of the chromophores. The reaction center consists of chlorophyll (Chl *a*) and pheophytin (Pheo *a*) molecules bound to the D1 and D2 proteins (see Fig. 2). Excitation of the reaction center results in charge separation within picoseconds, generating the radical cation, $P680^{\bullet+}$, and a radical anion, $Pheo_{D1}^{\bullet-}$. $P680^{\bullet+}$ has an estimated reduction potential of +1.2 V (vs SHE), making it one of the most oxidizing species known in biology [28, 29]. Following the formation of the initial charge-separated state, the electron is transferred from $Pheo_{D1}^{\bullet-}$ to a tightly bound plastoquinone molecule (Q_A) and then through a non-heme Fe(II) to the final acceptor Q_B. The rapid and efficient multistep electron transfer over this distance of more than 30 Å is critical for suppressing charge recombination and results in the high quantum efficiency of biological water splitting. After two reduction and protonation steps the formed plastoquinol Q_BH_2 diffuses into the thylakoid membrane and is replaced by another plastoquinone.

On the opposite side of the enzyme, the electron hole at $P680^{\bullet+}$ is filled by an electron donated by the redox-active Tyr161 residue (or Y_Z) of the D1 protein. Oxidation of Tyr161 is coupled to its deprotonation; the nearby His190 is the residue accommodating the proton on its imidazole ring upon formation of the tyrosyl radical. The $Y_Z^{\bullet}$ subsequently oxidizes the manganese cluster of the OEC, which can accumulate up to four oxidizing equivalents before it eventually claims back these four electrons by oxidizing two water molecules, forming molecular dioxygen.

2.2 *Channel Architecture*

In addition to satisfying the requirement of precise spatial organization of redox-active components involved in excitation energy and electron transfer, the folding and structural arrangement of the PS-II proteins must also serve the need for tight control of accessibility and water delivery at the OEC, proton transfer, and product release. Although precise understanding of this type of regulation is still missing, analysis of crystallographic models [18, 20, 30–32], noble gas studies [33], pK_a calculations [34], and, most importantly, molecular dynamics studies that take into account the dynamic nature of the channel architecture [35, 36] have already identified possible channels within PS-II that may be involved in water or dioxygen transport and proton transfer.

In this respect, it is important to realize that water plays at least four distinct roles: as the surrounding medium on each side of the thylakoid membrane, as structural water, as ordered chains involved in proton transfer, and as the reactant to be oxidized at the active site of the enzyme. Existing information indicates the presence of five solvent-accessible channels originating at the OEC. It is not currently possible to assign the role of each individual channel (water delivery, proton transfer, or dioxygen release), but it appears likely that at least one of the channels associated with a terminal Mn ion and one channel associated with the Ca^{2+} ion may serve as water delivery channels, another channel that involves calcium may be a dioxygen release pathway [36], and two channels associated with a proximal chloride ion may facilitate proton transfer away from the OEC [37].

2.3 Photoprotection and Repair

A critical aspect of natural photosynthesis and in particular PS-II is its efficient protection and repair machinery, an aspect that may not lend itself easily to biomimetic approaches in synthetic systems [38]. Being tasked with catalyzing one of the most challenging reactions in biology using sunlight, PS-II has to achieve tight control on how to regulate light absorption and how to use the absorbed energy to drive the electron transport reactions. Excess light energy that cannot be put to use would result in damage through unwanted side reactions and generation of deleterious reactive oxygen species. PS-II has a series of photoprotective mechanisms that serve to dissipate excess energy or quench photoexcited states under such conditions. Without going into detail, these mechanisms may be distinguished in those that involve recirculating electrons within the components of the reaction center, redirecting the electrons to alternative acceptors, or disposing of the absorbed light energy as heat by reconfiguration of the antenna system [39–42].

Nevertheless, photoinhibition cannot be entirely avoided. The D1 chain is the part of PS-II most vulnerable to oxidative damage – of the electron transfer components, amino acid residues, or the Mn cluster itself. There is still controversy regarding how D1 exchange is triggered but it is known that the protein has to be replaced approximately every half hour of operation. A process intimately related to self-repair of PS-II is the assembly of the Mn_4Ca cluster: remarkably, the biosynthesis of the inorganic cluster does not rely on chaperones but is a light-driven process which builds the oxo-bridged metal cluster in place by sequential oxidations of Mn^{2+} ions by $Y_Z^{\bullet}$.

3 Oxygen Evolving Complex

3.1 Catalytic Cycle and Redox Leveling

As described above, the tetramanganese cluster of the OEC couples the one-electron process of photoexcitation and charge separation at the reaction center with the four-electron chemistry of water oxidation by storing oxidizing equivalents. At a phenomenological level, four photo-driven charge separation events lead to successive removal of four electrons from the Mn ions of the OEC, via the redox-active Tyr161 (Y_Z), at which point O_2 is evolved and the OEC is reset to its most reduced state. These observations [43] have been encoded in a model of the catalytic cycle consisting of five intermediates (Fig. 3) [44], known as the S_i states, where the subscript indicates the number of stored oxidizing equivalents ($i = 0$–4). The S_0 is the most reduced state of the cluster, but S_1 is the dark-stable resting state [4]. The S_4 is a transient and so far unobserved state that decays spontaneously to S_0 by releasing O_2.

As indicated in Fig. 3, the removal of electrons and protons follows a strictly alternating pattern [45–48], with all transitions involving proton loss except $S_1 \rightarrow S_2$ [2, 28, 49, 50]. Considerable information from electron paramagnetic resonance (EPR) spectroscopy is available on many of the intermediate $S_iY_Z^{\bullet}$ states, and evidence also exists regarding additional transient intermediates in the $S_0 \rightarrow S_1$ and $S_2 \rightarrow S_3$ transitions [46, 51], and a deprotonated intermediate immediately prior to the final electron transfer after formation of the $S_3Y_Z^{\bullet}$ state [52–54].

The critical importance of the coupled removal of protons and electrons from the OEC is that it avoids the build-up of excess charge which would quickly render the Mn cluster impossible to oxidize by the Y_Z radical. This is known as a "redox-leveling" effect and allows for four successive oxidations of the cluster to take place within a narrow potential range. This becomes apparent in the reduction potentials of the transitions, which are ca. 0.85, 1.10, 1.15, and 1.0 eV for states S_0–S_3 [4]. It

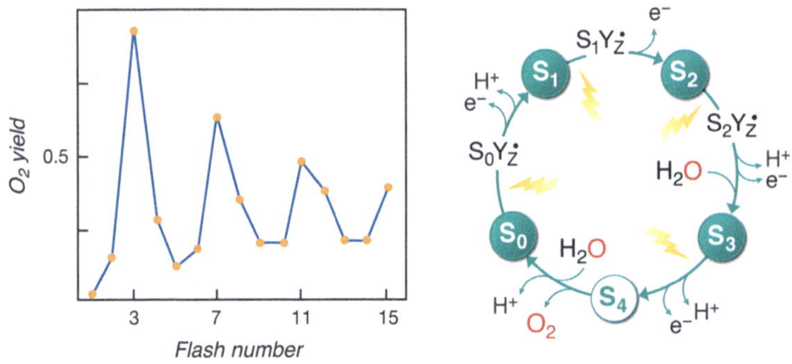

Fig. 3 *Left*: dioxygen evolved by PS-II with a series of flashes, demonstrating the four-flash periodicity in the maximum yield. *Right*: the catalytic cycle of five S_i states, including Y_Z radical intermediates

should be noted that although the electron and proton transfers are coupled, they are not necessarily concerted. Obviously, the highly organized and intricately balanced hydrogen bonding network around the OEC is of fundamental importance in regulating the transfer of protons [55–58].

3.2 Structural Information

The Mn–Mn distances obtained from Extended X-ray Absorption Fine Structure (EXAFS) spectroscopy represented the first insight into the geometry of the Mn_4CaO_5 core of the OEC. The 1981 study by Klein and coworkers was also the first direct observation of the Mn ions contained in chloroplasts [59]. Already at that time, even before the number of Mn ions per active site was known, the authors concluded that the Mn ions should be connected in a similar way to the bis-μ-oxo-bridged model dimers they also investigated. EXAFS remains a powerful tool and has provided metal–metal and metal–ligand distances for all stable catalytic intermediates [60–63], but it is unlikely that it can decipher structurally heterogeneous states of the OEC. The relative orientation of the metal–metal vectors can in principle be revealed by the use of polarized EXAFS [64], but the extraction of a unique three-dimensional structure remains challenging.

X-Ray diffraction studies of improving resolution have progressively refined our view of the three-dimensional topology and connectivity of the OEC up to the point where an atomic-resolution model (1.9 Å resolution) was reported in 2011 [22]. These studies are typically carried out on dark-adapted samples that have accumulated in the S_1 state. However, the employed X-ray doses resulted in radiation damage in the form of partial reduction of the Mn ions [65–67], leading to significant amounts of Mn(II), as opposed to Mn(III) and Mn(IV) which are present in the S_1 state. This also explains the significant differences in Mn–Mn distances between the XRD models and those observed by EXAFS or predicted computationally [61, 68, 69]. Most recently, by employing X-ray Free Electron Laser (XFEL) pulses, the three-dimensional structure of the inorganic cluster in the S_1 state was obtained at 1.95 Å resolution with probably no radiation damage [23]. Comparison of the XFEL Mn–Mn distances with those of the previous XRD structure [22] confirms the consensus reached by most researchers in the field, that a significant elongation of up to 0.20 Å for the Mn–Mn distances in the XRD structure can result from radiation-induced reduction of Mn ions in the cluster. Importantly, this XFEL structural model has now largely converged to EXAFS with respect to Mn–Mn distances and is in agreement with detailed computational models for the S_1 state (see Table 1), although some inconsistencies remain with respect to oxo-bridge positions [70, 71].

The inorganic core of the OEC has been described as taking the shape of a "distorted chair" [22], with the base formed by a heterometallic Mn_3CaO_4 cuboidal unit and the backrest by an Mn–O linkage connected to one of the Mn ions and one

Table 1 Mn–Mn distances (in Å) in the S_1 state of the OEC derived from experiment and from selected computational models

	Mn1–Mn2	Mn2–Mn3	Mn3–Mn4	Mn1–Mn3	Reference
S_1					
XRD	2.80	2.90	2.94	3.30	Umena et al. [22] (average)
XFEL-XRD	2.68	2.70	2.87	3.20	Suga et al. [23] (average)
EXAFS	2.71	2.71	2.79	3.28	Glöckner et al. [62]
DFT	2.79	2.80	2.76	3.34	Krewald et al. [70]
DFT-QM/MM	2.80	2.80	2.75	3.38	Luber et al. [68]
R-QM/MM	2.76	2.71	2.76	3.25	Luber et al. [68]

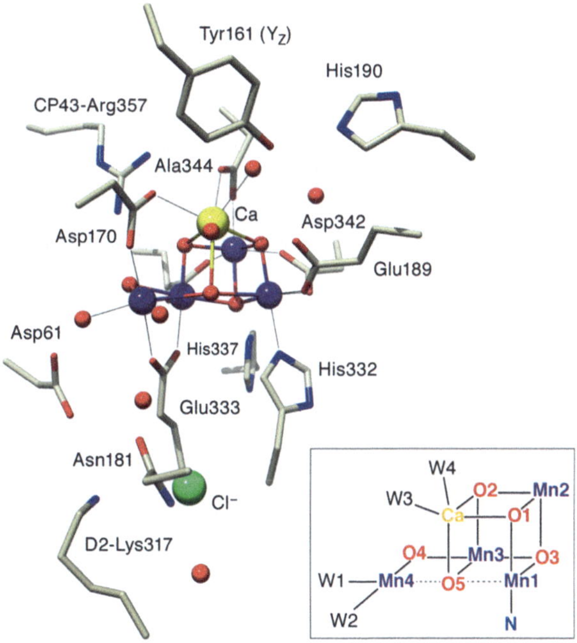

Fig. 4 A view of the OEC with the Mn_4CaO_5 cluster and its immediate environment (crystallographic coordinates are taken from Umena et al. [22]); a common labeling scheme for the inorganic core is indicated in the *inset*

of the oxo bridges of the cubane (see Fig. 4). Four water molecules, or water-derived ligands, are directly ligated to metal ions of the cluster, two at Mn4 and two at Ca. This unit is embedded in the D1 protein and connected to the CP43 protein by one direct ligand. Carboxylate residues bridge Mn1–Mn2 (Asp342), Mn2–Mn3 (CP43-Glu354), Mn3–Mn4 (Glu333), and Ca^{2+} with Mn2 (Ala344) and Mn4 (Asp170). Only one N-donor, His332, is coordinating the cluster, at Mn1, which is also mono-coordinated by Glu189. Important residues in the second coordination sphere include the Tyr161–His190 couple (electron transfer gate to the special pair), Asp61 (probably important for proton transfer), CP43-Arg357, and His337 (hydrogen bond to O3).

Remarkably, the precise nuclearity and bonding connectivity of the inorganic cluster and its immediate ligand environment had already been predicted quite accurately in computational studies by Siegbahn, based on energetic considerations, before the advent of the first atomic-resolution XRD structure [72, 73]. It should be noted that XRD as yet reports, ideally if not in practice, on the dark-adapted S_1 state only. The structural changes occurring as the catalytic cycle progresses can be traced from EXAFS data as follows. In the S_1 and S_2 states, there exist three short Mn–Mn distances of 2.7–2.8 Å and a longer Mn–Mn distance of 3.28 Å [62]. The very small structural changes between these two states are consistent with the fact that this transition represents only an oxidation event [45, 74, 75]. The S_0 state appears to be characterized by an elongation of one of the short Mn–Mn distances to 2.85 Å, which may arise from the protonation of one of the oxo bridges. The reduction by ca. 0.1 Å between S_0 and S_1 is in line with observations of geometrical changes in model systems upon oxo bridge protonation [76, 77]. Oxidation of the S_2 to the S_3 state also leads to elongation of at least one Mn–Mn distance by ca. 0.1 Å. Before the unequivocal assignment of the Mn oxidation states in the S_3 state by coupled EPR and theoretical studies [78], this elongation upon oxidation had given weight to arguments in favor of a ligand-centered rather than a metal-centered oxidation in the $S_2 \rightarrow S_3$ transition.

Despite this valuable structural information, the three-dimensional topology and the precise connectivity of the OEC cannot be directly deduced experimentally at the atomic level *in each S-state*, and information on the protonation states of bridging and terminal oxygen ligands is out of reach. However, all these aspects and a detailed view of the orientation of Mn(III) Jahn–Teller axes in each state can be obtained by experimentally constrained computational models, as described in the following.

3.3 Mn Oxidation States and Detailed Structural Models

Although the relative oxidation levels of the OEC are defined by the Kok cycle (Fig. 3), the absolute oxidation states of the Mn ions are not. Two schemes had historically evolved, which accommodated several structural constraints and spectroscopic observations [79–90]. Since the early 1980s it has been known that the four Mn ions in oxidation states III and IV composing the active site gave rise to an EPR signal with a spin ground state of $S_{GS} = 1/2$ in the S_2 state [79]. This finding could be explained equally well with the oxidation state combinations Mn(III)$_3$Mn(IV) and Mn(III)Mn(IV)$_3$, where in both cases one Mn(III) and one Mn(IV) magnetically couple to $S_{dimer} = 1/2$ and the remaining two Mn(III) or two Mn(IV) couple to yield $S_{dimer} = 0$, so that in both cases $S_{GS} = 1/2$ can be obtained. These two proposals are known as the "low oxidation state" and "high oxidation state" schemes, or equivalently, "low valent" (LV) and "high valent" (HV) schemes, and they differ by two in the total number of electrons in all states. Specifically, the LV scheme would correspond to Mn oxidation states of (II, III, III, III) in S_0 up to (III, III, IV, IV) in S_3, whereas the HV scheme requires the S_0 state to be (III, III, III, IV) instead. Answering this question is central to understanding the principles of biological water oxidation and places immediate restrictions on the mechanism of O–O bond formation, because, for example, in the HV scheme either oxo-oxyl coupling or nucleophilic attack mechanisms can be imagined, whereas in the LV scheme only the latter appears to be a chemically reasonable option.

Attempts to "count" the number of redox equivalents needed to photoactivate the OEC [91, 92] or to disassemble a specific S state [93], as well as experiments that probe the super-reduced S states [94] are promising, as they both involve measurable quantities of Mn(II). However, these experiments as well as $H_2{}^{16}O$/$H_2{}^{18}O$ substrate water exchange experiments [95–97] have been interpreted controversially [7, 98–100]. EPR and electron nuclear double resonance (ENDOR) studies have contributed a wealth of information on the water oxidizing cluster and have also addressed the question of overall oxidation level. Most importantly, the ground state spin multiplicities [101] of all stable intermediates have been established: $S_{GS} = 1/2$ for the S_0 state [102–105], $S_{GS} = 0$ for the S_1 state [106, 107], $S_{GS} = 1/2$ and $\geq 5/2$ for the S_2 state ($g = 2$ and $g \geq 4.1$ signals) [79, 108–110], and $S_{GS} = 3$ for the S_3 state [78, 111, 112]. All of these are compatible with the two oxidation state schemes introduced above. The same is true for the ^{55}Mn hyperfine coupling constants in the S_2 state, which have been used both to support the low valent scheme [81] and to rule it out [113, 114], particularly because ENDOR studies of the S_0 state were interpreted as indicating the absence of Mn(II) in that state [114], which would exclude the LV scenario.

X-Ray absorption and emission spectroscopies [115, 116] have probed the relative changes in Mn oxidation states, but the limits of interpretation in terms of absolute oxidation states are still being debated. Although in principle the absolute oxidation state of a metal can be extracted from the edge position, the situation for a multinuclear complex in which only one out of four ions changes valence upon oxidation of the cluster is more complicated: the shift of ca. 1 eV per oxidation event for a mononuclear site is diminished to ca. 0.25 eV for a tetranuclear cluster. Additionally, rearrangements of the coordination sphere, which are plausible upon oxidation of a multinuclear system, are known to influence the edge position. XANES data has been interpreted as consistent with both the HV [82–85, 117] and the LV [118–120] schemes, the latter mostly based on comparison of experimental with calculated spectra.

Computational studies have typically been carried out under the assumption of either the LV [119–123] or the HV scheme [72, 124–140]. The first comparative study of the two schemes was recently presented, including all stable intermediates (S_0–S_3) and combining the widest possible collection of experimental information as evaluation criteria, ranging from EPR properties and EXAFS or XFEL structural information to Mn K pre-edge spectroscopy. After construction of computational models valid for the S_0–S_3 states in either the LV or the HV schemes, i.e., ranging

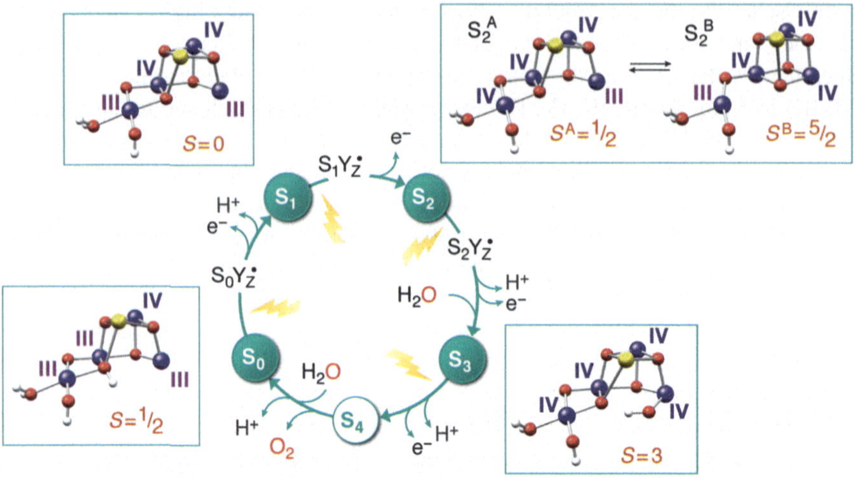

Fig. 5 Computationally determined and experimentally consistent models of the inorganic core of the OEC for all S-state intermediates. The distribution of Mn oxidation states and the ground state spin *S* are shown for each model. Coordinating amino acid ligands are not shown for clarity

from Mn(II)Mn(III)$_3$ to Mn(IV)$_4$, and systematic variation of the overall protonation level, protonation sites, orientation of Jahn–Teller axes for Mn(III) and other such permutations, the relative energies, geometric characteristics, spin ground states (see next section for details), ^{55}Mn hyperfine coupling constants and Mn K pre-edge signatures of the models were compared with experimental data. The study relied in large part on a series of methods developed over the years for the accurate calculation of magnetic and spectroscopic properties [130, 141–146]. Although agreement with isolated experimental observations could be reached for some low-valent models, S_0 and S_1 models based on the LV scheme were found to be incompatible with all experimental constraints. Therefore, this study established beyond doubt that a consistent catalytic cycle is only possible with the high-valent scheme, i.e., Mn(III)$_2$(IV)$_2$ in the S_0 state to Mn(IV)$_4$ in the S_3 state. Simultaneously, the protonation state and pattern in each S-state were uniquely determined. Figure 5 summarizes the sequence of all stable S-state catalytic intermediates resulting from this comprehensive study.

A somewhat different debate, independent of the overall oxidation level of the cluster, concerned the correct interpretation of XANES data with respect to the nature of the S_2 to S_3 transition [83, 147–149]. The two suggestions were (1) ligand-centered oxidation, as implied by the absence of a K-edge shift in the S_2 to S_3 transition and (2) metal-centered oxidation with concomitant binding of a ligand to a five-coordinate Mn(III) in S_2 to a six-coordinate Mn(IV) in S_3, such that the change in coordination environment compensates for the K-edge shift produced by metal center oxidation [83, 149]. A recent combined EPR/ENDOR and DFT study of the S_3 state conclusively supports the latter interpretation [78], showing that all four Mn centers are present as octahedrally coordinated Mn(IV) ions before the final oxidation event leading to spontaneous formation and release of dioxygen. Proposed mechanisms for the O–O bond formation step are discussed in a later section.

3.4 Spin States and Structural Flexibility

As noted above, EPR spectroscopy has been central in determining the spin states of the OEC cluster in the S_0–S_3 states of the catalytic cycle, but also in probing the local electronic properties of the Mn ions, especially through ^{55}Mn-ENDOR [113, 114, 150–152]. In terms of spin states, the data suggest that the Mn cluster evolves from low-spin states in the early stable intermediates (S_0 and S_1) to coexistence of low-spin and higher-spin states in S_2, progressing to a high-spin state in S_3, in preparation for the final O–O bond formation step (see Fig. 5).

The S_2 state is the catalytic intermediate studied the most by EPR/ENDOR, not only because of its easy accessibility through a single light flash from dark-adapted samples, but also because of the intriguing phenomenology observed for this state. It has also been the state targeted in pioneering studies aimed at providing theoretically supported topological interpretations of spin state data [153, 154] before the availability of crystallographic information. The S_2 state is associated with two EPR signals that are quantitatively interconvertible, implying that the structural changes must be small. The two signals correspond to spin ground states of $S_{GS} = 1/2$ at $g = 2.0$ ("multiline signal") and $S_{GS} = 5/2$ at $g \geq 4.1$ ("$g4$ signal"). Building on the first atomic-resolution XRD structure, Pantazis et al. constructed two models in the HV scheme that fully explained the phenomenology [137]. The structure giving rise to the $g4$ signal can be viewed as an Mn(IV)$_3$CaO$_4$ cubane, to which the fourth Mn is attached via a μ-oxo bridge providing a link to an Mn ion of the cubane (see Figs. 5 and 6). This structure is termed the "closed cubane" form, similar to a structure proposed by Barber and Murray [155]. It is known that Mn(IV)$_3$Ca cubanes are associated with spin ground states of $S_{GS} = 9/2$ [156]; therefore, the coupling of the trimeric $S_{cubane} = 9/2$ unit with the outer Mn(III) with $S_{Mn(III)} = 2$ leads to the observed spin ground state of $S_{GS} = 5/2$. The conversion into the almost isoenergetic structure associated with the multiline signal is achieved through the low-barrier translocation of a μ-oxo bridge of the cluster and a valence exchange so that the Mn(III) in the closed cubane structure becomes Mn(IV) in the open cubane structure (see Fig. 5). The open cubane form is similar to that put forward originally by Siegbahn [72, 73]. The thus altered connectivity of the Mn ions in the core leads to a magnetic coupling topology drastically different from that in the closed cubane structure, ultimately rendering the $S_{GS} = 1/2$ ground state. Two antiferromagnetic coupling pathways (Mn1–Mn2 and Mn3–Mn4) and one ferromagnetic coupling (Mn2–Mn3) dominate the magnetic interactions in the tetranuclear cluster. Importantly, the magnetic coupling topology places restrictions on the pattern of ^{55}Mn hyperfine coupling constants, which have been used as a powerful criterion to

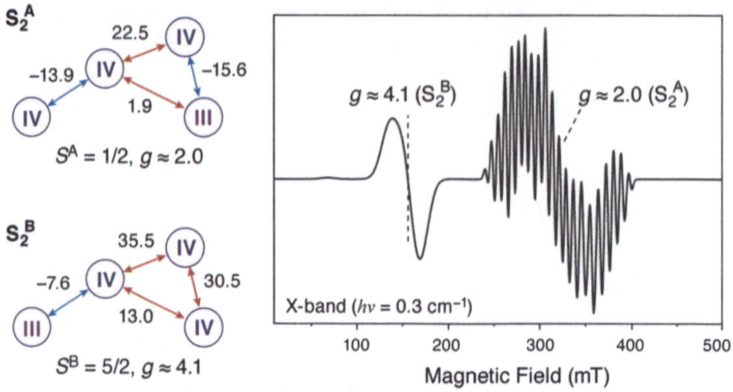

Fig. 6 Exchange coupling constants (in cm^{-1}) for the two interconvertible structural models of the S_2 state shown in Fig. 5, along with a first-derivative EPR simulation of the two configurations, the open cubane ($S = 1/2$) $g = 2.0$ multiline and the closed cubane ($S = 5/2$) $g = 4.1$ S_2-state OEC signals [137]

discriminate between different electronic configurations, structural possibilities, and protonation states of terminal and bridging ligands [70, 134, 136].

These studies represent important showcases of how a consistent picture of spectroscopic observations and structural information is yielded through the combination of experimental facts and a theoretical approach oriented towards prediction of observable properties. Structural polymorphism had previously been invoked for the S_1 state by Kusunoki [128], motivated by the quest to rationalize the radiation-damaged crystal structure using a superposition of structural forms, but it now appears that no such drastic rearrangement is present in the S_1 state. Several studies suggest that heterogeneity of the geometric and electronic structures of the S_2 state [132, 137, 157] may be absent in the S_3 state [70, 78, 124], but this point requires further investigation. It is not clear whether structural flexibility in S_2 serves a fundamental mechanistic purpose, such as facilitating or controlling water binding [132, 158]. Nevertheless, it suggests that the manganese cluster in PS-II is not merely a small rigid piece of rock embedded in a polypeptide. This structural flexibility, and the progression from low-spin to high-spin states along the catalytic cycle, are two intriguing aspects of the OEC with potential mechanistic implications requiring further experimental and theoretical work to be understood fully.

3.5 The Role of Calcium

Calcium is an integral part of the water-splitting catalyst, demonstrated, for instance, by the fact that the only ion that can replace it while maintaining the water-splitting capability of the OEC is Sr^{2+}, although the efficiency is diminished [159–162]. As other metals of the same charge and similar ionic radii are easily available in nature, the role of Ca^{2+} must be more than purely structural [163]. In line with the early, redox-equivalent accumulating S-states displaying structural heterogeneity, which the later S-states do not, the role of Ca^{2+} has also been suggested to change as the catalytic cycle progresses. Although Dy^{3+} and Cd^{2+} cause a loss of the O_2-evolution capability of the cluster, the OEC substituted with these cations can advance from S_1 to S_2, giving weight to a dominantly structural role of Ca^{2+} in the early S states [164]. Ca^{2+} has also been suggested to bind substrate water [3, 165] and to be involved in maintaining a hydrogen-bond network for proton transfer [166]. The possible involvement of calcium in the delivery of water to the Mn cluster has also been suggested on the basis of ab initio molecular dynamics studies [132].

The similar Lewis acidity of Sr^{2+} and Ca^{2+} has been considered to be an important factor for the fully functional substitution [159, 163, 164, 167]. Synthetic studies on MMn_3O_4 and MMn_3O_2 model complexes that are structural subunits of the OEC have established a linear correlation between the pK_a of $M(aquo)^{n+}$ ions of redox-inactive metals as a measure of their Lewis acidity and their reduction potentials [168]. Although the two series investigated differ in Mn oxidation states and ancillary ligands, the slopes of the linear correlations are similar. Thus, more negative reduction potentials are achieved by incorporation of redox-inactive metals with high Lewis acidities and by incorporation of more oxo-bridges into the metal-oxo core. The redox potentials of these models shifted distinctively relative to the Ca^{2+}-containing model for ions such as Zn^{2+}, Y^{3+}, Sc^{3+}, and Mn^{3+}, but not for Sr^{2+}, which is attributed to the similar Lewis acidities of Sr^{2+} and Ca^{2+} [168, 169]. Arguing by analogy, it might be expected that one reason for Nature's selectivity for Ca^{2+} would be the optimization of redox potential range for the Mn_4CaO_5 cluster, an effect that was also implied by DFT calculations on an OEC model [170].

Additionally, experimental evidence in the form of peptide carbonyl frequency shifts upon Ca^{2+} substitution by other cations imply that hydrogen bonding interactions are affected by this type of modification of the OEC cluster [171]. Moreover, recent work on the higher states of the OEC suggests that Ca/Sr substitution perturbs the distribution of the conformational microstates in the S_3–S_0 transition, highlighting the role of calcium in structuring the environment of Y_Z in the S_3 state [54]. Given the proximity of Ca^{2+} to the redox-active tyrosine (Tyr161, Y_Z), and because a sequence of balanced redox potentials is necessary for efficient energy conversion, it is pertinent to ask how precisely Ca^{2+} influences the properties of Y_Z, besides its effect on the Mn cluster itself. A recent computational study examined the effect of Ca^{2+} on the acidity, arrangement, and ordering of the water molecules between Ca^{2+} and Y_Z [140]. The two major factors are: (1) influence of Ca^{2+} on the electronic structure and therefore the redox potential of Y_Z by the organization of hydrogen bonds towards the tyrosine and (2) modulation of the communication between Y_Z and the cluster through these water molecules. The absence or presence of water molecules, or equivalently the "structured polarity" of the hydrogen-bonding surrounding the phenolic oxygen, leads to significant changes in predicted spectroscopic parameters (*g* tensor of the tyrosyl radical) and large shifts in the

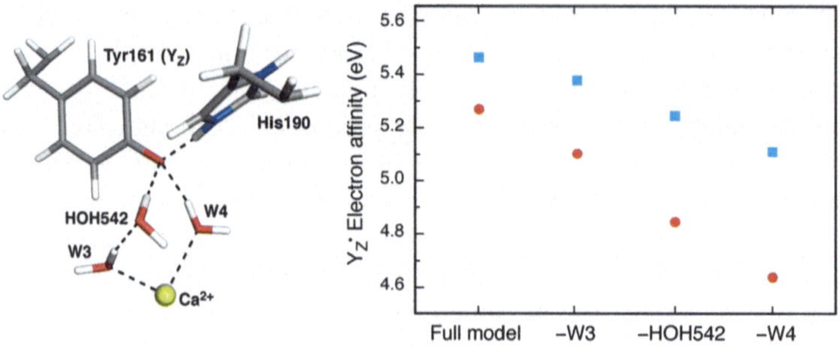

Fig. 7 Depiction of the Ca^{2+} ion of the OEC and the water molecules involved in the interaction of the ion with the redox-active tyrosine. The structure is taken from an optimized model of the $S_2Y_Z^{\bullet}$ state. The diagram shows the computed change in the electron affinity (*squares*: adiabatic; *circles*: vertical) of the tyrosyl radical upon successive loss of interaction with each ordered water molecule (adapted from Retegan et al. [140])

electron affinity of the residue of ca. 0.46/0.28 eV (vertical/adiabatic values). This demonstrates that Ca^{2+} plays a combined structural and electronic role by controlling the hydration environment of Y_Z, orienting the tyrosine ring, ordering the hydrogen-bonding network, and regulating the acidities of bound waters and the redox potential of the Y_Z residue in both its reduced and oxidized forms (Fig. 7) [140].

3.6 *Role of Chloride*

Two sites of Cl^- ions in close proximity to the active site have been identified in numerous crystal structures, again by substitution with the heavier and thus easier to identify Br^- and I^- ions [21, 172]. The role of Cl^- can be partially substituted by other negatively charged ions such as Br^-, NO_3^-, and NO_2^- at the cost of a lower efficiency, but probably not with iodide [172]. The binding sites are ca. 6.55 Å and 7.45 Å from the closest Mn ion of the OEC. Most studies have focused on the closer binding site, known as "binding site 1" (see Fig. 4). It is surrounded by Lys317, Asn181, and the backbone N of Glu333. Two water molecules (PDB IDs 517 and 518 in 4UB8) fill the space between this binding site, Asp61 and the W2 ligand to Mn4. A QM cluster study assigned the role of Cl^- mainly to its charge, with the effect that the last steps before product release are lowered in energy relative to Cl^--containing models [173]. In molecular dynamics studies, structural flexibility of the Cl^- ion and the amino acid residues in its vicinity was observed, with a salt bridge forming between Asp61 and Lys317 upon Cl^- depletion [37]. Because Asp61 is considered essential for proton removal from the catalytic site, chloride would serve to stabilize this function and would therefore be essential for the function of the proton relay network [174, 175]. Furthermore, a role of chloride in fine tuning the pK_a of Asp61 has been suggested [176].

The close connection to the Mn4 ligand W2 via one water molecule and the position in the immediate vicinity of the Mn1–Mn4 bridging entity Glu332-His333 are reminiscent of the organizational role that Ca^{2+} plays. Glu333, the D1 residue bridging Mn3 and Mn4, is suggested to interact with D2-Lys317 in the absence of Cl^-, so that Cl^- depletion would lead to a distorted geometry and hence altered magnetic interactions in the Mn_4Ca cluster [176–178].

4 Biological Water Oxidation Mechanism

4.1 *Substrate Identification*

The origin of the substrates and the evolution of their protonation states in the consecutive steps of the catalytic cycle determine the details of the water oxidation mechanism at the atomic level. Because the substrate waters are present close to the OEC and water molecules play structural roles or participate in proton channels, it is impossible to identify the substrates directly from any source of structural information.

As complex as the OEC is to study experimentally and computationally, one immediate advantage over other bioinorganic catalysts is that the product it forms is gaseous. By exposing the active site to isotopically labeled water ($H_2{}^{18}O$) the composition of the final O_2 ($^{16}O^{16}O$, $^{16}O^{18}O$, $^{18}O^{18}O$) can be analyzed by mass spectrometry. Invaluable advances have been achieved by membrane-inlet mass spectrometry (MIMS) studies [2, 100], in which the exposure time of the protein sample to the labeled water is varied, from which the exchange rates of the substrate waters can be determined [95, 97]. Two substrate binding sites with different exchange rates have thus been identified. Their exchangeability varies as the catalytic cycle progresses: the slowly exchanging water, "W_{slow}", exchanges fastest in S_0 and slowest in S_1, and accelerates its exchange rate again as the cluster is oxidized to the S_2 state. Although no molecular model studies exist for direct comparison of $H_2O/OH^-/O^{2-}$ exchange rates to the natural system, it is unexpected to observe an increase in exchange rate by several orders of magnitude upon metal oxidation. The other water, "W_{fast}", cannot be detected in S_0 and S_1 [179], either because of kinetic limitations in the setup or because it is not yet bound to the cluster. As shown in Table 2, W_{fast} has a slower exchange rate in the S_3 state although still exchanging much more rapidly than W_{slow} in S_3. Combining the MIMS experiment with Ca^{2+}/Sr^{2+} exchange, it was found that W_{slow} is accelerated by substitution of Ca^{2+} with Sr^{2+}, whereas W_{fast} is unaffected [180]. It should be noted that MIMS experiments are not sensitive to the precise location of the water exchange sites and the protonation state of the exchanging O-entity. This requires input from other experimental techniques and theoretical modeling.

Table 2 Approximate exchange rates of substrate water molecules (s^{-1}) from spinach thylakoid membranes at 10 °C

	S_0	S_1	S_2	S_3
W_{fast}	–	–	120	35
W_{slow}	14	0.02	2	2

To observe changes that affect water molecules between S states, FTIR difference spectra in the frequency region of OH stretching vibrations are a valuable tool. Weakly H-bonded OH groups are easier to detect than strongly H-bonded ones, as the latter overlap with NH stretching vibrations [181]. By subtracting the spectra of two S states, the protein background is removed, revealing those changes associated with the S-state transition, e.g., the hydration of the OEC. The effect of dehydration of OEC samples was found to be largest for the S_2–S_3 and S_3–S_0 transitions, which was interpreted as caused by water binding to the active site during these transitions [181, 182]. The difference in one vibrational mode seen in the S_2–S_1 difference spectrum upon suspension of PS-II in an $H_2{}^{18}O$ buffer and the effect on the mode of Ca/Sr substitution also suggested that a candidate for W_{slow} could be an Mn–O–Ca bridge of the inorganic core [183, 184].

A unique contribution to the effort towards substrate identification at the atomic level comes from EPR techniques. Specifically, high-field ^{17}O-ELDOR-detected NMR (EDNMR) and ^{17}O-ENDOR has been used to probe exchangeable ^{17}O species [184, 185]. Three types of ^{17}O nuclei were observed (μ-oxo bridge, terminal Mn–OH or Mn–OH_2, and Ca-bound or second-shell water). Experimentally, either O4 or O5 could correspond to the exchangeable μ-oxo bridge, but DFT models favored the assignments shown in Fig. 8, that is, with O5 as the exchangeable bridge. Further studies relied on the use of ammonia, a substrate analogue that can access the OEC and binds directly to the Mn_4Ca cluster by ligand substitution. Although the replacement of a μ-oxo bridge by a deprotonated form of ammonia is still considered as a viable scenario [186], experimental and theoretical studies have come to the conclusion that NH_3 most likely binds to the cluster by displacing W1 [139, 187, 188], stabilized at this position by hydrogen bonding (Fig. 8). By observing the response of the EDNMR signal of the exchangeable bridge to the presence of both NH_3 binding and Ca/Sr replacement, these studies strongly support the assignment of O5 as the slowly exchanging water binding site.

An exchange rate similar to that of W_{slow} is still much faster than exchange rates observed for μ-oxo bridges in high-valent Mn systems [185, 189]. This may be because there exists so far no synthetic model that mimics all structural features of the OEC and its environment. The structural flexibility observed in the S_2 state, enabled through the dynamic nature of the bonding between O5 and the terminal Mn ions, may be relevant in this respect [137], and recent DFT simulations of water exchange have produced computed barriers consistent with observed exchange rates [190].

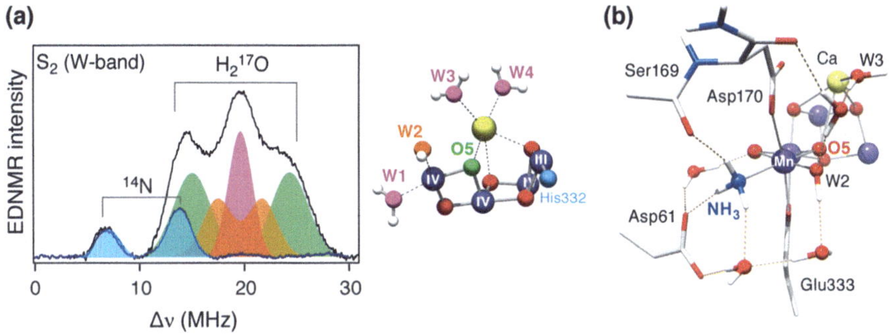

Fig. 8 (**a**) S_2-state ^{17}O-ELDOR-detected NMR spectra (94 GHz) of the OEC exchanged in $H_2{}^{17}O$ and color-coded assignments of exchangeable H_2O/OH^- ligands (adapted with permission from Cox et al. [5], copyright 2013 American Chemical Society); (**b**) DFT model for the binding of NH_3 by W1 displacement in the S_2 state (adapted from Lohmiller et al. [139])

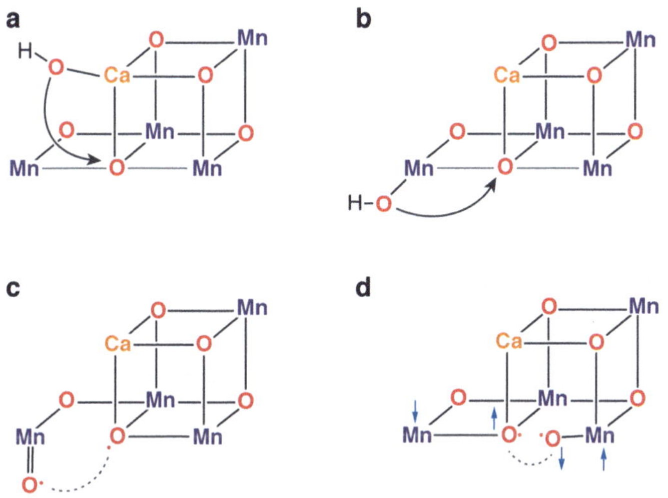

Fig. 9 (a–d) Possible schemes for O–O bond formation in PS-II

4.2 *O–O Bond Formation*

It would be impossible to review here all the different mechanistic proposals that have appeared over the years and all the variants of strictly hypothetical or computationally defined models for the critical steps in biological water oxidation. We refer the interested reader to the numerous reviews dealing with parts of this enormous subject, e.g., [3, 15, 99]. Here we focus exclusively on some of the possibilities regarding the O–O bond formation step, as constrained by the information presented in previous sections. The assignment of O5 as a substrate considerably limits the number of models that can be imagined, as shown in Fig. 9. The four models depicted here do not exclude variations in the bonding topology for each scheme, for example, in whether a terminal OH group is associated with Mn1 or Mn4. However, mechanisms that can be excluded if the O5 bridge is a substrate are those that would involve nucleophilic attack of a Ca-bound OH/H_2O on a terminal Mn4-oxo, and those that would involve direct coupling of terminal Mn4-oxo groups.

The models in Fig. 9a, b involve nucleophilic attack on O5. Nucleophilic attack of a Ca-bound OH or H_2O on a terminal high-valent oxo group, Mn(V)–O (oxo) or Mn(IV)–O$^•$ (oxyl) [191] has been discussed extensively in the past [127, 165, 192, 193], but the assignment of O5 as a substrate is not consistent with this idea as it requires this bridge and not a terminal oxo group to play the role of electrophile. A variation of the O–O bond formation scenario depicted in Fig. 9a, b would involve attack on O5 not by an Mn or Ca-bound OH/H_2O ligand but by a second-sphere water molecule positioned appropriately by hydrogen bonding interactions. We are not aware of any theoretical studies that explicitly evaluate and compare all these proposals in terms of their electronic, thermodynamic, and kinetic feasibility using realistic OEC models. Of these possibilities, an Mn-bound attacking group (Fig. 9b) would appear to be more consistent with the exchange kinetics described above.

The models in Fig. 9c, d involve radical mechanisms: in the first case the O–O bond is formed between a terminal Mn4(IV)–O$^•$ radical and the O5 bridge that is supposed to acquire radical character in the S_4 state [194]. The last scheme (Fig. 9d) corresponds to a model elaborated by Siegbahn in a long process of mechanistic refinement aimed at yielding the lowest possible energy barriers. After examining several possible models and O–O bond formation pathways, the most favorable scenario was identified as a direct coupling of an Mn1(IV)–O$^•$ radical [195] with the μ-oxo bridge (O5) bound to Mn4 (Fig. 9d) [72, 73, 196]. This structure is entirely consistent with the S_3 structure shown to be a best match with experimental observations (see Fig. 5) by independent calculations of spectroscopic properties [70, 78]. In this scenario, the oxyl radical on Mn1 originates from water that binds as OH^- upon formation of the S_3 state. An important requirement for the O–O bond formation to proceed with a low barrier is an alternate-spin alignment of the two oxygen atoms and the two terminal Mn ions, as shown in Fig. 9d.

The above coupling model represents a convincing and self-consistent hypothesis. However, several aspects of the mechanism of water oxidation at the later stages of the catalytic cycle still require more extensive studies to be sufficiently understood and, most important, to receive solid experimental support.

5 Summary

Water oxidation in biological photosynthesis relies on a complex arrangement of specialized components and finely orchestrated processes. Photosystem II is the "engine of life" that couples sunlight collection with charge separation and water oxidation. The catalytic oxygen evolving complex is composed of four oxo-bridged high-valent Mn ions. Although the cluster resembles a piece of Mn oxide "rock," it features structural flexibility which might be important for mechanistic regulation in the form of substrate inclusion or electronic control. The catalytic cluster self-assembles from Mn(II) ions by a photo-driven process, and the PS-II enzyme itself has a number of mechanisms both to avoid damage as long as possible and to repair its sensitive components when damage does occur. The Mn ions have localized unpaired electrons and local high-spin configurations in all catalytic states that have been probed experimentally, and the magnetic coupling between the Mn ions leads to a characteristic evolution of ground spin states from low-spin early in the cycle to high-spin at the later stages. The protein matrix plays an important functional role by optimizing accessibility of reactants and removal of products, but also by maintaining an efficient proton-removal network crucial for achieving an almost constant redox potential for four consecutive oxidations of the cluster. Integral (calcium) and peripheral (chloride) cofactors optimize the electronic properties and control the reactivity of the cluster. An all-Mn(IV) state is reached before the final catalytic step, implying formation of an Mn(V) oxo or an Mn(IV) oxyl species in the final step of the catalytic cycle leading to evolution of dioxygen. According to quantum chemical calculations, O–O bond formation most likely proceeds by a radical process. However, experimental proof of the final stages of biological water oxidation is still hard to obtain. Structural aspects of the biological water oxidizing complex, such as the cuboidal Mn_3Ca motif, are already partly realized in heterogeneous Mn/Ca oxides which show water oxidizing behavior [197], but without displaying catalytic performance comparable to that of the biological system. It remains to be seen whether features of the natural catalyst such as the structural flexibility, the switching from low-spin to activated high-spin states, and the exquisite control of proton transfer can be, or indeed need to be, implemented in synthetic manganese-based systems for efficient water oxidation.

References

1. Blankenship RE (2001) Molecular mechanisms of photosynthesis. Blackwell, Oxford
2. Hillier W, Messinger J (2005) Mechanism of photosynthetic oxygen production. In: Wydrzynski T, Satoh K (eds) Photosystem II. The light-driven water:plastoquinone oxido-reductase, vol 22, Advances in photosynthesis and respiration. Springer, Dordrecht, pp 567–608
3. McEvoy JP, Brudvig GW (2006) Chem Rev 106:4455–4483
4. Messinger J, Renger G (2008) Photosynthetic water splitting. In: Renger G (ed) Primary processes of photosynthesis, part 2: principles and apparatus, vol 9. The Royal Society of Chemistry, Cambridge, pp 291–349
5. Cox N, Pantazis DA, Neese F, Lubitz W (2013) Acc Chem Res 46:1588–1596
6. Pantazis DA, Cox N, Lubitz W, Neese F (2014) Oxygen-evolving photosystem II. In: Scott RA (ed) Encyclopedia of inorganic and bioinorganic chemistry. Wiley. doi:10.1002/9781119951438.eibc2166
7. Vinyard DJ, Ananyev GM, Dismukes GC (2013) Annu Rev Biochem 82:577–606
8. Lubitz W, Reijerse EJ, Messinger J (2008) Energy Environ Sci 1:15–31

9. Faunce TA, Lubitz W, Rutherford AW, MacFarlane D, Moore GF, Yang P, Nocera DG, Moore TA, Gregory DH, Fukuzumi S, Yoon KB, Armstrong FA, Wasielewski MR, Styring S (2013) Energy Environ Sci 6:695–698
10. Lewis NS, Nocera DG (2006) Proc Natl Acad Sci U S A 103:15729–15735
11. Nocera DG (2012) Acc Chem Res 45:767–776
12. Schlögl R (2010) ChemSusChem 3:209–222
13. Cogdell RJ, Gardiner AT, Cronin L (2012) Philos Trans R Soc A 370:3819–3826
14. Dau H, Zaharieva I (2009) Acc Chem Res 42:1861–1870
15. Dau H, Limberg C, Reier T, Risch M, Roggan S, Strasser P (2010) ChemCatChem 2:724–761
16. Zouni A, Witt HT, Kern J, Fromme P, Krauss N, Saenger W, Orth P (2001) Nature 409: 739–743
17. Kamiya N, Shen J-R (2003) Proc Natl Acad Sci U S A 100:98–103
18. Ferreira KN, Iverson TM, Maghlaoui K, Barber J, Iwata S (2004) Science 303:1831–1838
19. Biesiadka J, Loll B, Kern J, Irrgang K-D, Zouni A (2004) Phys Chem Chem Phys 6: 4733–4736
20. Loll B, Kern J, Saenger W, Zouni A, Biesiadka J (2005) Nature 438:1040–1044
21. Guskov A, Kern J, Gabdulkhakov A, Broser M, Zouni A, Saenger W (2009) Nat Struct Mol Biol 16:334–342
22. Umena Y, Kawakami K, Shen J-R, Kamiya N (2011) Nature 473:55–60
23. Suga M, Akita F, Hirata K, Ueno G, Murakami H, Nakajima Y, Shimizu T, Yamashita K, Yamamoto M, Ago H, Shen J-R (2014) Nature 517:99–103
24. Becker K, Cormann KU, Nowaczyk MM (2011) J Photochem Photobiol B 104:204–211
25. Shi L-X, Hall M, Funk C, Schröder WP (2012) Biochim Biophys Acta Bioenerg 1817:13–25
26. Fagerlund RD, Eaton-Rye JJ (2011) J Photochem Photobiol B 104:191–203
27. Wydrzynski T, Satoh K (eds) (2005) Photosystem II. The light-driven water:plastoquinone oxidoreductase, vol 22. Springer, Dordrecht
28. Rappaport F, Diner BA (2008) Coord Chem Rev 252:259–272
29. Diner BA, Rappaport F (2002) Annu Rev Plant Biol 53:551–580
30. Murray JW, Barber J (2007) J Struct Biol 159:228–237
31. Ho FM, Styring S (2008) Biochim Biophys Acta Bioenerg 1777:140–153
32. Ho FM (2008) Photosynth Res 98:503–522
33. Gabdulkhakov A, Guskov A, Broser M, Kern J, Müh F, Saenger W, Zouni A (2009) Structure 17:1223–1234
34. Ishikita H, Saenger W, Loll B, Biesiadka J, Knapp E-W (2006) Biochemistry 45:2063–2071
35. Vassiliev S, Comte P, Mahboob A, Bruce D (2010) Biochemistry 49:1873–1881
36. Vassiliev S, Zaraiskaya T, Bruce D (2012) Biochim Biophys Acta Bioenerg 1817:1671–1678
37. Rivalta I, Amin M, Luber S, Vassiliev S, Pokhrel R, Umena Y, Kawakami K, Shen JR, Kamiya N, Bruce D, Brudvig GW, Gunner MR, Batista VS (2011) Biochemistry 50: 6312–6315
38. Najafpour MM, Fekete M, Sedigh DJ, Aro E-M, Carpentier R, Eaton-Rye JJ, Nishihara H, Shen J-R, Allakhverdiev SI, Spiccia L (2015) ACS Catal 1499–1512
39. Cardona T, Sedoud A, Cox N, Rutherford AW (2012) Biochim Biophys Acta Bioenerg 1817:26–43
40. Niyogi KK (2000) Curr Opin Plant Biol 3:455–460
41. Pascal AA, Liu Z, Broess K, van Oort B, van Amerongen H, Wang C, Horton P, Robert B, Chang W, Ruban A (2005) Nature 436:134–137
42. Derks A, Schaven K, Bruce D (2015) Biochim Biophys Acta Bioenerg 1847:468–485
43. Joliot P, Barbieri G, Chabaud R (1969) Photochem Photobiol 10:309–329
44. Kok B, Forbush B, McGloin M (1970) Photochem Photobiol 11:457–475
45. Dau H, Haumann M (2008) Coord Chem Rev 252:273–295
46. Klauss A, Haumann M, Dau H (2012) Proc Natl Acad Sci U S A 109:16035–16040
47. Dau H, Haumann M (2007) Biochim Biophys Acta Bioenerg 1767:472–483
48. Klauss A, Haumann M, Dau H (2015) J Phys Chem B 119:2677–2689

49. Lavergne J, Junge W (1993) Photosynth Res 38:279–296
50. Rappaport F, Lavergne J (2001) Biochim Biophys Acta Bioenerg 1503:246–259
51. Klauss A, Krivanek R, Dau H, Haumann M (2009) Photosynth Res 102:499–509
52. Haumann M, Liebisch P, Müller C, Barra M, Grabolle M, Dau H (2005) Science 310: 1019–1021
53. Gerencsér L, Dau H (2010) Biochemistry 49:10098–10106
54. Rappaport F, Ishida N, Sugiura M, Boussac A (2011) Energy Environ Sci 4:2520–2524
55. Bao H, Dilbeck P, Burnap R (2013) Photosynth Res 116:215–229
56. Service RJ, Hillier W, Debus RJ (2014) Biochemistry 53:1001–1017
57. Bondar A-N, Dau H (2012) Biochim Biophys Acta Bioenerg 1817:1177–1190
58. Vogt L, Vinyard DJ, Khan S, Brudvig GW (2015) Curr Opin Chem Biol 25:152–158
59. Kirby JA, Robertson AS, Smith JP, Thompson AC, Cooper SR, Klein MP (1981) J Am Chem Soc 103:5529–5537
60. Yano J, Pushkar Y, Glatzel P, Lewis A, Sauer K, Messinger J, Bergmann U, Yachandra V (2005) J Am Chem Soc 127:14974–14975
61. Grundmeier A, Dau H (2012) Biochim Biophys Acta Bioenerg 1817:88–105
62. Glöckner C, Kern J, Broser M, Zouni A, Yachandra V, Yano J (2013) J Biol Chem 288: 22607–22620
63. Yano J, Yachandra V (2014) Chem Rev 114:4175–4205
64. Yano J, Kern J, Sauer K, Latimer MJ, Pushkar Y, Biesiadka J, Loll B, Saenger W, Messinger J, Zouni A, Yachandra VK (2006) Science 314:821–825
65. Dau H, Liebisch P, Haumann M (2004) Phys Chem Chem Phys 6:4781–4792
66. Grabolle M, Haumann M, Müller C, Liebisch P, Dau H (2006) J Biol Chem 281:4580–4588
67. Yano J, Kern J, Irrgang K-D, Latimer MJ, Bergmann U, Glatzel P, Pushkar Y, Biesiadka J, Loll B, Sauer K, Messinger J, Zouni A, Yachandra VK (2005) Proc Natl Acad Sci U S A 102: 12047–12052
68. Luber S, Rivalta I, Umena Y, Kawakami K, Shen JR, Kamiya N, Brudvig GW, Batista VS (2011) Biochemistry 50:6308–6311
69. Galstyan A, Robertazzi A, Knapp EW (2012) J Am Chem Soc 134:7442–7449
70. Krewald V, Retegan M, Cox N, Messinger J, Lubitz W, DeBeer S, Neese F, Pantazis DA (2015) Chem Sci 6:1676–1695
71. Askerka M, Vinyard DJ, Wang J, Brudvig GW, Batista VS (2015) Biochemistry 54: 1713–1716
72. Siegbahn PEM (2008) Chem Eur J 14:8290–8302
73. Siegbahn PEM (2009) Acc Chem Res 42:1871–1880
74. Zimmermann JL, Rutherford AW (1984) Biochim Biophys Acta Bioenerg 767:160–167
75. Brudvig GW, Casey JL, Sauer K (1983) Biochim Biophys Acta Bioenerg 723:366–371
76. Baldwin MJ, Stemmler TL, Riggs-Gelasco PJ, Kirk ML, Penner-Hahn JE, Pecoraro VL (1994) J Am Chem Soc 116:11349–11356
77. Krewald V, Lassalle-Kaiser B, Boron TT, Pollock CJ, Kern J, Beckwith MA, Yachandra VK, Pecoraro VL, Yano J, Neese F, DeBeer S (2013) Inorg Chem 52:12904–12914
78. Cox N, Retegan M, Neese F, Pantazis DA, Boussac A, Lubitz W (2014) Science 345:804–808
79. Dismukes GC, Siderer Y (1981) Proc Natl Acad Sci U S A 78:274–278
80. Penner-Hahn JE, Fronko RM, Pecoraro VL, Yocum CF, Betts SD, Bowlby NR (1990) J Am Chem Soc 112:2549–2557
81. Zheng M, Dismukes GC (1996) Inorg Chem 35:3307–3319
82. Roelofs TA, Liang W, Latimer MJ, Cinco RM, Rompel A, Andrews JC, Sauer K, Yachandra VK, Klein MP (1996) Proc Natl Acad Sci U S A 93:3335–3340
83. Iuzzolino L, Dittmer J, Dörner W, Meyer-Klaucke W, Dau H (1998) Biochemistry 37: 17112–17119
84. Dau H, Liebisch P, Haumann M (2003) Anal Bioanal Chem 376:562–583
85. Ono T-A, Noguchi T, Inoue Y, Kusunoki M, Matsushita T, Oyanagi H (1992) Science 258: 1335–1337

86. Hansson O, Andreasson LE (1982) Biochim Biophys Acta 679:261–268
87. De Paula JC, Brudvig GW (1985) J Am Chem Soc 107:2643–2648
88. De Paula JC, Beck WF, Brudvig GW (1986) J Am Chem Soc 108:4002–4009
89. de Paula JC, Beck WF, Miller A-F, Wilson RB, Brudvig GW (1987) J Chem Soc Faraday Trans 83:3635–3651
90. Pace RJ, Jin L, Stranger R (2012) Dalton Trans 41:11145–11160
91. Dasgupta J, Ananyev GM, Dismukes GC (2008) Coord Chem Rev 252:347–360
92. Kolling DRJ, Cox N, Ananyev GM, Pace RJ, Dismukes GC (2012) Biophys J 103:313–322
93. Kuntzleman T, Yocum CF (2005) Biochemistry 44:2129–2142
94. Messinger J, Seaton G, Wydrzynski T, Wacker U, Renger G (1997) Biochemistry 36: 6862–6873
95. Messinger J, Badger M, Wydrzynski T (1995) Proc Natl Acad Sci U S A 92:3209–3213
96. Beckmann K, Messinger J, Badger M, Wydrzynski T, Hillier W (2009) Photosynth Res 102: 511–522
97. Cox N, Messinger J (2013) Biochim Biophys Acta Bioenerg 1827:1020–1030
98. Hillier W, Wydrzynski T (2001) Biochim Biophys Acta Bioenerg 1503:197–209
99. Messinger J (2004) Phys Chem Chem Phys 6:4764–4771
100. Hillier W, Wydrzynski T (2008) Coord Chem Rev 252:306–317
101. Haddy A (2007) Photosynth Res 92:357–368
102. Åhrling KA, Peterson S, Styring S (1997) Biochemistry 36:13148–13152
103. Messinger J, Nugent JHA, Evans MCW (1997) Biochemistry 36:11055–11060
104. Messinger J, Robblee JH, Yu WO, Sauer K, Yachandra VK, Klein MP (1997) J Am Chem Soc 119:11349–11350
105. Åhrling KA, Peterson S, Styring S (1998) Biochemistry 37:8115–8120
106. Koulougliotis D, Hirsh DJ, Brudvig GW (1992) J Am Chem Soc 114:8322–8323
107. Hsieh W-Y, Campbell KA, Gregor W, Britt RD, Yoder DW, Penner-Hahn JE, Pecoraro VL (2004) Biochim Biophys Acta Bioenerg 1655:149–157
108. Zimmermann JL, Rutherford AW (1986) Biochemistry 25:4609–4615
109. Haddy A, Lakshmi KV, Brudvig GW, Frank HA (2004) Biophys J 87:2885–2896
110. Horner O, Rivière E, Blondin G, Un S, Rutherford AW, Girerd J-J, Boussac A (1998) J Am Chem Soc 120:7924–7928
111. Ioannidis N, Petrouleas V (2000) Biochemistry 39:5246–5254
112. Boussac A, Sugiura M, Rutherford AW, Dorlet P (2009) J Am Chem Soc 131:5050–5051
113. Kulik LV, Epel B, Lubitz W, Messinger J (2005) J Am Chem Soc 127:2392–2393
114. Kulik LV, Epel B, Lubitz W, Messinger J (2007) J Am Chem Soc 129:13421–13435
115. Sauer K, Yano J, Yachandra VK (2005) Photosynth Res 85:73–86
116. Sauer K, Yano J, Yachandra VK (2008) Coord Chem Rev 252:318–335
117. Yachandra VK, DeRose VJ, Latimer MJ, Mukerji I, Sauer K, Klein MP (1993) Science 260: 675–679
118. Gatt P, Stranger R, Pace RJ (2011) J Photochem Photobiol B Biol 104:80–93
119. Jaszewski AR, Petrie S, Pace RJ, Stranger R (2011) Chem Eur J 17:5699–5713
120. Jaszewski AR, Stranger R, Pace RJ (2011) J Phys Chem B 115:4484–4499
121. Petrie S, Stranger R, Pace RL (2008) Chem Eur J 14:5482–5494
122. Gatt P, Petrie S, Stranger R, Pace RJ (2012) Angew Chem Int Ed 51:12025–12028
123. Petrie S, Gatt P, Stranger R, Pace RJ (2012) Phys Chem Chem Phys 14:11333–11343
124. Siegbahn PEM (2013) Biochim Biophys Acta Bioenerg 1827:1003–1019
125. Kusunoki M (2007) Biochim Biophys Acta Bioenerg 1767:484–492
126. Sproviero EM, Gascon JA, McEvoy JP, Brudvig GW, Batista VS (2008) J Am Chem Soc 130:6728–6730
127. Sproviero EM, Gascon JA, McEvoy JP, Brudvig GW, Batista VS (2008) J Am Chem Soc 130:3428–3442
128. Kusunoki M (2011) Photochem Photobiol B 104:100–110

129. Saito T, Yamanaka S, Kanda K, Isobe H, Takano Y, Shigeta Y, Umena Y, Kawakami K, Shen JR, Kamiya N, Okumura M, Shoji M, Yoshioka Y, Yamaguchi K (2012) Int J Quantum Chem 112:253–276
130. Pantazis DA, Orio M, Petrenko T, Zein S, Lubitz W, Messinger J, Neese F (2009) Phys Chem Chem Phys 11:6788–6798
131. Yamaguchi K, Isobe H, Yamanaka S, Saito T, Kanda K, Shoji M, Umena Y, Kawakami K, Shen JR, Kamiya N, Okumura M (2012) Int J Quantum Chem 113:525–541
132. Bovi D, Narzi D, Guidoni L (2013) Angew Chem Int Ed 52:11744–11749
133. Schinzel S, Schraut J, Arbuznikov AV, Siegbahn PEM, Kaupp M (2010) Chem Eur J 16: 10424–10438
134. Ames W, Pantazis DA, Krewald V, Cox N, Messinger J, Lubitz W, Neese F (2011) J Am Chem Soc 133:19743–19757
135. Cox N, Rapatskiy L, Su J-H, Pantazis DA, Sugiura M, Kulik L, Dorlet P, Rutherford AW, Neese F, Boussac A, Lubitz W, Messinger J (2011) J Am Chem Soc 133:3635–3648
136. Su J-H, Cox N, Ames W, Pantazis DA, Rapatskiy L, Lohmiller T, Kulik LV, Dorlet P, Rutherford AW, Neese F, Boussac A, Lubitz W, Messinger J (2011) Biochim Biophys Acta Bioenerg 1807:829–840
137. Pantazis DA, Ames W, Cox N, Lubitz W, Neese F (2012) Angew Chem Int Ed 51:9935–9940
138. Retegan M, Neese F, Pantazis DA (2013) J Chem Theory Comput 9:3832–3842
139. Lohmiller T, Krewald V, Pérez Navarro M, Retegan M, Rapatskiy L, Nowaczyk MM, Boussac A, Neese F, Lubitz W, Pantazis DA, Cox N (2014) Phys Chem Chem Phys 16: 11877–11892
140. Retegan M, Cox N, Lubitz W, Neese F, Pantazis DA (2014) Phys Chem Chem Phys 16: 11901–11910
141. Zein S, Kulik LV, Yano J, Kern J, Pushkar Y, Zouni A, Yachandra VK, Lubitz W, Neese F, Messinger J (2008) Philos Trans R Soc B 363:1167–1177
142. Orio M, Pantazis DA, Neese F (2009) Photosynth Res 102:443–453
143. Pantazis DA, Orio M, Petrenko T, Zein S, Bill E, Lubitz W, Messinger J, Neese F (2009) Chem Eur J 15:5108–5123
144. Neese F, Ames W, Christian G, Kampa M, Liakos DG, Pantazis DA, Roemelt M, Surawatanawong P, Ye SF (2010) Adv Inorg Chem 62:301–349
145. Neese F (2009) Coord Chem Rev 253:526–563
146. DeBeer George S, Petrenko T, Neese F (2008) J Phys Chem A 112:12936–12943
147. Kusunoki M, Ono T, Noguchi T, Inoue Y, Oyanagi H (1993) Photosynth Res 38:331–339
148. Messinger J, Robblee JH, Bergmann U, Fernandez C, Glatzel P, Visser H, Cinco RM, McFarlane KL, Bellacchio E, Pizarro SA, Cramer SP, Sauer K, Klein MP, Yachandra VK (2001) J Am Chem Soc 123:7804–7820
149. Haumann M, Müller C, Liebisch P, Iuzzolino L, Dittmer J, Grabolle M, Neisius T, Meyer-Klaucke W, Dau H (2005) Biochemistry 44:1894–1908
150. Peloquin JM, Campbell KA, Randall DW, Evanchik MA, Pecoraro VL, Armstrong WH, Britt RD (2000) J Am Chem Soc 122:10926–10942
151. Peloquin JM, Britt RD (2001) Biochim Biophys Acta Bioenerg 1503:96–111
152. Kulik L, Lubitz W (2009) Photosynth Res 102:391–401
153. Hasegawa K, Ono T, Inoue Y, Kusunoki M (1999) Bull Chem Soc Jpn 72:1013–1023
154. Hasegawa K, Ono T-A, Inoue Y, Kusunoki M (1999) Chem Phys Lett 300:9–19
155. Barber J, Murray JW (2008) Philos Trans R Soc B 363:1129–1137
156. Krewald V, Neese F, Pantazis DA (2013) J Am Chem Soc 135:5726–5739
157. Isobe H, Shoji M, Yamanaka S, Umena Y, Kawakami K, Kamiya N, Shen JR, Yamaguchi K (2012) Dalton Trans 41:13727–13740
158. Narzi D, Bovi D, Guidoni L (2014) Proc Natl Acad Sci U S A 111:8723–8728
159. Ghanotakis DF, Babcock GT, Yocum CF (1984) FEBS Lett 167:127–130
160. Boussac A, Rutherford AW (1988) Biochemistry 27:3476–3483

161. Ishida N, Sugiura M, Rappaport F, Lai T-L, Rutherford AW, Boussac A (2008) J Biol Chem 283:13330–13340
162. Boussac A, Rappaport F, Carrier P, Verbavatz J-M, Gobin R, Kirilovsky D, Rutherford AW, Sugiura M (2004) J Biol Chem 279:22809–22819
163. Brudvig GW (2008) Philos Trans R Soc B 363:1211–1219
164. Lee C-I, Lakshmi KV, Brudvig GW (2007) Biochemistry 46:3211–3223
165. Vrettos JS, Limburg J, Brudvig GW (2001) Biochim Biophys Acta Bioenerg 1503:229–245
166. McEvoy JP, Brudvig GW (2004) Phys Chem Chem Phys 6:4754–4763
167. Vrettos JS, Stone DA, Brudvig GW (2001) Biochemistry 40:7937–7945
168. Tsui EY, Agapie T (2013) Proc Natl Acad Sci U S A 110:10084–10088
169. Tsui EY, Kanady JS, Agapie T (2013) Inorg Chem 52:13833–13848
170. Siegbahn PEM (2014) Phys Chem Chem Phys 16:11893–11900
171. Polander BC, Barry BA (2013) J Phys Chem Lett 786–791
172. Kawakami K, Umena Y, Kamiya N, Shen J-R (2009) Proc Natl Acad Sci U S A 106: 8567–8572
173. Siegbahn PEM (2009) Dalton Trans 10063–10068
174. Olesen K, Andréasson L-E (2003) Biochemistry 42:2025–2035
175. Amin M, Vogt L, Szejgis W, Vassiliev S, Brudvig GW, Bruce D, Gunner MR (2015) J Phys Chem B 119:7366–7377
176. Pokhrel R, McConnell IL, Brudvig GW (2011) Biochemistry 50:2725–2734
177. Pokhrel R, Service RJ, Debus RJ, Brudvig GW (2013) Biochemistry 52:4758–4773
178. van Vliet P, Rutherford AW (1996) Biochemistry 35:1829–1839
179. Hillier W, Wydrzynski T (2004) Phys Chem Chem Phys 6:4882–4889
180. Hendry G, Wydrzynski T (2003) Biochemistry 42:6209–6217
181. Noguchi T (2008) Philos Trans R Soc B 363:1189–1195
182. Noguchi T, Sugiura M (2002) Biochemistry 41:15706–15712
183. Chu H-A, Sackett H, Babcock GT (2000) Biochemistry 39:14371–14376
184. Rapatskiy L, Cox N, Savitsky A, Ames WM, Sander J, Nowaczyk MM, Rögner M, Boussac A, Neese F, Messinger J, Lubitz W (2012) J Am Chem Soc 134:16619–16634
185. McConnell IL, Grigoryants VM, Scholes CP, Myers WK, Chen P-Y, Whittaker JW, Brudvig GW (2012) J Am Chem Soc 134:1504–1512
186. Pokhrel R, Brudvig G (2014) Phys Chem Chem Phys 16:11812–11821
187. Pérez Navarro M, Ames WM, Nilsson H, Lohmiller T, Pantazis DA, Rapatskiy L, Nowaczyk MM, Neese F, Boussac A, Messinger J, Lubitz W, Cox N (2013) Proc Natl Acad Sci U S A 110:15561–15566
188. Schraut J, Kaupp M (2014) Chem Eur J 20:7300–7308
189. Tagore R, Chen H, Crabtree RH, Brudvig GW (2006) J Am Chem Soc 128:9457–9465
190. Siegbahn PEM (2013) J Am Chem Soc 135:9442–9449
191. Yamaguchi K, Takahara Y, Fueno T (1986) Ab-initio molecular orbital studies of structure and reactivity of transition metal-oxo compounds. In: Smith VH Jr, Scheafer HF III, Morokuma K (eds) Applied quantum chemistry. Reidel, Boston, pp 155–184
192. Pecoraro VL, Baldwin MJ, Caudle MT, Hsieh W-Y, Law NA (1998) Pure Appl Chem 70: 925–929
193. Barber J, Ferreira K, Maghlaoui K, Iwata S (2004) Phys Chem Chem Phys 6:4737–4742
194. Yamaguchi K, Yamanaka S, Isobe H, Tanaka K, Ueyama N (2012) Int J Quantum Chem 112: 3849–3866
195. Siegbahn PEM, Crabtree RH (1999) J Am Chem Soc 121:117–127
196. Siegbahn PEM (2006) Chem Eur J 12:9217–9227
197. Shevela D, Koroidov S, Najafpour MM, Messinger J, Kurz P (2011) Chem Eur J 17: 5415–5423

Top Curr Chem (2016) 371: 49–72
DOI: 10.1007/128_2015_634

Published online: 16 May 2015

Biomimetic Water-Oxidation Catalysts: Manganese Oxides

Philipp Kurz

Abstract The catalytic oxidation of water to molecular oxygen is a key process for the production of solar fuels. Inspired by the biological manganese-based active site for this reaction in the enzyme Photosystem II, researchers have made impressive progress in the last decades regarding the development of synthetic manganese catalysts for water oxidation. For this, it has been especially fruitful to explore the many different types of known manganese oxides MnO_x.

This chapter first offers an overview of the structural, thermodynamic, and mechanistic aspects of water-oxidation catalysis by MnO_x. The different test systems used for catalytic studies are then presented together with general reactivity trends. As a result, it has been possible to identify layered, mixed $Mn^{III/IV}$-oxides as an especially promising class of bio-inspired catalysts and an attempt is made to give structure-based reasons for the good performances of these materials.

In the outlook, the challenges of catalyst screenings (and hence the identification of a "best MnO_x catalyst") are discussed. There is a great variety of reaction conditions which might be relevant for the application of manganese oxide catalysts in technological solar fuel-producing devices, and thus catalyst improvements are currently still addressing a very large parameter space. Nonetheless, detailed knowledge about the biological catalyst and a solid experimental basis concerning the syntheses and water-oxidation reactivities of MnO_x materials have been established in the last decade and thus this research field is well positioned to make important contributions to solar fuel research in the future.

Keywords Artificial photosynthesis · Heterogeneous catalysis · Manganese · Oxides · Water oxidation

---

P. Kurz (✉)
Institut für Anorganische und Analytische Chemie, Albert-Ludwigs-Universität Freiburg, Albertstraße 21, 79104 Freiburg, Germany
e-mail: philipp.kurz@ac.uni-freiburg.de

Contents

1 Introduction

It is hardly necessary to mention that one of the challenges – if not the greatest challenge – facing researchers in chemistry today is the development of the compounds needed to make a large scale use of renewable energy by mankind possible [1]. Within this research field, the problem of finding a suitable way to store energy from abundant sources such as sunlight, wind or geothermal heat is central. When activities in this area gained momentum globally around the year 2000, a number of research teams rediscovered the concept of "artificial photosynthesis" [2, 3]. It had already been suggested in the 1970s (and even earlier) that the biological processes of photosynthesis provided a very good starting point for the development of systems to store renewable energy, namely in the form of high-energy chemical substances [4]. For oxygenic photosynthesis, the converted energy form is sunlight (the most abundant source of renewable energy on Earth) and the substances produced for energy storage are reduced forms of carbon [5]. Such "$C^{<+IV}$-compounds" make up the bulk of the enormous amounts of biomass found on the surface of the Earth, e.g. as marine algae, forests or grasslands. To represent this very complex and diverse process by a simple chemical equation, the reduction of carbon dioxide to glucose is often given (1) [5], although, of course, the chemical composition of entire organisms is much more complicated.

$$6C^{+IV}O_2 + 12H_2O^{-II} \rightarrow C_6{}^{0}H_{12}O_6 + 6O_2{}^{0} + 6H_2O \quad (1)$$

Energetically, reaction (1) is a significantly uphill process ($\Delta H \sim +3\ \mathrm{MJ \cdot mol^{-1}}$) and this energy is obtained by photosynthesizing organisms from sunlight via an amazingly optimized sequence of light-harvesting reactions of photosynthetic dyes. As also indicated in (1), the chemical reaction used in biology to store energy is a multi-electron redox reaction where the two half-reactions are the reduction of carbon dioxide and the oxidation of water. In consequence, efficient electron transfer and redox catalysis are essential.

As simple as the overall reaction of (1) might look, it has been revealed by decades of scientific work that an extremely complex enzymatic machinery is at work in the cells to turn over the rather inert substrates CO_2 and H_2O [5]. Thus, it was already clear for the first scientists suggesting bio-inspired energy-conversion systems that, in a device built up of synthetic compounds, it would only be possible and realistic to mimic general features but never construct a "1:1 copy" of the biological reaction sequence [2–4].

By now, there are a number of different proposed designs for artificial photosynthesis systems, but in general it seems to be agreed that a bio-inspired system for solar energy conversion should be built up according to the following guidelines [2, 3, 6–8]:

- The aim is a device that uses the energy of sunlight for the synthesis of high-energy compounds, so-called "solar fuels"
- The conversion of solar energy occurs via light-driven redox reactions
- Similar to biological photosystems, the energy of more than one visible-light photon (two or three) per electron is used to drive charge separation and redox catalysis
- As in biology, the processes of light absorption and the oxidation/reduction catalyses are each carried out by different entities

Furthermore, to make the scale-up for global impact possible, many in the field additionally agree that:

- Water should be the reaction medium of choice
- The use of non-abundant elements has to be avoided
- The devices should be operated at moderate temperatures

To visualize this concept, a much generalised depiction of the components, reactions and electron flows of a bio-inspired artificial photosynthetic device is shown in Fig. 1.

In the following sections, this chapter only deals with recent developments concerning a single component of the scheme shown in Fig. 1, the water-oxidation catalyst Cat_{ox}. Additionally, this review is limited to a single class of compounds used as Cat_{ox}: manganese oxides. Even so, it is still only possible to offer a general introduction to this field here. Nevertheless, several reasons make this topic especially interesting in the author's opinion: (1) a manganese-oxido cluster has evolved

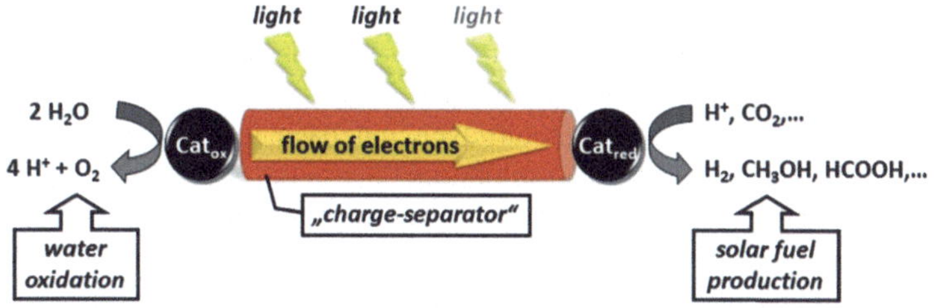

Fig. 1 Concept for an artificial photosynthesis system emphasising the importance of catalysts for water oxidation (Cat_{ox}) and substrate reduction (Cat_{red}). Depending on the design, charge separation normally requires the energy of two or three absorbed visible-light photons per transferred electron

over billions of years of evolution as biology's one and only catalyst for water oxidation, (2) manganese is an abundant, non-toxic element and thus Mn-based catalysts could indeed be suitable for technological use on a large scale and (3) because of a number of recent research projects, the synthesis and understanding of manganese oxide catalysts for water oxidation has seen very encouraging progress in recent years.

2 Key Features of the Water-Oxidation Reaction (WOR)

The half-reaction of the water-oxidation reaction (WOR), the process taking place on the left hand side of the scheme shown in Fig. 1, is given in (2):

$$2H_2O^{-\mathbf{II}} \rightarrow O_2{}^{\mathbf{0}} + 4H^+ + 4e^- \tag{2}$$

The standard redox potential at which the WOR can occur according to thermodynamics is $E^0 = +1.23$ V (*note*: all redox potentials in this chapter are given vs the normal hydrogen electrode, NHE) [9]. However, because of the complicated mechanism of this four-electron redox process, water oxidation is not observed at redox potentials close to E^0 in the absence of catalysts and significant overpotentials η of at least hundreds of millivolts are needed to achieve useful reaction rates, even in the presence of good WOR catalysts. Furthermore, as the oxidation of water yields protons as well as electrons (2), the thermodynamic potential of the WOR is strongly pH-dependent (3):

$$E(O_2/H_2O) = +1.23\,\text{V} - 59\,\text{mV}\,\text{pH} \tag{3}$$

Even without any further knowledge about the mechanistic details of the WOR, some important consequences for the properties of a water-oxidation catalyst can be drawn from (2) and (3):

- The transfer of four electrons has to be mastered by the catalyst to yield one O_2 product molecule
- Despite the redox chemistry alone, the catalyst most likely also has to facilitate the formation of the oxygen–oxygen bond
- Four protons are generated per turnover and have to be removed from the catalytic site to prevent "product inhibition"
- The WOR takes place at potentials where most organic but also many inorganic compounds are not stable (> +1 V for most pH values)
- The pH of the aqueous reaction system is a key parameter influencing the thermodynamic requirements

3 Biology's Manganese-Based Catalyst for the WOR

As shown in Fig. 1, a number of possible reduction reactions yielding different solar fuels are currently explored. On the other hand, the only realistic electron source for these reduction processes appears to be the water-oxidation reaction (WOR). Alternative reactions yielding reducing equivalents are either too energy intensive (e.g. the oxidation of H_2O to H_2O_2 or $OH^{\bullet}$-radicals require potentials of +1.76 or +2.85 V, respectively) [9] or require globally scarce substrates such as H_2S or Cl^-.

In consequence, photosynthesizing organisms also had to develop a WOR-catalyst to succeed on a global scale and the result is the oxygen-evolving complex (OEC) of the enzyme complex Photosystem II (PSII) [10]. The OEC is able to catalyse reaction (2) at a rate of about 100 turnovers per second. It first loses electrons to its primary oxidation agent, an oxidized side-chain of a tyrosine amino acid that is part of PSII's protein sequence and which has a redox potential of just over +1 V. After fourfold oxidation, the OEC regains the four lost electrons by oxidizing two water molecules at near-neutral pH [11]. Using (3), one can calculate the overpotential of water oxidation for the enzymatic process to be $\eta \sim 150$–200 mV.

An incredible amount of work and resources has been spent in the last 50 years to elucidate how the OEC manages to achieve its outstanding catalytic rate and efficiency in water-oxidation catalysis [10–12]. As a result, many details of the OEC's mode of action have been revealed and it is probably fair to say that no catalyst for the WOR is currently as well understood as the OEC. Central aspects of this knowledge are presented in a separate chapter of this issue by Dimitrios Pantazis [13]. Nevertheless, I want to summarize some key points relevant for the development of water-oxidation catalysts inspired by the OEC in the following list. These elementary steps are also depicted in the simplified catalytic cycle for the OEC shown in Fig. 3:

- The OEC contains four manganese ions and one calcium ion (Fig. 2) interconnected by bridging water-derived ligands (WDLs, i.e. H_2O, OH^- and O^{2-}).
- In addition to these μ-WDLs, the metal centres are almost exclusively coordinated to terminal WDLs and carboxylate moieties of the protein chain. As a result, all four manganese centres have "oxygen-only" coordination spheres with five or six O^- ligands ($[MnO_{5/6}]$).
- The OEC can exist in five redox states (the so-called S-states S_0 to S_4), of which four (S_0–S_3) have long enough lifetimes to allow spectroscopic investigation.
- The total oxidation state of the S-states differs by one electron from one state to the next. In this way, four oxidation equivalents have to be extracted from the most reduced S_0 state to reach the most oxidized S_4 state. These are the four electron/hole pairs needed for one turnover of the WOR.

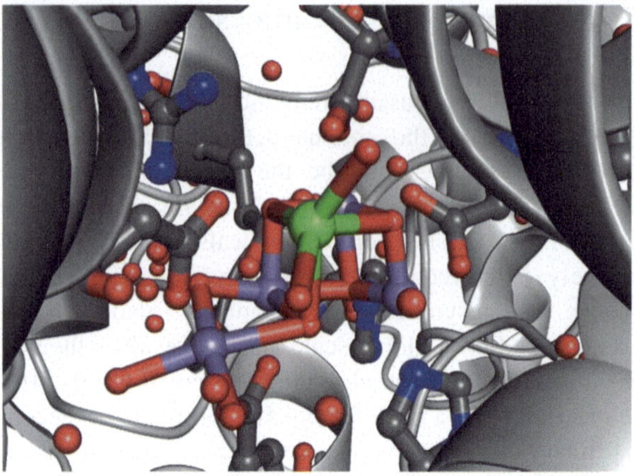

Fig. 2 Architecture of the oxygen-evolving-complex (OEC) of Photosystem II, the active site of the only enzyme catalysing the water-oxidation reaction. The structure from X-ray data most likely represents the S_0 state and was plotted using atom coordinates from Suga et al. [16] (PDB dataset 4UB8). Atom colour code: Mn: *purple*; Ca: *green*; O: *red*; C: *grey* and N: *blue*. As hydrogen atom coordinates could not be directly detected, *red spheres* might indicate the O positions for O^{2-}, OH^- or H_2O

- Even though some disagreement remains concerning oxidation state details, most S-state transitions are best described as equivalent to single-electron oxidations of the type $Mn^{III} \rightarrow Mn^{IV}$.
- In addition to being the redox active part of the OEC, manganese ions are also the binding sites for water, the substrate molecules for the reaction. As a water molecule coordinated to an Mn^{III}- or Mn^{IV}-ion is much more acidic than a water molecule in aqueous solution, the manganese centres also facilitate the important reaction steps of proton removal.
- Many suggestions for the last, O–O bond-forming step of water-oxidation catalysis by the OEC involve the reaction between one μ-WDL and at least one partially oxidized O ligand in form of an oxido-radical ($O^{\bullet -}$, oxidation state O^{-I}, marked in red in Fig. 3).

In an attempt to visually summarize our current knowledge about the OEC, Fig. 2 shows the structure of the $CaMn_4O_5$ cluster and its immediate environment as found in a recent X-ray diffraction study [16]. Additionally, one of the currently much discussed models for the OEC's catalytic cycle from spectroscopic data and theoretical calculations is presented in Fig. 3 [12, 15]. Readers wishing to learn more about mechanistic aspects of water-oxidation catalysis by the OEC are referred to some excellent recent overview articles [11, 12, 14, 15] and the chapter by Dimitrios Pantazis of this issue [13].

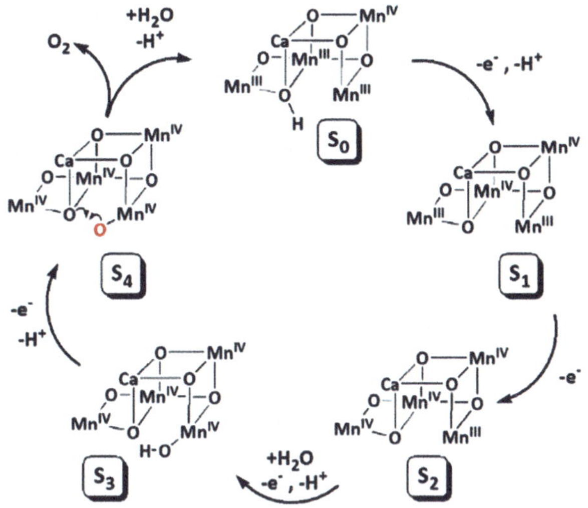

Fig. 3 Possible catalytic cycle for water oxidation by the OEC in accordance with currently available experimental and theoretical data [11, 12, 14]. Assignments of the individual Mn oxidation states as well as details of water binding, proton removal and μ-O^{2-} bridge formation are shown as described in Krewald et al. [15], but these are still topics of ongoing research. The oxygen atom marked in *red* for S_4 represents an oxyl radical ($O^{\bullet -}$, oxidation state O^{-I})

4 Manganese Oxide Phases and Their Stabilities in an Aqueous Environment

As described in the previous section, an advanced picture of nearly atomic resolution exists by now for the structure of the OEC and water-oxidation catalysis within the PSII enzyme. In "classical" bioinorganic model chemistry, one would now proceed with the synthesis of molecular coordination compounds of related structures and thus aim to develop a synthetic *molecular* Mn-based catalyst for the WOR [17, 18]. Indeed, this has been tried over many decades in a number of laboratories. However, success along this route has been very limited as far as developing manganese catalysts for a possible technological application are concerned. Instead, slow to negligible catalytic rates and poor compound stabilities at water-oxidation conditions have been the predominant observations for manganese, and instead the "champions" of *molecular* water-oxidation catalysis today are complexes of ruthenium and not of manganese [6, 19]. The central difference, in the view of this author, between Ru and Mn is that none of the so far synthesized polynuclear manganese coordination compounds is able to accumulate four oxidation equivalents at the high redox potentials needed to drive the WOR [19].

Despite these hardly encouraging outcomes from manganese coordination chemistry, the impressive performance of the OEC clearly demonstrates that fast and energy-efficient water-oxidation catalysis by a catalyst containing manganese as the only redox-active component is possible. Thus, research in the last 5 years or so has picked up (and by now greatly improved) a completely different bio-inspired approach to water-oxidation catalysis dating from the 1970s and 1980s: the use of manganese oxide materials as *heterogeneous* catalysts for the WOR. To set the stage for this compound class, some general remarks on manganese oxides seem

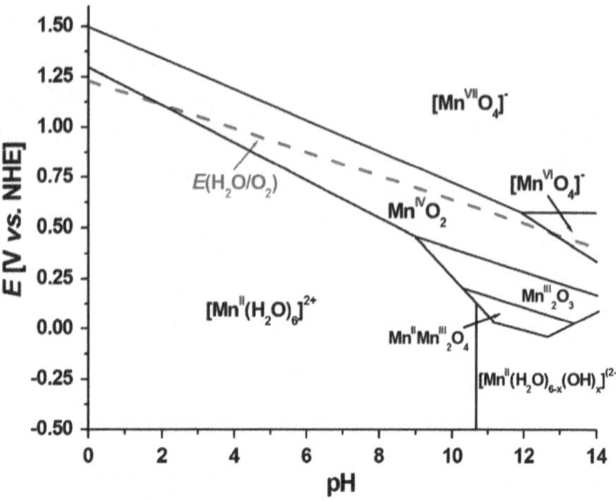

Fig. 4 Pourbaix diagram for manganese with the most important Mn species for different values of pH and oxidation potential E. The *grey dashed line* indicates the thermodynamic potential necessary for water oxidation. The diagram was drawn using data from Takeno [21] for a total Mn concentration of $[Mn]_{tot} \sim 10^{-10}$ M

appropriate (*note*: if meant in general, manganese oxides are simply abbreviated as "MnO_x" in the following).

Depending on the reaction partners and conditions, manganese shows one of the largest variations of oxidation states of all elements. For example, six of the eight possible oxidation states between Mn^{0} and Mn^{VII} are observed for the simple reaction system comprising manganese and water, the reaction medium of importance in the context of water oxidation [9, 20]. For a rough orientation concerning the dominant species of each element in an aqueous environment, Pourbaix diagrams show the most stable compounds as function of both pH and redox potential. There is a certain degree of uncertainty concerning the data used to draw such diagrams and it is especially difficult to determine redox potentials for solid oxides and hydroxides. Thus, slightly differing versions of the Mn Pourbaix diagram exist and Fig. 4 is an attempt to present a good consensus between the different available datasets [21].

As clearly visible in Fig. 4, one species with a large stability range is the hydrated Mn^{2+} ion, present in solution as the hexaaqua complex $[Mn(OH_2)_6]^{2+}$. The protons of this complex are not very acidic, so a first deprotonation only occurs above pH ~ 10.5 to obtain the hydroxido complex $[Mn(OH)(OH_2)_5]^{+}$. When the redox potential is increased (especially at neutral to basic pH), manganese oxidation occurs at potentials below $E = +1$ V. For most pH values, the first oxidation products of Mn^{2+} in water are insoluble manganese oxides or hydroxides of various, sometimes also mixed oxidation states between Mn^{II} and Mn^{IV}. Taken together, the MnO_x solids represent a second large zone in the manganese Pourbaix diagram. As any synthetic chemist who has tried to prepare manganese coordination compounds knows, these MnO_x are a thermodynamic "sink" for any manganese chemistry taking place at moderately oxidizing potentials in the presence of water. Only at high potentials, especially $E > +1$ V, are the highest accessible Mn oxidation states Mn^{VI} and Mn^{VII} reached. As the corresponding $Mn^{6+/7+}$ ions are very small, their coordination number, even with small WDLs, decreases to four. Additionally, the

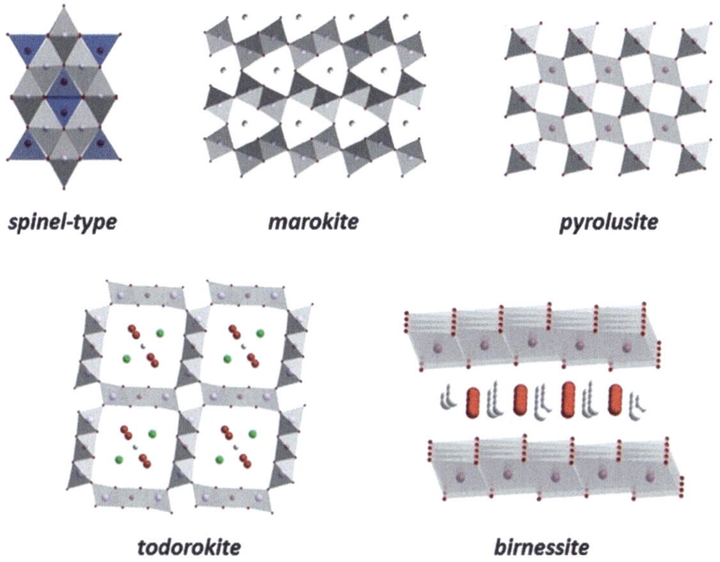

Fig. 5 Examples of structural motifs found in typical manganese oxides of network, tunnelled and layered structures. In each case, the common $[MnO_6]$-building blocks are given in *grey*

coordinated water is so acidic that only the fully deprotonated forms with O^{2-} ligands are found in the forms of the manganate and permanganate anions $MnO_4^{2-/-}$.

In a simplified Pourbaix diagram similar to that shown here, the solids of the "oxide zone" are typically indicated as phase pure materials with stoichiometries Mn_3O_4, Mn_2O_3 or MnO_2. These phases indeed exist as stable, crystalline materials [9, 20], but in reality the chemistry of manganese oxides and hydroxides of oxidation states + II, +III and + IV is much more diverse. Up to 30 different forms of MnO_x solids might be formed depending on subtle differences in oxidation potential, redox agent, pH, and also the concentration of manganese and other cations present in the reaction system [22, 23].

Fortunately, decades of research, especially by geochemists and mineralogists, have established solid foundations of MnO_x chemistry. It has been found that the dominant building unit of most $Mn^{II/III/IV}$-oxides is the MnO_6-octahedron, but that many variations exist for how these MnO_6-units might be arranged and interconnected in the solid compounds. To simplify, the wealth of solid state structures is commonly grouped into three classes: network, tunnelled and layered structures [23]. Examples from each family are shown in Fig. 5 and are briefly described in the following section.

In network structures, dense packings of Mn^{n+} cations and O^{2-} anions characterize the solid state arrangement, resulting in, for example, structures with corner-sharing (MnO, NaCl type) or edge-sharing (Mn_3O_4, spinel type) arrangements of the MnO_6-octahedra. Mixed oxides of the network type are also found: as an example, the structure of the calcium manganese oxide marokite ($CaMn_2O_4$) is built up of $Mn^{III}O_6$-octahedra linked by eight-coordinate Ca^{2+}-ions. Important characteristics of synthetic oxides from this "network family" are rigid, dense, well-ordered structures. The synthetic routes leading to these oxides usually require solid state reactions at high temperatures (>500 °C) and typically yield non-porous materials of low surface areas (few $m^2 \cdot g^{-1}$). Network oxides have well-defined Mn oxidation states which sometimes contain a significant contribution of Mn^{2+} (e.g. hausmannite: $Mn^{II}Mn^{III}{}_2O_4$).

This is different for the manganese oxides containing tunnels in their structures. Such tunnels are formed by arranging chains of edge-sharing MnO_6 units at near perpendicular angles towards each other. Depending on the width of the chains, very narrow tunnels such as the 1 × 1-motif found in pyrolusite (the most stable form of MnO_2) might be formed. As another extreme, the chains linked in the structure of todorokite each consist of three edge-sharing MnO_6-octahedra and in consequence the channels of its 3 × 3-tunnel structure are about 10 Å wide. These cavities are normally not empty, but contain – depending on the origin of the tunnel oxide – cations of other metals and water. As a result, typical elemental compositions of todorokites are given as $(Ca, Na, K)_{0.3–0.5}[Mn^{III/IV}, Mg]_6O_{12} \cdot (3–4.5\ H_2O)$. A similar diversity of compositions is also found for hollandite ($(Ba, Na, K)_{0.8–1.5}[Mn^{III/IV}, Mg]_8O_{16}$), the oxide of intermediate tunnel diameter formed by a 2 × 2 arrangement of MnO_6-chains. As a general summary, manganese oxides of tunnelled structures differ significantly from their network cousins: the structures contain large cavities, are much less ordered, and their external and internal surface areas are often higher (>25 $m^2 \cdot g^{-1}$). The average Mn oxidation states of tunnelled oxides vary depending on their compositions (see formulae above), but are usually well above +3.0, so that Mn^{2+} cations are at best minor components. Synthetically, such oxides are mostly obtained by precipitation reactions from aqueous solutions at temperatures above 50 °C, sometimes also using hydrothermal conditions.

Finally, layered oxides contain sheets of edge-sharing MnO_6-octahedra separated by 7- to 10-Å interlayer spaces. Similar to the oxides with wide tunnels, this interlayer space is usually filled with varying amounts of additional cations and water. Because of manganese vacancies in the sheets and differences in sheet stacking or sheet-to-sheet distances, the layered oxides are the least well-defined MnO_x group and rarely form well-ordered ion arrangements as drawn on the bottom right of Fig. 5. Even the names used for the layered oxides are often confusing. The most common group of layered oxides are mainly referred to as "birnessite type", an oxide class characterized by interlayer distances of about 7 Å, an average Mn oxidation state of 3.5–4.0, and a fairly large content of intercalated alkali or alkaline earth cations and water. A typical birnessite composition might be written as $(Na, K, Ca)_{0.2–0.7}MnO_2 \cdot (1.0–2.0\ H_2O)$. Variations of the birnessite theme exist where the interlayer space is wider (~10 Å, so-called buserites) or where Mn^{2+} is one of the main intercalated cation species (vernadite or "δ-MnO_2").

In general, layered MnO_x display many of the properties mentioned before for tunnelled oxides, as they are also porous materials of low order, characterised by a varying, Mn^{IV}-rich oxidation state and high surface areas (commonly $>100\ m^2 \cdot g^{-1}$). Synthetically, the preparation of these oxides is least demanding, because birnessite-type materials are often the dominant product when manganese (hydr)oxides are formed in precipitation reactions from water. In general, moderately oxidizing conditions, manganese ions in aqueous solution and a significant concentration of alkaline (earth) cations result in birnessite formation and the dried materials retain their layered structures unless heated to more than 400 °C.

The previous paragraphs might indicate that, even though a great structural variety exists, an individual manganese oxide sample can be easily assigned to a certain family or phase. For the well-ordered, crystalline network oxides this is indeed mainly correct and an identification of the MnO_x phase by, for example, standard X-ray powder diffraction experiments (XRD) is often possible. However, for the tunnelled or layered oxides, XRD or even X-ray absorption spectroscopy (XAS) at best allows for an identification of the oxide family (e.g. hollandite or birnessite) and further analytical tools have to be used to characterize the sample. In addition or in combination with XRD and XAS, X-ray photoelectron spectroscopy (XPS) or redox titrations have been successfully used to determine Mn redox states. IR spectroscopy also proved useful to detect water as part of the oxide structure, and ion-chromatography or energy dispersive X-ray spectroscopy (EDX) have enabled scientists to measure metal ion contents of different MnO_x samples. Additional analyses with electron microscopy or sorption methods are essential to identify morphological aspects of the oxide particles.

Fortunately, a number of examples of the state-of-the-art of MnO_x characterisation can be found in the recent literature [24–30], but a closer study of these publications reveals that a thorough characterization of a given manganese oxide sample is a rather tedious affair. This is further complicated by the fact that the water molecules and the secondary cations of tunnelled or layered oxides are exchangeable once the material is exposed to an aqueous solution, resulting in a very dynamic situation. As a result, the MnO_x samples used for water-oxidation catalysis (described in the next sections) and especially their fate during the catalytic reactions are so far mainly rather ill-defined. However, as manganese oxides currently gain large attention in the context of the WOR, the author is very confident that this situation is about to improve greatly in the coming years.

5 Early Examples of MnO_x Catalysts for Water Oxidation: Reports from the 1980s

From a look at the manganese Pourbaix diagram of Fig. 4, one can see that the manganese oxidation state closest to the line indicating the pH-dependent water-oxidation potential is Mn^{IV}. This area is marked in Fig. 4 simply as "MnO_2", but –

as described in the previous section – this could stand for various Mn^{IV}-containing oxides depending on the actual reaction conditions. This correlation between Mn oxidation states and the thermodynamic potential for the WOR has at least two important consequences:

1. The Mn Pourbaix diagram offers a very simplistic explanation for why the $Mn^{III} \rightarrow Mn^{IV}$ oxidation steps observed for the S-state transitions of the OEC are a good choice to build up oxidation equivalents for water oxidation at neutral pH: for MnO_6-units, this is apparently the least energy intensive way to generate holes that are oxidizing enough to drive water oxidation
2. The "$Mn^{III/IV}$-oxide zone" in the Pourbaix diagram is quite large so that these materials might be used at a variety of *pH*/*E*-combinations without entering areas of corrosion where the oxides dissolve into soluble forms ($[Mn^{II}(H_2O)_6]^{2+}$ or $Mn^{VI/VII}O_4^{2-/-}$)

Pioneering and largely forgotten work where Mn oxides were successfully used as bio-inspired catalysts was published in the former USSR. As early as 1968, Glikman and Shchegoleva from Kiev reported on oxygen evolution from aqueous suspensions of MnO_2 in the presence of the strong oxidation agent Ce^{4+} [31]. In 1981, the group of Alexander Shilov from Chernogolovka continued work on this system and realized its model character for water oxidation in biological photosynthesis [32]. Long before details of the WOR catalysed by PSII were known, Shilov et al. in their publication already rightly assumed that manganese-centred redox chemistry in a Mn^{IV}-rich environment might be a very suitable process for a water-oxidation catalyst mimicking PSII's mode of action. Having failed to observe significant amounts of O_2-formation for WOR catalysis by Mn^{IV}-complexes, they followed Glikman and Shchegoleva and turned to manganese (IV) oxides as potential catalysts. Indeed, catalytic water oxidation could be observed for a wide pH range ($2 < pH < 8$) when MnO_x were exposed to $[Ru(bipy)_3]^{3+}$ acting as chemical oxidation agent. In their report, the scientists concluded that "*the results for* [...] *$Ru(bpy)^{3+}$ and MnO_2 permit consideration of this system as a functional model of PSII in plant photosynthesis*" [32].

However, the materials used as catalysts in both studies were not characterized in any detail and just mentioned as "MnO_2" without further specifications. The MnO_x for the 1981 study were prepared as suspensions or colloids by the reduction of MnO_4^- by Mn^{2+} or H_2O_2 in aqueous buffer solutions. Their real composition is unknown, but from the general trends described in the previous section, it seems unlikely that phase-pure MnO_2, as assumed by Shilov et al., might have been formed under such conditions. Nevertheless, this work is a remarkable paper, drawing many correct conclusions concerning the catalysis of the WOR by manganese oxides well ahead of later work. Unfortunately, biomimetic water-oxidation chemistry was mainly concerned with manganese coordination compounds at that time and therefore Shilov's "PSII model" was little cited in the 1980s and soon forgotten.

More influential but far away from any link to enzymatic water-oxidation chemistry were the studies by electrochemists also carried out around 1980 in

Osaka, Japan and Milan, Italy [33, 34]. In both laboratories, it was demonstrated that manganese oxides acted as electrocatalysts for the WOR either in the form of massive MnO_x discs/rods or as coatings on conductive surfaces. These electrolysis reactions using MnO_x as anode materials were carried out at the extreme pH conditions of acidic or alkaline electrolysers (pH ~0 or ~14, respectively), but even here manganese oxides proved to be relatively stable. By using MnO_x as the anode material, it was possible to reach water-oxidation current densities comparable to those observed for the other "hot candidates" amongst MO_x-materials for the catalysis of the WOR, the oxides of Ru, Ir, Co and Ni. The Japanese team at this time went furthest concerning both synthetic variations and characterizations of the oxide materials and found that best catalytic results for MnO_x anodes could be obtained if a mixture of MnO_2 and Mn_2O_3 (as detected by XRD and chemical analysis) was employed [34]. The authors recorded large currents when the freshly prepared Mn(III,IV) oxide mixture was exposed to potentials of ~ +1.5 V in acidic solution and concluded that this oxidation somehow represents a transformation of the original material (at least at the electrode surface) to a more Mn^{IV}-rich form. This material is then catalytically active for the WOR and current densities of $i \sim 1\ mA \cdot cm^{-2}$ at overpotentials of $\eta \sim 400–500$ mV for both pH ~0 and ~14 were reported. However, in all cases these early electrocatalytic studies did not investigate for how long such current densities can be maintained and what kind of transformations occur in the MnO_x materials during extended electrolyses.

As a third set of influential work, the studies by Okuno (Sapporo, Japan, 1983) [35] and especially Harriman (London, UK, 1988) [36] should be mentioned, as they established that water-oxidation catalysis by MnO_x could (1) also be driven by visible light using $[Ru(bipy)_3]^{2+}$ as photosensitizer and appropriate electron acceptors, and (2) was indeed very much dependent on the phase of the manganese oxide used. Phase-pure, crystalline MnO_2 generally performed much worse than Mn_2O_3 or especially "mixed MnO_x oxides" of unknown detailed composition and structure. In the Harriman study on an oxide series of 15 different metals, Mn_2O_3 again ranked among the best catalysts. Additionally, Harriman et al. also determined catalyst surface areas and found that, as expected for heterogeneous catalysis, these were important factors and the possibility of enhancing catalysis by preparing small oxide particles was already indicated [36].

As these reports show, WOR catalysis by manganese oxides has been known for quite a long time. Interestingly, the early studies also provide first examples of the three commonly used reaction systems to determine catalytic WOR activities still used today. These are: (1) **Ce**, the use of Ce^{4+}-solutions as chemical oxidation agent at pH ~ 1–3 with an oxidation potential of about +1.6 V; (2) **Ru**, a system where the MnO_x catalyst reacts with the strongly oxidizing complex $[Ru^{III}(bipy)_3]^{3+}$($E \sim +1.3$ V) at pH ~4–7; and (3) **ecat**, electrochemical set-ups, in which manganese oxides act as electrocatalysts at the anode surfaces. Cartoons representing the flow of electrons in these systems are shown in Fig. 6. For **Ce** and **Ru**, one usually measures the amount of O_2 product formed over time to compare catalytic rates. For **ecat**, the current detected for a certain externally applied potential normalized to the geometric area of the anode (current density, *i*) is

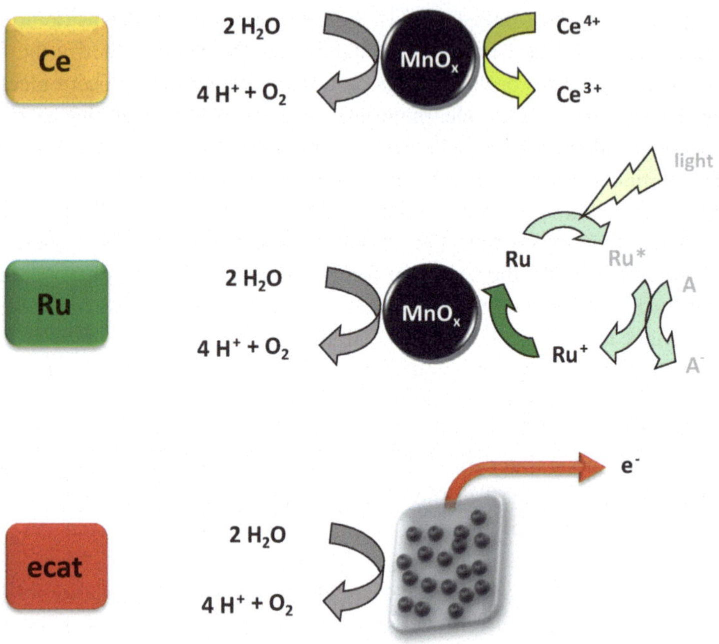

Fig. 6 The three most commonly used systems to determine rates for the WOR catalysed by manganese oxides. *Top*: chemical oxidation by Ce^{4+}. *Middle*: chemical oxidation by $[Ru^{III}(bipy)_3]^{3+}$, which has either been prepared before or is generated in situ by photochemical reaction of $[Ru^{II}(bipy)_3]^{3+}$ (**Ru**) with an appropriate electron acceptor A. *Bottom*: electrochemical water oxidation using MnO_x particles as electrocatalyst on the anodic side of an electrolyser

most often taken as an indication for the rate at which electrochemical water oxidation occurs. Even as the catalytic reaction is the same, Fig. 6 illustrates how diverse these catalyst test systems are and I return to this topic at the end of the next section.

Because of very limited analytical data, the authors of the publications from the 1970s/1980s had to remain rather vague on possible reasons for the good performance of certain manganese oxides. The fact that the materials contain the same metal as the active site of PSII was mentioned of course [32], but that really did not explain too much. Binding energies for metal-oxido bonds were analysed (they should be neither too weak nor too strong) [33], but the accuracy of those values at the time was low. The best rationale from the last century thus seems to be based on the aforementioned very suitable position of the redox potential for the reaction $Mn^{III} \rightarrow Mn^{IV}$ in some Mn oxide materials. In his review of the topic from 1984, Tseung from London, UK even went as far as to write that the value of the potential for the transition going from lower → higher oxidation state in an oxide was *the* key parameter to explain the differences in electrocatalytic efficiencies [37]. However, convincing experimental proof for these claims hardly existed at that time. Thus, by about 1985, water-oxidation catalysis by manganese oxides was a firmly established but not well understood phenomenon.

6 On the Way to "Solar Fuels": Optimization of MnO_x-Materials for Water-Splitting Devices Since the Year 2005

The promising first results on MnO_x-based water oxidation described in the previous section were not followed any further from 1985 onwards, probably because the two oil crises of the 1970s had passed and the first burst of activity (and funding) in the field was over. Some projects, for example the Swedish Consortium for Artificial Photosynthesis founded in 1993, kept the research field of bio-inspired WOR catalysis alive, but it was not until about 2005 that intensive research in this area really picked up again, most likely as a consequence of the greater urge to develop renewable energy systems to avoid further global warming. A very influential review paper from 2006 [38] and the first international conferences with extensive program sections on artificial photosynthesis such as IPS-16 in Uppsala, Sweden in 2006 or the first Gordon Conference on Renewable Energy: Solar Fuels in Ventura, USA in 2007 marked the start of a new phase of research activities.

Immediately, the need for efficient and affordable water-oxidation catalysts became a central problem and, again, molecular approaches for homogeneous WOR catalysis dominated many discussions. The very influential (re)discovery of cobalt oxides for anodic water oxidation by the group of Daniel Nocera from Cambridge, USA changed this in 2008 [39] and, since then, research in the use of heterogeneous metal-oxide catalysts for the WOR has developed into a very large research field within just over 5 years.

As part of these activities, a number of groups also revitalized the bio-inspired approach of using manganese oxides. The 2010 publications by the group of Heinz Frei from Berkeley, USA [40] and this author's team from (then) Kiel, Germany [41] might be seen as starting points for these new research activities. The group in the USA showed that MnO_x nanoparticles prepared on the surfaces of a porous silica support acted as fast WOR catalysts [40]. Materials that had been calcined at different temperatures were used and it was found that the catalysts heated to 600–700 °C performed best. An XAS analysis was carried out to determine the Mn oxidation state for these nanoparticles and Frei et al. concluded that the fastest catalysts were characterized by a mixed $Mn^{III/IV}$-state rich in Mn^{IV}.

In our own work, we tried to be "even more biomimetic" and synthesized mixed calcium-manganese(III) oxides in order imitate both the Mn oxidation state and the presence of Ca^{2+}. By that time Ca^{2+} was firmly known to be part of the structure of the biological catalyst (the first crystal structures of PSII had been published in 2004 [10]). We chose marokite ($CaMn_2O_4$) as the target oxide phase for our study and prepared this compound using a comproportionation reaction between Mn^{2+} and MnO_4^- in water and in the presence of Ca^{2+}, followed by heating of the precipitate to 1,000 °C. Such a synthesis indeed yields phase-pure marokite (as detected by XRD), but this oxide showed little catalytic activity for the WOR [41]. However, the initially formed precipitates, used either simply as dried powders or preferentially after heating to 400 °C, proved to be remarkably good catalysts for water oxidation. Both the "600 °C nanoparticles" of Frei et al. and our "marokite precursors" outperformed crystalline oxides such as Mn_2O_3 or MnO_2 in WOR catalysis by a wide margin [40, 41].

A common problem for both catalysts was the difficulty, as described in Sect. 4, in finding out what kind of MnO_x was really acting as catalyst. In the 2010 publication, we described the X-ray amorphous "marokite precursors" as "hydrated $CaMn_2O_4$" for lack of much analytical data (no information could be obtained from XRD, but the FTIR-spectra showed pronounced water-related bands). We only learned more about the interesting new catalysts after teaming up with the group of Holger Dau in Berlin, Germany. At Berlin's BESSY synchrotron, XAS measurements at the K-edges of both Mn and Ca were carried out and an elaborate analysis, especially of the EXAFS spectra, left little doubt that the most active materials were calcium-containing layered oxides from the birnessite family (Fig. 5) [42]. Their average Mn-oxidation state was ~+3.7 and thus much higher than the Mn^{III}-state expected for marokite. Additionally, vacancies in the MnO_6-layers and a low degree of stacking order characterized the synthetic oxides. All these features of MnO_x precipitated in such synthesis conditions were known to mineralogists long before (see Sect. 4 above), but we (and others) first had to "rediscover" this chemistry in the context of heterogeneous water-oxidation catalysis by MnO_x.

In the last few years, it has been striking how dominant the "birnessite theme" has become for WOR catalysis by Mn-compounds [25, 43–45]. It has been shown that the MnO_x materials obtained (1) from molecular Mn_2- or Mn_4-precursors in membranes or adsorbed on clays [46, 47], (2) from the comproportionation reaction of Mn^{2+} and MnO_4^- [48], (3) from electrochemical deposition starting from Mn^{2+} [28, 49], (4) from $Mn^{II}O$ operated as WOR catalyst [24], (5) from the hydrothermal reaction of $KMnO_4$ with water [27] or even (6) from the calcium-manganese oxides covering the cell walls of certain photosynthesizing algae [50] all belong to the birnessite family of layered MnO_x and all show catalytic activity for the WOR (Fig. 7).

Thus, birnessites seem to possess important properties for efficient water-oxidation catalysis. Additionally, in water this oxide type apparently represents some kind of local energy minimum for systems comprising Mn/water/alkali(ne earth) metal at the oxidation potentials relevant for water oxidation over the entire pH-range. As a result, layered oxide structures are formed from very different precursors at "WOR conditions". What are these important oxide properties and how can the synthesis of birnessite-type oxides be improved to render even better catalyst materials? The following list is an attempt to represent an overview of currently discussed points based on the by now quite extensive experimental data in this field:

- Birnessites represent the "*right Mn oxidation state*": it has been mentioned above that both the Mn Pourbaix diagram and the Mn oxidation states found for the OEC indicate that an Mn^{IV}-rich mixed Mn^{III}/Mn^{IV} oxidation state might be ideal for Mn-based WOR catalysis. It has been confirmed especially through

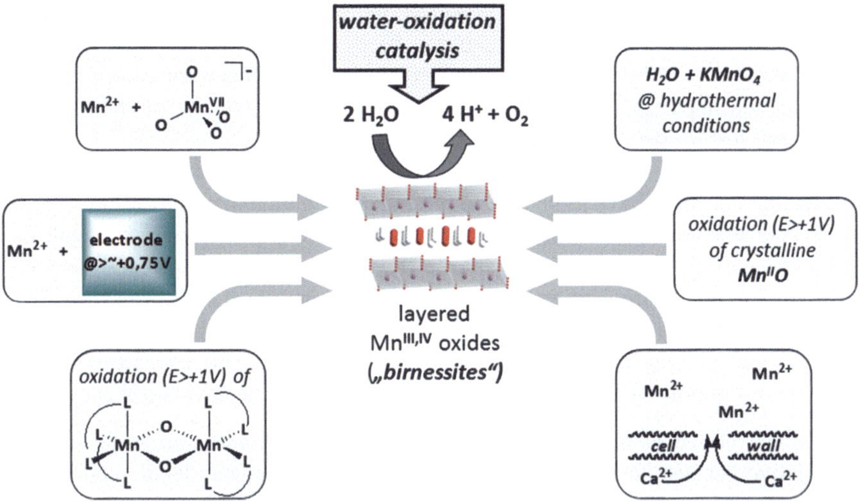

Fig. 7 Overview of the large variety of recently reported reaction conditions resulting in the formation of birnessite type oxide materials, all of which are catalytically active for the WOR

XAS of catalytically active birnessites that these are "$Mn^{+3.5–3.9}$-oxides" and thus fit this requirement perfectly [28, 42, 47].

- Studies of the bulk oxidation state changes observed for birnessites indicate that the *crucial redox transition* ($Mn^{III} \rightarrow Mn^{IV}$) occurs within some hundred millivolts above the thermodynamic $E(O_2/H_2O)$ [28, 49, 51] and is thus very well-suited for the WOR according to the theory of Tseung (see above) [37].
- Synthetic birnessites show a *low degree of structural order* [26, 48]: as a result, the environment of individual manganese ions is rather flexible and thus allows for the structural rearrangements known to be necessary for WOR catalysis from the S-state transitions of PSII (Fig. 3) [43, 48].
- Birnessites contain the "*right ions*": from studies of the OEC, it is known that Mn^{3+}/Mn^{4+} ions linked by OH^-/O^{2-} ligands form a catalytic site for a fast WOR. Additionally, the presence of the alkaline earth metal ions Ca^{2+} or Sr^{2+} as part of the OEC's structure is essential for catalytic activity. The exact role of these ions in the process has not been elucidated with certainty, but it is very likely that they are involved in the activation of the water substrate (in the form of some water-derived ligand coordinated to the M^{2+} ion; see Fig. 3). Additionally, the presence of the M^{2+} ions might tune the Mn^{III}/Mn^{IV} redox potential in a beneficial way [52]. Birnessites are amongst the few MnO_x where these large, catalytically important cations can be incorporated to carry out both potential functions and this might therefore be a crucial advantage [43, 48].
- Birnessites are *porous* materials with high outer and internal *surface areas*: because WOR catalysis by oxides is a heterogeneous process, a large contact area between catalyst and solution is a general advantage. Furthermore, the ~7 Å wide interlayer space already contains mobile, exchangeable H_2O molecules, the

substrate of the reaction. In combination, the porosity and layered architecture seem to result in a material which can be described as a "catalytic sponge", where $Mn^{III} \rightarrow Mn^{IV}$ oxidations, water binding (and deprotonation), and O_2 formation apparently occur in a large fraction of the volume of the particles and not just on the outer surfaces [26, 28].

- Birnessites show significant *electric conductivity*: during the catalytic process, electrons or holes have to be able to migrate through the material from the point of hole injection to the actual active sites for the WOR. For electrocatalysis, charge carriers also have to be able to traverse the material to reach the back contact in an electrochemical cell. From measurements on solid MnO_x samples [25] and through their use as battery materials, it is known that layered manganese oxides show reasonable (but definitely not excellent) electric conductivities, an important property for their application in water-oxidation catalysis.

The list above is most probably incomplete and certainly, to some degree, speculative because the experimental evidence on mechanistic details concerning WOR catalysis by birnessites is so far very limited. Nevertheless, there have been recent successes to improve the catalytic performance of birnessites for the WOR with the points listed above in mind – and this indicates that they are probably not entirely unimportant:

- Different catalyst screenings of MnO_x resulted in the identification of layered Mn oxides as "best" MnO_x materials (however, see text below concerning general problems of such rankings) [25, 26, 53].
- Synthetic routes to preparing birnessites of especially low order were developed and indeed resulted in an increase of catalytic rates [26, 49].
- The type and concentration of alkaline earth metal ions in synthetic birnessites was varied and a clear optimum was found for Ca^{2+} with a Ca:Mn ratio of ~0.2:1 (both remarkably similar to PSII) [48].
- Attempts to improve catalytic rates through the preparation of nanoparticles were generally successful [54–56], but did not result in a steep performance increase. This seems to indicate that the model of the "catalytic sponge" is probably quite appropriate for WOR catalysis by birnessites.

As convincing as the "birnessite story" presented so far might look, it is still far from proven that layered Mn(III,IV) oxides are truly the generally best type of MnO_x for WOR catalysis. A number of people (including this author) are of the opinion that birnessites are indeed the most promising Mn-based materials for water oxidation known today. However, it should be mentioned that others have observed different trends in reactivity and have thus focused on other types of MnO_x catalysts, namely from the family of crystalline network oxides [27, 30, 57]. The most important reason for this disagreement is the fact that the methodology for the determination of catalytic rates is very diverse. Also, the "target zone" for technological applications is still quite ill-defined.

It has already been mentioned above that established test systems for WOR catalysis exist which are either purely chemical, photochemical or electrochemical

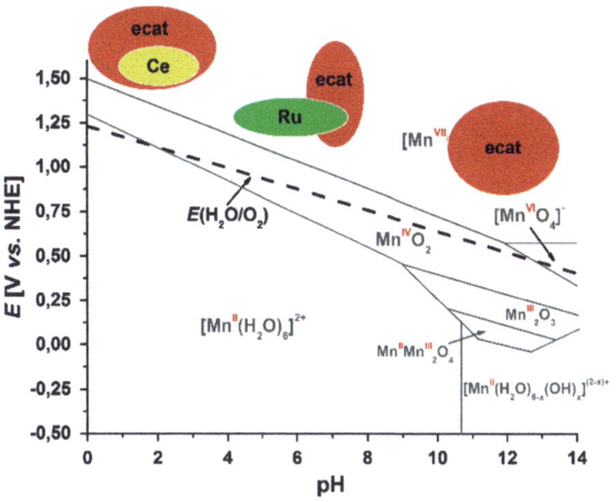

Fig. 8 Illustration of the typical E/pH domains where catalytic WOR studies using the test systems **Ce**, **Ru** or **ecat** introduced in Fig. 6 are typically carried out. The thermodynamic potential for water oxidation (*dashed line*) and the Mn Pourbaix diagram (*grey*) are also shown

in nature (Fig. 6). In consequence, the factors influencing reaction rates differ very much for each case. For example, the concentration and diffusion rate of Ce^{4+} are important for the **Ce**-system, whereas the efficiency of the photochemical generation of $[Ru(bipy)_3]^{3+}$ largely influences rates in **Ru** and the quality of the electric contact between the MnO_x particles and the anode material is a key issue in electrocatalysis (**ecat**).

Additionally, it is important to note that the three systems probe WOR catalysis at very different areas of the *E*/pH parameter space. To illustrate this, Fig. 8 again shows the Pourbaix diagram already introduced in Sect. 4, but this time with an emphasis on the typical conditions for the test systems **Ce**, **Ru** and **ecat** in relation to $E(O_2/H_2O)$ and the different MnO_x phases. As can be seen, the chemical systems **Ce** and **Ru** are limited to a rather small area in such a plot because of their fixed redox potentials and pH-limitations (precipitation of cerium hydroxides and side reactions of the Ru^{III}-complex, respectively). In contrast, electrochemical studies can in theory address the entire *E*/pH-space above the line marking the thermodynamic water-oxidation potential. In the practice of **ecat**-set-ups, it has been found that an overpotential of at least $\eta = 300$ mV is required to detect significant current densities ($i \sim 100\ \mu A \cdot cm^{-2}$) and so *E* has usually been set significantly above the thermodynamic $E(O_2/H_2O)$ in electrocatalytic studies. Furthermore, experiments on the electrocatalytic WOR have so far mainly concentrated on three pH-ranges: strongly acidic (pH ~ 0–3) [58] and strongly basic conditions (pH ~ 10–13) [28, 59, 60] with the aim to profit from the high conductivities of such electrolytes, and near-neutral pH [49, 53, 57, 60], where less corrosion problems during longer operation times and better model character of the observed processes in relation to PSII are expected.

Interestingly, the "overpotential" for the three common experimental conditions **Ce**, **Ru** and **ecat** are at least similar and always in the range 300–500 mV (though even higher values are sometimes used in electrocatalysis to reach current densities well above 1 $mA \cdot cm^{-2}$).

However, despite this similar thermodynamic driving force, Fig. 8 illustrates quite well why a comparison of catalytic results – and thus an identification of the "best" MnO_x catalyst – is hardly possible: the test conditions are far apart in the *E*/pH-space. Additionally, experimental protocols differ. For example, results for **Ce**- or **Ru**-experiments have been reported for different concentrations of the oxidation agents and different electron acceptors have been used in photochemical versions of set-up **Ru**. More general, reaction rates have variably been normalized per MnO_x catalyst mass, per mole Mn, or per surface area. Likewise, electrolysis experiments have been carried out using a wide variety of electrolytes, anode geometries, electrochemical measurement protocols, etc. To summarize, a lot of catalytic data on widely varying catalytic systems exists today, but comparisons of different datasets is largely impossible – a situation which has to be improved in the coming years to make concerted, target-oriented improvements of MnO_x catalysts possible.

However, such a "standardisation" is easier said than done because it is far from clear today how commercial devices for solar fuels production are to be constructed in detail. As a consequence, the reaction conditions for these devices are unclear. The general layout shown at the start (Fig. 1) allows many technological implementations. Among them, the following two each seem to be favoured by a large number of research groups:

- "*Photovoltaic (PV) cell + electrolyser*": in such a set-up, the light-driven charge separator at the centre of Fig. 1 would be constructed by operating two or three photovoltaic units in series to generate an electrolysis voltage of about 2 V. This is then coupled to a "classical" electrolysis cell, for example alkaline or PEM electrolysers [61]. To minimize energy losses caused by cell resistivity, the electrolyser is most probably run at extreme pH values as in the commercial electrolysers available today. As a result, a MnO_x electrocatalyst used in such devices has to be optimized at the conditions shown in Fig. 8 as the two **ecat** areas are at very low or very high pH values.
- "*Artificial leaf*": a very influential device design for solar fuel production of the last years has been "artificial leaves" [62, 63]. In this, the catalysts Cat_{ox} and Cat_{red} are directly attached to the surfaces of a two- to three-junction photovoltaic device to allow "wireless" electron transfer from the PV-unit to the catalysts through a thin interface between catalyst and semiconductor. This could make such leaves more compact and thus cheaper to produce than the PV-electrolyser design. Concerning the operating conditions, it could very well be advantageous to immerse artificial leaves into a neutral electrolyte to avoid an exposure of the PV/catalyst interfaces to the very corroding conditions of standard electrolyser media. Thus, a catalyst for artificial leaves most likely has to show especially good performances at near-neutral pH, the central **ecat** area in Fig. 8.

Both lines of argument have been followed and manganese oxides have been operated in electrocatalytic set-ups in acidic, neutral and alkaline electrolytes. Sometimes comparative studies were carried out where identical anodes were investigated at different pH values [51, 53, 59, 60]. From the results, it has been encouraging to see that MnO_x electrocatalysts can be used successfully over the entire pH range and thus show potential for many different cell set-ups. On the other hand, our knowledge concerning the importance of issues such as MnO_x type, electrolyte composition or pH, oxide conductivity, O_2 bubble formation, etc. for electrocatalytic water oxidation by manganese oxides is currently still very limited and many interesting results await us here in the future.

7 Conclusions and Outlook

The presented overview has hopefully been able to show that the field of manganese oxide catalysts for the WOR has developed rapidly in the last 5 years. Starting from the few examples of MnO_x catalysts known from the last century, impressive synthetic activity in this field has resulted in the discovery and improvement of many catalytically active MnO_x materials. Additionally, the effort put into analysing interesting oxides, also under "operating conditions", has increased tremendously.

On the basis of what we know today, it seems probable that the manganese-based material of choice for WOR catalysis is a mixed $Mn^{III/IV}$-oxide of low crystallinity and high surface area. Additionally, it might be beneficial to incorporate secondary metal ions into the structure to activate the water substrate, tune $Mn^{III/IV}$-redox potentials and/or increase oxide conductivities. Manganese oxides of layered architectures are currently often the "winners" of catalyst screenings, but more systematic work is needed to arrive at convincing structure-activity relationships for the many types of possible MnO_x materials. The extremely detailed knowledge about the OEC, biology's catalyst for water oxidation, puts researchers in this field in a very good position to arrive finally at a high level of mechanistic understanding regarding how the WOR might occur on the surface of an MnOx particle.

Catalytic test reactions, especially using the three chemical and electrochemical systems **Ce**, **Ru** and **ecat**, have shown that manganese oxides are able to catalyse water oxidation with significant reaction rates at overpotentials of ~400 mV over the entire pH range. This is unusual as most other metal oxides suffer from corrosion problems, especially under acidic conditions.

In the rather short time since their "rediscovery" as WOR catalyst materials, manganese oxides have emerged as very promising materials to be used in devices for the production of solar fuels. A number of "recipes" for the synthesis of well-performing MnO_x catalyst materials exist. The next steps in this research should now be to use these compounds in prototypes of solar-fuel-producing devices and test their efficiencies, stabilities and modes of action under conditions relevant for a possible technological application. Only if they also prove to be usable in such "real world" set-ups, one can rightly conclude in some years from now that it was a good idea to look at biology as inspiration for water-oxidation catalysts.

Acknowledgements First and foremost I would like to thank the Ph.D. students of my research group who have worked very hard in recent years to broaden our understanding of MnO_x WOR catalysis: Carolin E. Frey, Seung Y. Lee, M. Mahdi Najafpour, and Mathias Wiechen. In Freiburg special thanks go to Jann Sonnenfeld for preparing Fig. 2. Additionally, the close collaboration with Holger Dau and Ivelina Zaharieva at Freie Universität Berlin has been a great source of inspiration and results since its start in 2008. Finally, I would like to acknowledge generous financial support by the Fonds der Chemischen Industrie (FCI) and the German Science Foundation (DFG).

References

1. Armaroli N, Balzani V, Serpone N (2013) Powering planet Earth: energy solutions for the future. Wiley-VCH, Weinheim
2. Balzani V, Credi A, Venturi M (2008) Photochemical conversion of solar energy. ChemSusChem 1:26–58
3. Lubitz W, Reijerse EJ, Messinger J (2008) Solar water-splitting into H_2 and O_2: design principles of photosystem II and hydrogenases. Energy Environ Sci 1:15–31
4. Kirch M, Lehn JM, Sauvage JP (1979) Hydrogen generation by visible light irradiation of aqueous solutions of metal complexes. An approach to the photochemical conversion and storage of solar energy. Helv Chim Acta 62:1345–1384
5. Taiz L, Zeiger E (2010) Plant physiology. Sinauer Ass, Sunderland
6. Kärkäs MD, Verho O, Johnston EV et al (2014) Artificial photosynthesis: molecular systems for catalytic water oxidation. Chem Rev 114:11863–12001
7. Crabtree RH (ed) (2010) Energy production and storage: inorganic chemical strategies for a warming world. Wiley, Chichester
8. Thapper A, Styring S, Saracco G et al (2013) Artificial photosynthesis for solar fuels. Green 3:43–57
9. Holleman AF, Wiberg E (2007) Lehrbuch der Anorganischen Chemie. de Gruyter, Berlin
10. Barber J (2009) Photosynthetic energy conversion: natural and artificial. Chem Soc Rev 38: 185–196
11. Dau H, Limberg C, Reier T et al (2010) The mechanism of water oxidation: from electrolysis via homogeneous to biological catalysis. ChemCatChem 2:724–761
12. Cox N, Pantazis DA, Neese F et al (2013) Biological water oxidation. Acc Chem Res 46: 1588–1596
13. Krewald V, Retegan M, Pantazis DA (2015) Principles of natural photosynthesis. Top Curr Chem. doi:10.1007/128_2015_645
14. Siegbahn PE (2009) Structures and energetics for O_2 formation in photosystem II. Acc Chem Res 42:1871–1880
15. Krewald V, Retegan M, Cox N et al (2015) Metal oxidation states in biological water splitting. Chem Sci 6:1676–1695
16. Suga M, Akita F, Hirata K et al (2015) Native structure of photosystem II at 1.95 Å resolution viewed by femtosecond X-ray pulses. Nature 517:99–103
17. Rehder D (2014) Bioinorganic chemistry. Oxford University Press, Oxford
18. Kraatz H, Metzler-Nolte N (eds) (2006) Concepts and models in bioinorganic chemistry. Wiley-VCH, Weinheim
19. Llobet A (ed) (2014) Molecular water oxidation catalysis: a key topic for new sustainable energy conversion schemes. Wiley, Chichester
20. Greenwood NN, Earnshaw A (1998) Chemistry of the elements. Butterworth Heinemann, Oxford
21. Takeno N (2005) Atlas of Eh-pH diagrams. Geological Survey of Japan, National Institute of Advanced Industrial Science and Technology (Japan)

22. Tebo BM, Bargar JR, Clement BG et al (2004) Biogenic manganese oxides: properties and mechanisms of formation. Annu Rev Earth Planet Sci 32:287–328
23. Post JE (1999) Manganese oxide minerals: crystal structures and economic and environmental significance. Proc Natl Acad Sci U S A 96:3447–3454
24. Indra A, Menezes PW, Zaharieva I et al (2013) Active mixed-valent MnO(x) water oxidation catalysts through partial oxidation (corrosion) of nanostructured MnO particles. Angew Chem Int Ed 52:13206–13210
25. Meng Y, Song W, Huang H et al (2014) Structure–property relationship of bifunctional MnO_2 nanostructures: highly efficient, ultra-stable electrochemical water oxidation and oxygen reduction reaction catalysts identified in alkaline media. J Am Chem Soc 136:11452–11464
26. Frey CE, Wiechen M, Kurz P (2014) Water-oxidation catalysis by synthetic manganese oxides – systematic variations of the calcium birnessite theme. Dalton Trans 43:4370
27. Robinson DM, Go YB, Mui M et al (2013) Photochemical water oxidation by crystalline polymorphs of manganese oxides: structural requirements for catalysis. J Am Chem Soc 135: 3494–3501
28. Gorlin Y, Lassalle-Kaiser B, Benck JD et al (2013) In situ X-ray absorption spectroscopy investigation of a bifunctional manganese oxide catalyst with high activity for electrochemical water oxidation and oxygen reduction. J Am Chem Soc 135:8525–8534
29. Najafpour MM, Sedigh DJ, Pashaei B et al (2013) Water oxidation by nano-layered manganese oxides in the presence of cerium(IV) ammonium nitrate: important factors and a proposed self-repair mechanism. New J Chem 37:2448
30. Boppana VB, Jiao F (2011) Nanostructured MnO_2: an efficient and robust water oxidation catalyst. Chem Commun 47:8973–8975
31. Glikman TS, Shchegoleva IS (1968) The catalytic oxidation of water by quadrivalent cerium ions. Kinet Katal 1968:461–462
32. Shafirovich VY, Khannanov NK, Shilov AE (1981) Inorganic models of photosystem II of plant photosynthesis – catalytic and photocatalytic oxidation of water with participation of manganese compounds. J Inorg Biochem 15:113–129
33. Trasatti S (1980) Electrocatalysis by oxides – attempt at a unifying approach. J Electroanal Chem 111:125–131
34. Morita M, Iwakura C, Tamura H (1979) Anodic characteristics of massive manganese oxide electrode. Electrochim Acta 24:357–362
35. Okuno Y, Yonemitsu O, Chiba Y (1983) Manganese dioxide as specific redox catalyst in the photosensitized oxygen generation from water. Chem Lett 815–818
36. Harriman A, Pickering IJ, Thomas JM et al (1988) Metal-oxides as heterogeneous catalysts for oxygen evolution under photochemical conditions. J Chem Soc Faraday Trans I 84:2795–2806
37. Rasiyah P, Tseung AC (1984) The role of the lower metal-oxide higher metal oxide couple in oxygen evolution reactions. J Electrochem Soc 131:803–808
38. Lewis NS, Nocera DG (2006) Powering the planet: chemical challenges in solar energy utilization. Proc Natl Acad Sci U S A 103:15729–15735
39. Kanan MW, Nocera DG (2008) In situ formation of an oxygen-evolving catalyst in neutral water containing phosphate and Co^{2+}. Science 321:1072–1075
40. Jiao F, Frei H (2010) Nanostructured manganese oxide clusters supported on mesoporous silica as efficient oxygen-evolving catalysts. Chem Commun 46:2920–2922
41. Najafpour MM, Ehrenberg T, Wiechen M et al (2010) Calcium manganese(III) oxides ($CaMn_2O_4 \cdot xH_2O$) as biomimetic oxygen-evolving catalysts. Angew Chem Int Ed 49: 2233–2237
42. Zaharieva I, Najafpour MM, Wiechen M et al (2011) Synthetic manganese-calcium oxides mimic the water-oxidizing complex of photosynthesis functionally and structurally. Energy Environ Sci 4:2400–2408
43. Wiechen M, Najafpour MM, Allakhverdiev SI et al (2014) Water oxidation catalysis by manganese oxides: learning from evolution. Energy Environ Sci 7:2203

44. Wiechen M, Berends HM, Kurz P (2012) Water oxidation catalysed by manganese compounds: from complexes to 'biomimetic rocks'. Dalton Trans 41:21–31
45. Zhang M, Frei H (2015) Towards a molecular level understanding of the multi-electron catalysis of water oxidation on metal oxide surfaces. Catal Lett 145:420–435
46. Najafpour MM, Moghaddam AN, Dau H et al (2014) Fragments of layered manganese oxide are the real water oxidation catalyst after transformation of molecular precursor on clay. J Am Chem Soc 136:7245–7248
47. Hocking RK, Brimblecombe R, Chang LY et al (2011) Water-oxidation catalysis by manganese in a geochemical-like cycle. Nat Chem 3:461–466
48. Wiechen M, Zaharieva I, Dau H et al (2012) Layered manganese oxides for water-oxidation: alkaline earth cations influence catalytic activity in a photosystem II-like fashion. Chem Sci 3:2330–2339
49. Zaharieva I, Chernev P, Risch M et al (2012) Electrosynthesis, functional, and structural characterization of a water-oxidizing manganese oxide. Energy Environ Sci 5:7081–7089
50. Schöler A, Zaharieva I, Zimmermann S et al (2014) Biogenic manganese-calcium oxides on the cell walls of the algae *Chara Corallina*: elemental composition, atomic structure, and water-oxidation catalysis. Eur J Inorg Chem 2014:780–790
51. Bergmann A, Zaharieva I, Dau H et al (2013) Electrochemical water splitting by layered and 3D cross-linked manganese oxides: correlating structural motifs and catalytic activity. Energy Environ Sci 6:2745
52. Tsui EY, Tran R, Yano J et al (2013) Redox-inactive metals modulate the reduction potential in heterometallic manganese-oxido clusters. Nat Chem 5(4):293–299
53. Lee SY, González-Flores D, Ohms J et al (2014) Screen-printed calcium-birnessite electrodes for water oxidation at neutral pH and an "electrochemical Harriman series". ChemSusChem 7:3442–3451
54. Najafpour MM, Rahimi F, Amini M et al (2012) A very simple method to synthesize nano-sized manganese oxide: an efficient catalyst for water oxidation and epoxidation of olefins. Dalton Trans 41:11026–11031
55. Menezes PW, Indra A, Littlewood P et al (2014) Nanostructured manganese oxides as highly active water oxidation catalysts: a boost from manganese precursor chemistry. ChemSusChem 7:2202–2211
56. Elmaci G, Frey CE, Kurz P et al (2015) Water oxidation catalysis by birnessite@iron oxide core-shell nanocomposites. Inorg Chem 54:2734–2741
57. Mette K, Bergmann A, Tessonnier J et al (2012) Nanostructured manganese oxide supported on carbon nanotubes for electrocatalytic water splitting. ChemCatChem 4:851–862
58. Huynh M, Bediako DK, Nocera DG (2014) A functionally stable manganese oxide oxygen evolution catalyst in acid. J Am Chem Soc 136:6002–6010
59. Fekete M, Hocking RK, Chang SLY et al (2013) Highly active screen-printed electrocatalysts for water oxidation based on beta-manganese oxide. Energy Environ Sci 6:2222
60. Takashima T, Hashimoto K, Nakamura R (2012) Mechanisms of pH-dependent activity for water oxidation to molecular oxygen by MnO_2 electrocatalysts. J Am Chem Soc 134:1519–1527
61. Carmo M, Fritz DL, Mergel J et al (2013) A comprehensive review on PEM water electrolysis. Int J Hydrogen Energy 38:4901–4934
62. Joya KS, Joya YF, Ocakoglu K et al (2013) Water-splitting catalysis and solar fuel devices: artificial leaves on the move. Angew Chem Int Ed Engl 52:10426–10437
63. Nocera DG (2012) The artificial leaf. Acc Chem Res 45:767–776

Top Curr Chem (2016) 371: 73–104
DOI: 10.1007/128_2015_646

Published online: 2 July 2015

Solar Water Splitting Using Semiconductor Photocatalyst Powders

Kazuhiro Takanabe

Abstract Solar energy conversion is essential to address the gap between energy production and increasing demand. Large scale energy generation from solar energy can only be achieved through equally large scale collection of the solar spectrum. Overall water splitting using heterogeneous photocatalysts with a single semiconductor enables the direct generation of H_2 from photoreactors and is one of the most economical technologies for large-scale production of solar fuels. Efficient photocatalyst materials are essential to make this process feasible for future technologies. To achieve efficient photocatalysis for overall water splitting, all of the parameters involved at different time scales should be improved because the overall efficiency is obtained by the multiplication of all these fundamental efficiencies. Accumulation of knowledge ranging from solid-state physics to electrochemistry and a multidisciplinary approach to conduct various measurements are inevitable to be able to understand photocatalysis fully and to improve its efficiency.

Keywords Electrocatalysis · Hydrogen · Overall water splitting · Photocatalysis · Semiconductor

Contents

K. Takanabe (✉)
Division of Physical Sciences and Engineering, KAUST Catalysis Center (KCC),
King Abdullah University of Science and Technology (KAUST), 4700 KAUST,
Thuwal 23955-6900, Saudi Arabia
e-mail: kazuhiro.takanabe@kaust.edu.sa

1 Introduction

Solar energy is by far the most abundant renewable energy resource [1]. The total solar energy absorbed by the Earth is 3.85×10^{24} J $year^{-1}$, which is ~10^4 greater than the world energy consumption [2]. To compensate for increasing energy demand, solar energy conversion has to be at least partially implemented. To achieve extensive solar energy conversion, the large-scale collection of solar flux is essential. A simple calculation using the air-mass 1.5 global (AM 1.5G) (~1 kW m^{-2}) solar spectrum reported by the National Renewable Energy Laboratory (NREL) predicts that a collection area of ~750,000 km^2 is required to meet global energy demands [3]. Therefore, solar energy conversion technologies must have tremendous scalability, irrespective of the conversion method used. For easy storage and transport, producing chemicals and fuels in which the energy is in the form of chemical bonds is preferred. This technology requires water as the sole reactant and directly forms chemical energy (H_2) in a single reactor. The reactor contains water and photocatalysts as powders. It does not require any complicated parabolic mirrors or electronic devices. Overall, the simplicity of this type of solar hydrogen production plant makes the system economically advantageous [4].

The photocatalytic system for overall water splitting produces a mixture of H_2 and O_2 followed by the separation of these products [5]. It is critical to develop highly efficient photocatalysts made from abundant elements using a mass-production process. The overall water splitting reaction is

$$H_2O \rightarrow H_2 + \frac{1}{2}O_2 \quad \Delta G = 237\ \text{kJ mol}^{-1}. \tag{1}$$

The above reaction (two-electron reaction in this stoichiometry) has a Gibbs free energy of 237 kJ mol^{-1}. The photon energy, E, is expressed by

$$E = h\nu = \frac{hc}{\lambda}, \tag{2}$$

where h is Planck's constant, ν is the frequency of the photon, c is the speed of light, and λ is the wavelength of the photon. Thus, a photon energy of 1.23 eV is thermodynamically required to drive overall water splitting, which is equivalent to a wavelength of ~1,000 nm. A photocatalyst for water splitting thus requires a bandgap greater than 1.23 eV. It is also essential to consider the overpotential, i.e., the excess potential beyond the thermodynamic potential required to overcome the activation energy.

As far as solar energy conversion is concerned, analysis of the solar spectrum is essential to determine the important requirements for photocatalyst materials for overall water splitting. Figure 1 shows the solar to hydrogen (STH) energy and the number of photons as a function of wavelength according to the data from the standard AM 1.5G spectrum [3]. The STH efficiency can be calculated using the Gibbs free energy of reaction (1):

$$\mathrm{STH} = \frac{\text{Output energy}}{\text{Energy of incidence solar light}} = \frac{r_{\mathrm{H_2}} \times \Delta G}{P_{\mathrm{Sun}} \times A_{\mathrm{Geometric}}}, \tag{3}$$

where $r_{\mathrm{H_2}}$ is the hydrogen production rate, P_{Sun} is the energy flux of sunlight, and $A_{\mathrm{Geometric}}$ is the area of the reactor. The solar energy spectrum ($P_{\mathrm{Sun}} = 1{,}003$ W m^{-2}) has ~93 W m^{-2} in the ultraviolet (UV) region ($\lambda \leq 400$ nm; 9.3%), ~543 W m^{-2} in the visible region (400 nm $< \lambda \leq 800$ nm; 54.1%), and 367 W m^{-2} in the infrared (IR) region ($\lambda > 800$ nm; 36.6%). For overall water splitting, hydrogen is the sole product corresponding to an energy of 1.23 eV (or 237 kJ mol^{-1}) equivalent. The theoretically attainable STH efficiency and hydrogen production rate can thus be calculated from the number of photons in the spectrum from shortest wavelength to the respective wavelength at different quantum efficiencies (QE), e.g., 30%, 60%, or 100%:

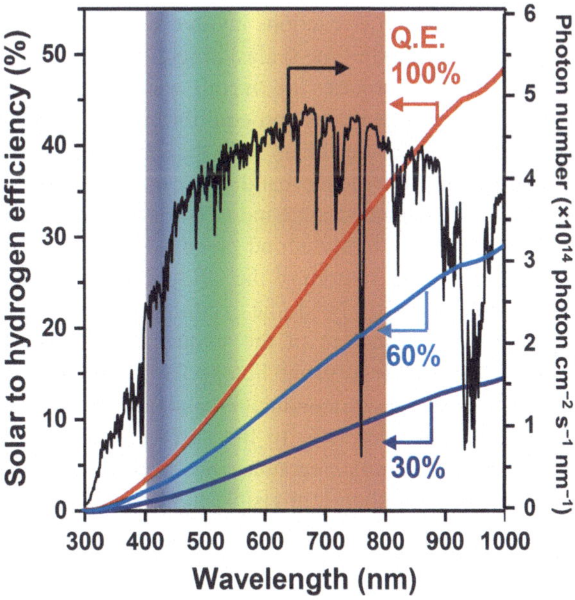

Fig. 1 Photon number of AM 1.5G as a function of wavelength and theoretical for solar-to-hydrogen efficiency integrated from a low wavelength to the respective wavelength at QEs of 30%, 60%, and 100%

$$QE(h\nu) = \frac{2 \times r_{H_2}}{I_0(h\nu)}. \tag{4}$$

By analyzing the solar irradiance, the theoretical maximum STH efficiency can be calculated to be ~48% (at 100% QE), which is integrated from the UV to a wavelength of ~1,000 nm (1.23 eV). Based on the definition of STH efficiency, the excess energy from photons greater than 1.23 eV has to be dissipated (mainly as heat). The energy loss becomes more apparent as the wavelength decreases. For UV photons, over half the energy is dissipated (only 1.23 eV can be utilized to generate H_2) and consequently the theoretical maximum STH efficiency using only UV light ($\lambda \leq 400$ nm) is only 3.3% (for a QE of 100% at each wavelength). In contrast, the target of STH efficiency is generally set to 10% to become competitive within the hydrogen market. Although the technology is simple, the low STH efficiency requires a large area for the solar reactor. This fact clearly shows why the development of a visible-light-responsive photocatalyst is essential for a high STH process.

As mentioned previously, the benchmark for the STH efficiency is set to 10%. This efficiency corresponds to a hydrogen production rate of ~154 μmol H_2 cm^{-2} h^{-1} and a photoelectrochemical current of ~8.3 mA cm^{-2}, i.e., a consumption rate of ~260 photons nm^{-2} s^{-1} on a flat surface. These values are practically useful for experiments in a laboratory scale. To achieve this efficiency, the development of materials that absorb wavelengths of light up to 600–700 nm (~1.8–2.0 eV) is essential. The challenge is that these materials have to have suitable band positions for water splitting, as discussed below.

The photocatalytic reactions involve various photophysical and electrocatalytic processes on different time scales. Figure 2 shows a general scheme for the photon-induced reaction process for overall water splitting using a solid photocatalyst. Photon absorption initiates non-equilibrium photophysical and photochemical processes. These processes begin with the generation of an exciton, i.e., excitation of an electron in the valence band (VB) or the highest occupied molecular orbital (HOMO) to the conduction band (CB) or the lowest unoccupied molecular orbital (LUMO) [6]. This femtosecond process is followed by relaxation of the electron and the hole to the bottom of the CB and to the top of the VB, respectively, on a similar time scale [6]. Next, the exciton (electron–hole pair) needs to be separated depending on the nature of the semiconductor. The electronic structure should guide the excited electron and hole (polaron) to move independently, fully utilizing the junctions at the semiconductor–catalyst and semiconductor–solution *interfaces*. Successful charge transfer to the surface and to the cocatalyst is followed by electrocatalytic redox reactions on a time scale longer than microseconds [6]. Because each photon in visible light possesses limited overpotential for water

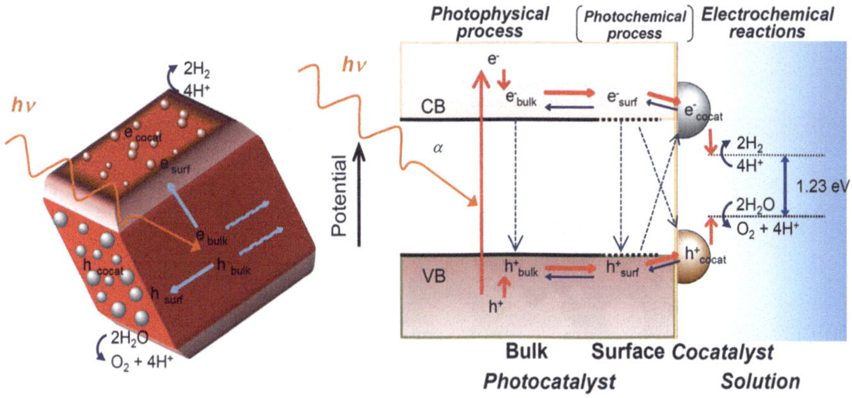

Fig. 2 Scheme of photon-induced reaction process for overall water splitting by solid photocatalyst

splitting, the presence of such electrocatalysts on the surface of semiconductor is essential.

The primary aim of this review is to focus on discussion of the fundamental parameters involved in photocatalytic overall water splitting. The processes involved in photocatalytic reactions, especially water splitting with powder photocatalysts suspended in the liquid phase, are of interest. The steps involved in the photocatalysis for water splitting are divided into the following six processes and the relevant parameters are described:

1. Photon absorption
2. Exciton separation
3. Carrier diffusion
4. Carrier transport
5. Catalytic efficiency
6. Mass transfer of reactants

Figure 3 summarizes the parameters that are highlighted according to the six processes. This review gives a broad and rather conceptual description only, rather than detailed discussion on the specific materials, which can be found in different reviews [7–26]. The author trusts that the understanding of these physicochemical properties leads to "photocatalyst materials by design" where the electronic structure of the semiconductor, *interface* development, and electrocatalytic properties are fully connected to achieve the complex sequential processes to achieve finally overall water splitting.

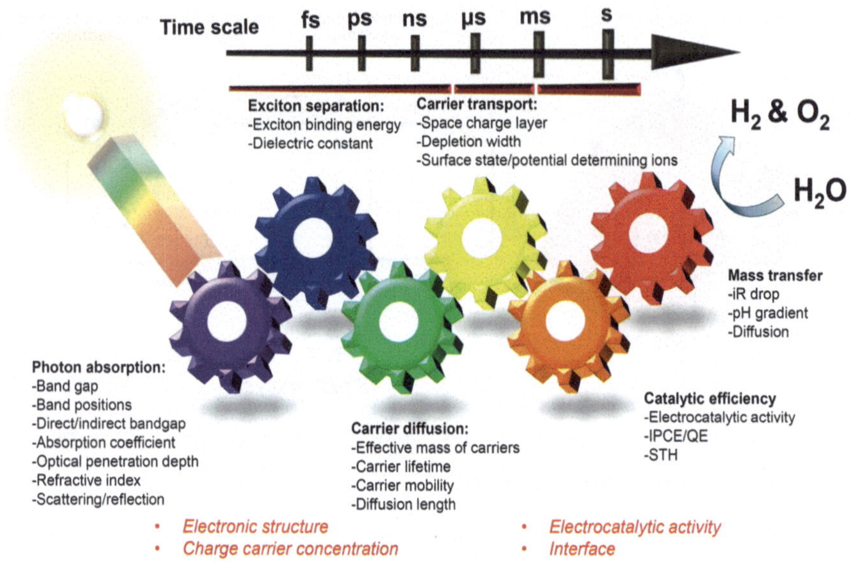

Fig. 3 Parameters requiring attention for efficient overall water splitting. Overall water splitting is only successful for high efficiencies at all six gears depicted in the scheme. Different time scales of the reactions are also displayed

2 Photocatalytic Processes

2.1 Photon Absorption and Exciton Generation

Photon absorption by semiconductors generally occurs via the excitation of electrons in the valence band into the conduction band, generating excitons (excited electron and hole pairs). At the macroscopic level, photon absorption by powder suspension systems includes intrinsic and extrinsic absorptions in addition to scattering, reflection, and transmission. A schematic image of the light absorption process by a semiconductor powder is shown in Fig. 4.

A spherical particle with a diameter of 100 nm is irradiated by nearly 10^7 photons (considering wavelengths <600 nm in the AM 1.5G spectrum) every second [27]. Among those photons, some of them experience scattering by the powder [28]. Scattering can be categorized into Rayleigh and Mie scattering, depending on the size of the particle. The size of a scattering particle is defined by the ratio of its characteristic dimension to the wavelength of the scattered light:

$$x = \frac{2\pi r}{\lambda}. \tag{5}$$

Scatter diameters much smaller than the wavelength result in Rayleigh scattering, whereas larger diameters result in Mie scattering. Based on the sizes of the photocatalyst particles (a few nanometers to 10 μm), the photons in the UV–vis

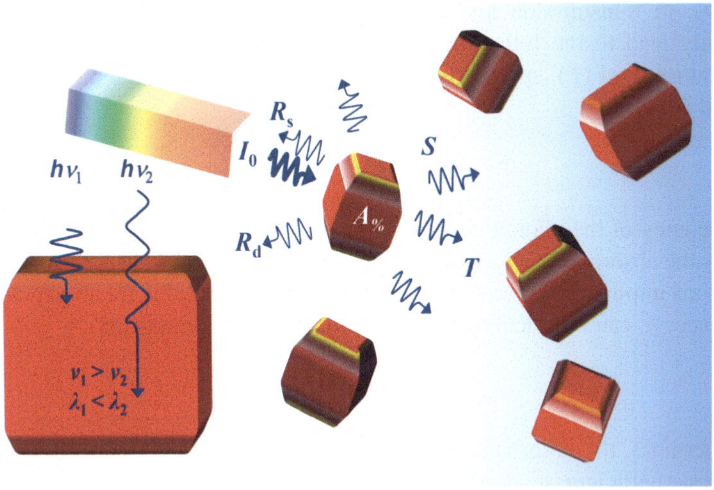

Fig. 4 Scheme of photon absorption process using semiconductor powder. See main text for explanation of the notations

range with wavelengths of 300–800 nm should be appropriately taken into consideration. The scattering coefficient for Rayleigh scattering varies for small particles inversely with the fourth power of the wavelength (~λ^{-4}), giving higher scattering for shorter wavelengths (i.e., UV > VIS > NIR).

Similarly, some of the photons undergo reflection and transmission through media with distinct refractive indices [29]. The description of light propagation in a complex photocatalyst system may be treated using the Fresnel equations. The reflection of light retains either the geometry or the energy of the incident light. Diffuse reflection can also occur, especially on a rough surface, which retains the energy and is indeed utilized to elucidate the absorption characteristics of the powder semiconductor.

In a photoreactor consisting of a semiconductor powder in suspension configuration, the scattered and reflected light can be further absorbed by other semiconductor particles. This is indeed beneficial for suspension systems compared with film configurations. It is very difficult to measure quantitatively the extent of the scattering and reflection in experiments, so the photocatalytic efficiency is generally associated with the incident photons (which can be separately measured) to the reactor or measured actinometrically [30].

The quantitative measurements of the absorption efficiencies of the photocatalytic powders provide useful information for the development of efficient photocatalyst materials and systems. The absorption spectra indicate the consequences of bandgap excitation, d-d transitions, phonon absorptions, and excitations associated with defect states [31]. The absorption coefficient, α (wavelength dependent), is an important parameter of semiconductors that provides their intrinsic characteristics. Ideally, absorption measurements using photon flux of incident light, Φ_0, in transmission and reflectance mode (using an integrating sphere) commonly lead to the following relationships with absorptance ($A_{\%}$), and light that is transmitted (T), specularly reflected (R_s), forward-scattered (S), and back-scattered (R_d) [31]:

$$\Phi_0 = A_{\%} + T + R_S + S + R_d. \tag{6}$$

The efficiency of the photon absorption process occurring within a sample is thus given by the absorptance $A_{\%}$, which is the fraction of photons absorbed out of all of the photons impinging on the sample. Using the resulting value for absorptance, the absorbance, A, can be calculated using the following relationship:

$$A_{\%} = 1 - 10^{-A}. \tag{7}$$

Furthermore, the absorbance can be normalized by the thickness to obtain the absorption coefficient, α, as

$$\alpha\left(\text{cm}^{-1}\right) = \frac{\ln(10) \times A}{l\,(\text{cm})}, \tag{8}$$

where l is the path length of the light through the sample. The light intensity, I, in the unit of power, as a function of distance, d, is given by

$$I(d) = \sum^{h\nu} \Phi_0(h\nu)e^{-\alpha d}. \tag{9}$$

The absorption coefficient, α, is preferably determined using absorptance, $A_{\%}$, which gives absorbance, A, though the contribution of scattering is excluded whenever possible. For a practical absorptance measurement, the thin film configuration of semiconductors provides a more precise description, compared to powder configurations, once the film thickness is more rigorously defined [32]. Different semiconductor synthesis techniques for different configurations are reviewed elsewhere [22]. For accurate measurement of the absorption coefficient, the absorption coefficient is linked with the measured transmittance (T) and reflectance (R) [32]:

$$T = \frac{(1 - R')^2 e^{-\alpha d}}{1 - R'^2 e^{-2\alpha d}}, \tag{10}$$

$$R = R'\left(1 + \frac{(1 - R')^2 e^{-2\alpha d}}{1 - R'^2 e^{-2\alpha d}}\right), \tag{11}$$

where R' is the single surface reflectance of the material. These equations lead to determination of the absorption coefficient as a function of measured T and R as follows [33]:

$$\alpha = \frac{1}{d}\ln\left(\frac{(1-R)^2}{2T} + \sqrt{\left(\frac{(1-R)^4}{4T^2} + R^2\right)}\right), \tag{12}$$

and, when R is negligible and/or the following inequality is satisfied [34],

$$Re^{-\alpha d} \ll 1, \tag{13}$$

then (12) is simplified to

$$T = (1-R)^2 e^{-\alpha d}, \tag{14}$$

and, thus,

$$\alpha = -\frac{1}{d}\ln\frac{T}{(1-R)^2}. \tag{15}$$

From the same spectra, analyzing interference fringes in a weaker absorption range would provide useful information. If the thickness d is uniform, the interference fringes can be used to obtain the position of the interference maxima λ [35]:

$$2nd = m\lambda, \tag{16}$$

where n is the refractive index and m is the order of the interference. When m is not known, the adjacent maxima can be used to eliminate m:

$$(2nd)^{-1} = \lambda_{m+1}^{-1} - \lambda_m^{-1}. \tag{17}$$

A complex refractive index n_c is described as

$$n_c = n + i\kappa, \tag{18}$$

where κ is the extinction coefficient. Once $\alpha(\lambda)$ is known, $\kappa(\lambda)$ can be calculated from the equation

$$\kappa(\lambda) = \frac{\alpha(\lambda)}{4\pi}. \tag{19}$$

Then the dielectric constant, a measure of the charge retention capacity of a medium, can be obtained. The complex dielectric constant, ε_r, is described as [36]

$$\varepsilon_r = \varepsilon_1 + i\varepsilon_2, \tag{20}$$

where ε_1 is the real part of the dielectric constant associated with dispersion:

$$\varepsilon_1 = n^2 - \kappa^2, \tag{21}$$

and ε_2 is dielectric loss, i.e., the imaginary part associated with the dissipative rate of the wave:

$$\varepsilon_2 = 2n\kappa. \tag{22}$$

These values lead to the dielectric loss tangent, $\tan \delta = \varepsilon_2/\varepsilon_1$, defined by the angle between the capacitor's impedance vector and the negative reactive axis, providing an important parameter to quantify the inherent dissipation of electromagnetic energy into heat. The dielectric constant, ε_r, can be divided into the contributions from the electronic density, ε_∞ and from the motion of ions constituting the material, ε_{vib} [37]:

$$\varepsilon_r = \varepsilon_\infty + \varepsilon_{vib} = \varepsilon_\infty + \frac{4\pi}{V}\sum_p \frac{Z_p^2}{v_p^2}, \tag{23}$$

where v_p is the phonon frequency of mode p, V is the unit cell volume, and Z_p is the mass-weighted mode effective born vector, which is proportional to IR absorbance, I_p.

Once the absorption coefficient is obtained, the absorption depth, which describes how far light can penetrate into a material before being absorbed, can be determined by simply taking the inverse of the absorption coefficient α (on a natural logarithm scale). These penetration depths are generally defined as the depths where the light intensity decreases by factors of $1/e$ (~36%) of the incident intensity. The absorption coefficient and the absorption depth vary with the incident light wavelength: typical density of states of the semiconductor suggests that shorter wavelength light is absorbed closer to the photocatalyst surface than longer wavelength light. Thus, before being absorbed, visible light travels further in a photocatalyst than UV light. Even if a long-wavelength photon is absorbed, it penetrates deep into the bulk of the photocatalyst, so the generated excitons have a long distance to travel to the surface (Fig. 4). Thus, there is a greater probability of excited electrons and holes recombining before they can participate in surface redox reactions. The absorption depth (together with scattering and reflection) also accounts for how many semiconductor particulates are required to absorb effectively the incident light in the photoreactor (the depth of the photoreactor).

The *electronic structure* of the semiconductor determines various critical characteristics relevant to photocatalysis. Regarding the absorption properties, the electronic structure of the semiconductor not only decides the bandgap and band positions but also the absorption coefficient and direct/indirect nature of the light absorption. The absorption parameters of a given semiconductor primarily depend on its crystal structure, which in turn determines its electronic structure. Essentially, direct bandgaps lead to high absorption coefficients, whereas indirect bandgaps give low absorption coefficients. The symmetry of the crystal structure gives

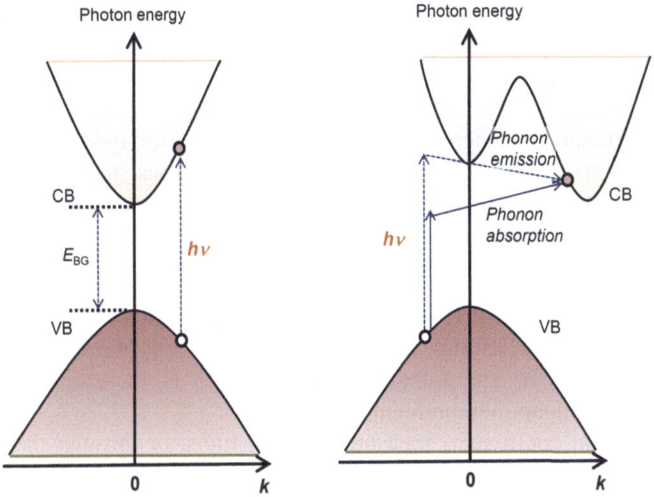

Fig. 5 Electronic structures for (*left*) direct and (*right*) indirect bandgap semiconductors and their excitations

allowed and forbidden energies. Indirect transitions involve both a photon and a phonon because the band edges of the conduction and valence bands are widely separated in k space, as described in Fig. 5. A typical indirect bandgap semiconductor, Si, possesses a typically low absorption coefficient of 1×10^3 to 1×10^5 cm^{-1}, giving absorption depths of up to a few micrometers for visible light (400–800 nm) [38]. Therefore, if Si nanoparticles (less than a micron in diameter) are synthesized for photocatalysis, there is a significant chance that a single photocatalyst particle may not absorb many photons per unit time. Typical absorption coefficients of direct bandgap semiconductors typically fall into the range of 1×10^4 to 1×10^6 cm^{-1}, equivalent to absorption depths of 1,000–10 nm. This range of absorption depths becomes comparable to the typical particle size of the photocatalysts. Therefore, it is important to consider the absorption coefficient to determine the efficiency of the photocatalytic materials and photocatalyst design.

For powder semiconductors it is often difficult to address the true absorption coefficient. Alternatively, the measurement of diffuse reflectance is a powerful tool to acquire information about light absorption [29]. Because of the anisotropic nature of the powder system, the Kubelka–Munk function $f(R)$ is frequently used to describe absorption/reflectance:

$$f(R) = \frac{(1-R)^2}{2R} = \frac{\alpha}{s}, \tag{24}$$

where s is the scattering coefficient. The Kubelka–Munk model is based on the following assumptions [39]:

- The sample is modeled as a plane layer of finite thickness but with infinite sheet approximation, so there are no boundary effects
- A perfectly diffuse and homogenous illumination is incident on the surface
- The only interactions of light with the medium are scattering and absorption; polarization and spontaneous emission (fluorescence) are ignored
- The sample is considered isotropic and homogeneous and contains optical heterogeneities
- No external or internal surface reflections occur
- The scattering coefficient, s, is constant for any layer thickness

Because of the lack of an s value, it is difficult to pin down the absolute value of α. Assuming that the scattering coefficient is independent of wavelength, $f(R)$ is then proportional to α. The Kubelka–Munk function is, therefore, effective for addressing the bandgap measurement.

The absorption coefficients measured in thin films, or through the Kubelka–Munk function for powdered samples, can be used to obtain the bandgap, E_g, of the semiconductor. When $\alpha > 10^4$ cm^{-1}, it often obeys the following relationship presented by Tauc and supported by Davis and Mott [40, 41]:

$$\alpha h\nu \propto \left(h\nu - E_g\right)^{\frac{1}{n}}, \tag{25}$$

where n can take values of 3, 2, 3/2, or 1/2, corresponding to indirect (forbidden), indirect (allowed), direct (forbidden), and direct (allowed) transitions, respectively [31]. Tauc plots, i.e., $(\alpha h\nu)^{\frac{1}{n}}$ as a function of $h\nu$ (n is the same as above), give E_g from the intersection of a tangent to the slope in the linear region of the absorption onset with the baseline. A more detailed description of how to utilize accurately the information of Tauc plots can be found in [31].

The generation rate, G, per photocatalyst particle is the number of electron–hole pairs generated per photon striking the particle with the depth, x. The assumption that each photon directly causes generation of an electron–hole pair can be first described using the Beer–Lambert law approximation [42]:

$$G = \alpha \Phi_{0,p} e^{-\alpha x}, \tag{26}$$

where $\Phi_{0,p}$ is the photon flux at the particle surface (photons (particle geometric area)$^{-1}$ s^{-1}). The photon flux density may be directly obtained from the irradiance measurements. As previously mentioned in (9), the light intensity, I, decays as it passes through the semiconductor materials because of light absorption and creates the generation of electron–hole pairs according to [43]

$$\frac{dI(x)}{dx} = -\alpha I(x) = -\sum^{h\nu} G(h\nu). \tag{27}$$

Recent advances in density functional theory (DFT) calculations give quite accurate and reliable estimates for the electronic structures of semiconductors from a given crystal structure, resulting in accurate descriptions of the densities of states (DOS). Essentially, the accurate DOS provides information about the bandgap energy, the band positions (vs vacuum), the connectivity of both the CB and VB within the crystal structure, and the absorption coefficient as a function of wavelength. As discussed later, DFT calculations are now able to describe fundamental parameters for semiconductors with high accuracy, such as the effective mass, exciton binding energy, and dielectric constant, etc. The effective depth (thickness) of the powder suspension can be theoretically estimated from the absorption depth, which is significant information because the concentration (amount) of the photocatalysts per given area and volume can be determined for a given semiconductor.

2.2 Exciton Separation

Upon absorption of light, excitons (electron–hole pairs) are generated [44]. For photocatalytic processes, the subsequent step includes exciton separation to generate excited electrons and holes (free carriers). When the dielectric screening potential and the exciton radius are large, the excitons are Mott–Wannier type, which are typical for bulk semiconductors. If the exciton radius is small, the exciton is Frenkel type, typical for the dielectric characteristics of molecules and organic polymers. Characteristics of these two types of exciton generation are summarized in Fig. 6.

The practical measurement of exciton binding energy involves photoemission spectroscopy, optical absorption spectroscopy, photoconductivity screening potential, spectroscopy, and magneto-optical spectroscopy [45]. Using a single dielectric constant ε_r and Bohr's quantum theory, the energy E_n of the exciton series for the Mott–Wannier type can be derived from [46]

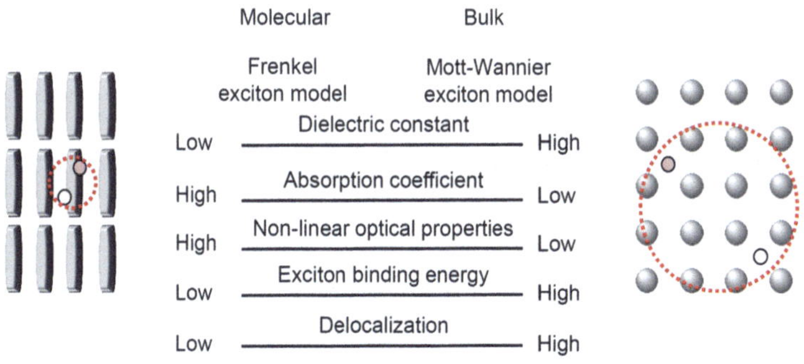

Fig. 6 Frenkel and Mott–Wannier exciton models and their characteristics

$$E_{\mathrm{n}} = E_{\infty} - \frac{R_{\mathrm{ex}}}{n^2}, \tag{28}$$

where E_{∞} is the series limit (a constant) and n is a quantum number $n = 1, 2, \ldots$. Thus, R_{ex} is the exciton binding energy, which represents the energy required to ionize an exciton in its lowest energy state, i.e., the energy separation between the lowest bound state ($n = 1$) and the series limit, and is much smaller than that for a hydrogen atom [47]. The 1 s state of the exciton is the exciton binding energy, R_{ex}, and can be described as

$$R_{\mathrm{ex}} = \frac{m^* e^4}{2h^2 \varepsilon_{\mathrm{r}}^2} = E_1 \frac{m^*}{\varepsilon_{\mathrm{r}}^2}, \tag{29}$$

where m^* is the reduced effective mass of the electron–hole system $\left(\frac{1}{m^*} = \frac{1}{m_{\mathrm{n}}^*} + \frac{1}{m_{\mathrm{p}}^*}\right)$, e is the elemental charge, and h is Planck's constant. From the equation, it is observed that the binding energy becomes small if the dielectric constant of the semiconductor is high. For Mott–Wannier excitons, the typical binding energy is less than 10 meV and the radius is ~10 nm. For Frenkel excitons, these values can be as high as 1 eV and ~1 nm. The energy should be compared with the thermal energy (25 meV at room temperature) [37], and efficient separation of excitons requires the binding energy to be lower than this value. For the high binding energies in Frenkel-type excitons, heterojunctions at the molecular levels are essential to separate charges and are prevalent in organic polymer photovoltaic cells. A high delocalization of the charge carriers occurs because of the low effective masses and large collision times associated with defects.

The anisotropic nature of exciton binding energy is also critical. The crystal and electronic structures determine the anisotropic nature in the different crystal orientations. These differences are currently possible to predict using DFT calculations, and it is therefore an effective method for this investigation because of its high accuracy. Inoue also discussed the correlation between distorted metal-oxygen crystal structures and photocatalytic activity; high distortion creates an anisotropic electronic field upon exciton generation which assists in separation [17].

2.3 Carrier Diffusion and Recombination

Once the exciton is separated, free charge carriers have to transfer to the surfaces for successful photocatalysis [48]. For the electrons and holes, the ability to move around in a material and transport charge is called mobility (electron mobility and hole mobility, respectively). The diffusion coefficient, D, and the mobility of the charge carrier, μ, are connected through the Einstein relations [37]:

$$D = \frac{k_B T}{e}\mu, \tag{30}$$

where k_B is the Boltzmann constant and e is the elemental charge. Thus, the mobility (in a specific direction) can be obtained as

$$\mu = e\frac{\tau_c}{m^*}, \tag{31}$$

where τ_c is the collision time of the charge carrier and m^* is the effective mass. Now the flow of electrons and holes are considered current in a given direction, e.g., the x-direction. There are two driving forces for the carrier movement: the concentration gradient of the carriers (change in concentration) and external electric fields [44]. When a concentration gradient of the carriers is present, they distribute by themselves from regions of high concentration to regions of low concentration only through thermal motion. The movement of charge results in a so-called diffusion current. In a gradient of electrons or holes, $\frac{dp_x}{dx}$ and $\frac{dn_x}{dx}$ are non-zero, generating diffusion currents described with diffusion coefficients, D_p and D_n, and charges, $+q$ and $-q$, according to Fick's law (p for holes, n for electrons, respectively). When an electric field is present, the potential gradient also causes a drift current. The drift and diffusion currents make up the total current in a semiconductor. The total current density (for one vector) is driven by the carrier gradients and the potential gradients:

$$\begin{aligned} \vec{J}_{\text{total}} &= \vec{J}_{\text{p,diffusion}} + \vec{J}_{\text{n,diffusion}} + \vec{J}_{\text{p,drift}} + \vec{J}_{\text{n,drift}} \\ &= -qD_p\frac{dp_x}{dx} + qD_n\frac{dn_x}{dx} + pq\mu_p\vec{E}_x + nq\mu_n\vec{E}_x. \end{aligned} \tag{32}$$

Under equilibrium conditions, the total current density should be zero. In powder semiconductor systems there are no external electric fields by choice. Hence the carrier mobility of the photocatalyst must be high because it strongly relies on the diffusion process of the excited carriers. However, if the doping or electronic structure is not uniform in terms of potential, there is a concentration gradient which can create an electric field within the semiconductor and result in non-zero current densities, which may cause electrons and holes to transfer. The anisotropic nature of the electronic structure within a semiconductor is also of great importance. The anisotropy guides electrons and holes to move to different crystal orientations, which allows them to avoid recombination. The gradient is even facilitated by junctions intentionally added between the metal, semiconductor, and electrolyte. This result is discussed in the following section.

Carrier mobility is one of the most important parameters that determines overall photocatalytic efficiency. The mobility is affected by temperature, doping concentration, and the magnitude of the applied field. It also depends on the effective masses of the electrons and holes, which are determined primarily by the electronic structure of the semiconductor. The carriers with small effective masses have large mobilities. As a result, the holes are significantly less mobile than the electrons. The doping concentration also has a significant influence on the mobility. When the doping concentration is low, the mobility can be considered independent of the doping concentration. When the concentration of dopants becomes high, the mobility of the carriers decreases monotonically. These factors are closely associated with the recombination process. Practically, the resistivity, charge carrier concentration, and resultant mobility of the semiconductors can be measured by the van der Pauw technique with the Hall measurement [49, 50], although this method requires a high quality semiconductor slab.

The minority carrier lifetime, a consequence of the *charge carrier concentrations* and charge mobilities, is the intrinsic indicator of whether a semiconductor material is an effective photocatalyst. The minority carrier lifetime is the average time a typical minority carrier exists before recombination. In other words, the lifetime from indirect transitions is inversely proportional to the trap density. It is denoted τ_n for electrons in a p-type material and τ_p for holes in an n-type material. The minority carrier lifetime is measured optically or electronically using spectroscopic techniques [51]. One of the methods is using the simultaneous measurement of the light-induced photoconductance of the sample and the corresponding light intensity. The transient mode using a short light pulse used to be the common method, but the recent alternative method allows for measurement under quasi-steady state or quasi-transient illumination. For quasi-steady-state and quasi-transient carrier lifetime measurements, effective carrier lifetimes τ_{eff} can be given in the following generalized form [51]:

$$\tau_{\text{eff}} = \frac{\Delta n(t)}{G(t) - \frac{d\Delta n(t)}{dt}}, \tag{33}$$

where G is the generation rate and Δn is the time-dependent values of the excess carrier density.

Once the carrier lifetime is measured, the minority carrier diffusion length, denoted L_{p} for holes in an n-type material and L_{n} for electrons in a p-type material, represents the average distance the excess minority carrier travels from where it was generated to where it is annihilated:

$$L_{\text{n}} \equiv \sqrt{D_{\text{n}}\tau_{\text{n}}},\ L_{\text{p}} \equiv \sqrt{D_{\text{p}}\tau_{\text{p}}}. \tag{34}$$

Examples of the lifetime as a function of carrier concentrations for Si (indirect bandgap) are shown in Fig. 7. Law et al. show that the minority carrier lifetime can be fitted numerically with the following equation [52]:

$$\tau = \frac{\tau_{\text{o}}}{1 + N_{\text{D}}/N_{\text{ref}} + \tau_{\text{o}} C_{\text{A}} N_{\text{D}}^2}. \tag{35}$$

The parameters used in the fit for n-type and p-type Si are ($\tau_{\text{o}} = 10$ μs, $N_{\text{ref}} = 1 \times 10^{17}$ cm^{-2}, and $C_{\text{A}} = 1.8 \times 10^{-31}$ cm^6 s^{-1}) and ($\tau_{\text{o}} = 30$ μs, $N_{\text{ref}} = 1 \times 10^{17}$ cm^{-2}, and $C_{\text{A}} = 8.3 \times 10^{-32}$ cm^6 s^{-1}), respectively. If the

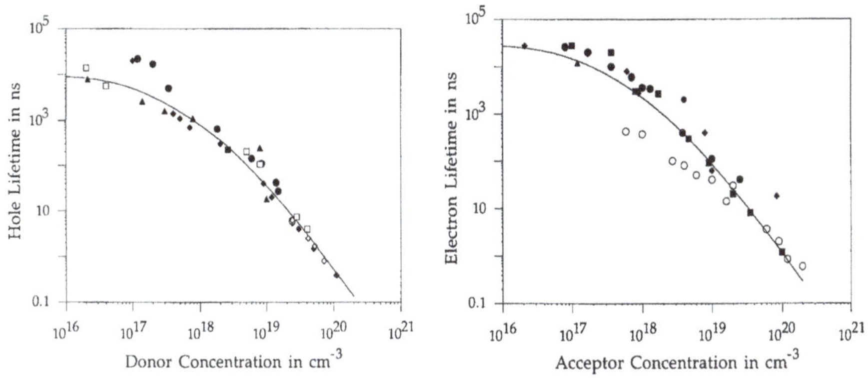

Fig. 7 Hole (*left*) and electron (*right*) lifetimes in heavily doped n-type and p-type silicon, respectively. Copyright 1991, IEEE. Reprinted, with permission, from [52]

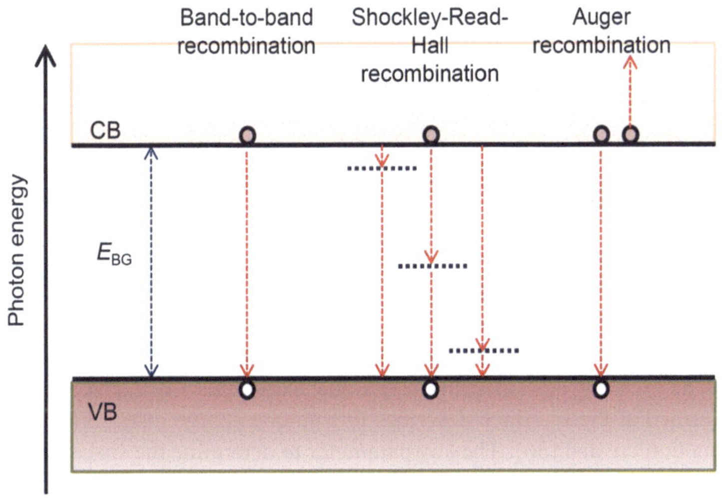

Fig. 8 Types of recombination processes

photocatalysis depends on the diffusion of excited carriers, the carrier lifetime and relevant diffusion length must be long for successful photocatalysis. This consideration is often associated with the semiconductor designed with “high crystallinity” and least defects [22].

The photocatalytic efficiency is decreased when the excitons or free carriers recombine: free carriers are no longer able to move because they are participating in covalent bonds in the crystal. The electron in the conduction band recombines by returning to the valence band whereas a hole in the valence band recombines when an electron annihilates it by falling from the conduction band. An electron can also be captured by a trap or recombination center. Recombination can be categorized into the following three types, as shown in Fig. 8:

- Band-to-band recombination
- Shockley-Read-Hall recombination (defects) [53, 54]
- Auger recombination [55]

The types of recombination strongly depends on the electronic structure of the semiconductor. For direct bandgap semiconductors, band-to-band recombination is dominant, generating radiative processes such as luminescence. This recombination depends proportionally on the density of available electrons and holes. For indirect bandgap semiconductors, the band-to-band recombination is negligibly low, but recombination through defect levels, so-called Shockley–Read–Hall recombination, is dominant because the recombination is facilitated by the exchange of thermal energy with a phonon. Surface recombination occurs in a very similar manner. The surfaces and interfaces often act as trap sites because they contain impurities and abrupt terminations (the presence of dangling bonds which are electronically active). Auger recombination is a process in which an electron and a hole recombine in a band-to-band transition, but with the resulting energy transferred to another electron or hole. The involvement of a third particle affects the recombination rate so the Auger recombination has to be treated differently from band-to-band recombination.

2.4 Carrier Transport

For efficient photocatalysis, the free carriers should transfer to active sites according to the type of semiconductor. For intact semiconductors, the carriers have to transfer via diffusion. Thus, the efficiency is associated with the carrier lifetime. For improved charge separation, the *interface* (surface modification) is essential to creating new effective electronic structures [48]. The space charge region formed at the interface describes this unique charge distribution according to the junction created [56]. The key parameter to determine the space charge layer is the carrier density. For extensive charge separation, one of the driving forces of charge transfer is creating and utilizing band bending, which is strongly driven by the newly-created electronic structure at the interface. The interfaces involve metal–semiconductor, semiconductor–semiconductor, and semiconductor–electrolyte pairs [57]. The electronic structure should have a slope or bending for the free carriers to be transferred.

Where band bending is achieved, the potential description relative to the reference level is drawn in three-dimensional space. Figure 9 depicts the band bending that occurs when an n-type semiconductor is in contact with a solution [58]. In this case, the majority charges (electrons) near the interface of the semiconductor transfer into the solution until the potentials are equilibrated, i.e., the Fermi level of the semiconductor ($E_{F,s}$) equals the redox potential (E^0) in the solution while preserving the band edge positions (pinning). This transfer of the majority carrier creates a depletion layer. Band bending can be described by the Poisson equation as follows [44]:

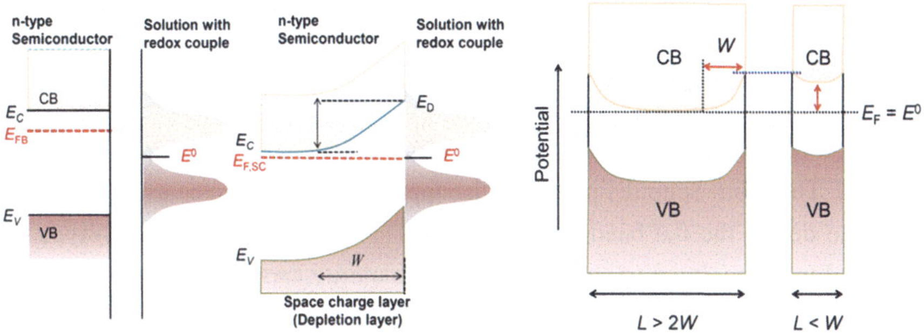

Fig. 9 Scheme of semiconductor–electrolyte junction and influence of particle size on depletion layer

$$\frac{d^2\Psi(x)}{dx^2} = -\frac{eN_D}{\varepsilon_0\varepsilon_r} \quad (0 \leq x \leq W), \tag{36}$$

where N_D is the majority carrier density, W is the depletion layer width, ε_0 is the static permittivity in vacuum, and ε_r is the static dielectric constants of the semiconductor. Solving this for Ψ gives [44]

$$\Psi(x) = -\frac{eN_D}{\varepsilon_0\varepsilon_r}\left(Wx - \frac{1}{2}x^2\right) - V_B. \tag{37}$$

When $x = W$,

$$W = \left\{\frac{2\varepsilon_0\varepsilon_r V_B}{eN_D}\right\}^{1/2}, \tag{38}$$

where e is the elementary charge and V_B is the potential drop in the space-charge layer or the band bending that originates from the potential barrier height for the majority carriers to pass over on transferring from the semiconductor to the solution.

The flat-band potential, E_{FB}, is the potential where band bending diminishes and the band becomes flat. The potential, E_{FB}, is used to estimate the intrinsic band edge of the majority carriers, assuming that the surface states do not affect the surface potential. The absolute position of the majority band edge of the semiconductor can be estimated by determining the majority carrier concentration from Mott–Schottky measurements [58]. This potential is expressed as the flatband potential, E_{FB}, where the band bending becomes zero. The space-charge capacity of the semiconductor, C, is given by the Mott–Schottky equation as follows [59]:

$$\frac{1}{C^2} = \frac{2}{e\varepsilon_0\varepsilon_r A^2 N_D}\left(E_{app} - E_{FB} - \frac{kT}{e}\right), \quad (39)$$

where A is the surface area and E_{app} is the applied potential. The capacity of the electrode with a target semiconductor is measured as a function of the applied potential. The Mott–Schottky plot, $\frac{1}{C^2}$ vs the applied potential, can be extrapolated to $\frac{1}{C^2} = 0$ to derive the flat-band potential E_{FB}. The slope not only gives n-type or p-type information, but also an estimate of the majority carrier concentrations, N_D, from knowing ε_r and A. Fabrication of high-quality electrodes using powder semiconductors is critical because the Mott–Schottky relation is only applicable to "ideal" semiconductors, where the bulk and surface are uniform with preferably known surface areas. The surface states also significantly affect the results of the plot because the pH effects follow the Nernstian relationship of -59 mV pH^{-1} for many semiconductors [60].

When metallic particles are deposited on the semiconductor surfaces, there is a new electronic structure at the metal–semiconductor interface [61]. The theory of metal–semiconductor interfaces has also been developed; the details of this theory can be also found in the work of Tung [62]. Simplified schematics for metal–semiconductor interfaces are shown in Fig. 10. The initial local electronic structure at the interface is strongly influenced by the relative positions between the work function of the metal and the Fermi level of the semiconductor. In short, there is the formation of a barrier, a so-called Schottky barrier, or of an energetically-smooth interface, a so-called Ohmic contact. The generalized criteria to generate such Ohmic contacts are $\phi_{SC,n\text{-}type} > \phi_m$ and $\phi_m > \phi_{SC,p\text{-}type}$, and the opposite is

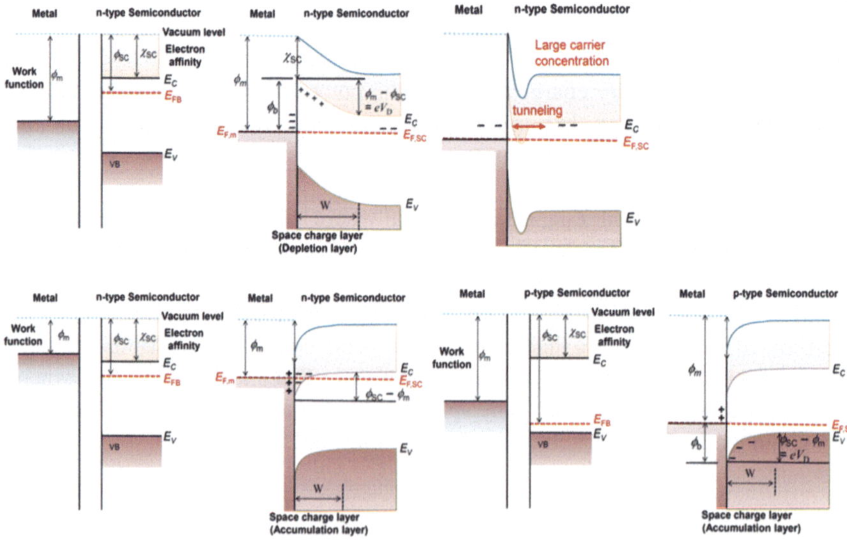

Fig. 10 Schematic images showing different metal–semiconductor interfaces

applicable to Schottky junctions. Additionally, a high concentration of dopants may lead to Ohmic contact because of tunneling effects. The Schottky barrier height is a complex problem because it depends on the atomic structure of the metal–semiconductor interface. The reduction of the metal and its hydride formation also causes a change, namely a reduction in the barrier height [62]. For interface driven potentials, the difference between the work function, ϕ_m, and the flatband potential of the semiconductor, ϕ_{SC}, gives the diffusion voltage, V_D:

$$eV_D = |\phi_m - \phi_{SC}|. \tag{40}$$

The maximum Schottky barrier height is described as the potential difference between the work function and the electron affinity of the semiconductor, χ_{SC}:

$$\phi_B = |\phi_m - \chi_{SC}| \text{ (Schottky limit)} \tag{41}$$

Each surface atom has a dangling bond, forming new surface state. When such surface states or interface states pin the Fermi level of the system, independent of the work function of the metal, the barrier height becomes smaller than the Schottky limit (another limit, called the Bardeen limit) [63]. In reality, the barrier comes between the Schottky and Bardeen limits. Using empirical equations, the barrier can be described as

$$\phi_B = S|\phi_m - \chi_{SC}| + AE_g \tag{42}$$

where S varies between 0 and 1, A is a constant, and E_g is the bandgap of the semiconductor. The previously discussed Schottky barrier model only describes semiconductor–metallic ideal contacts. The classic model fails to describe realistic porous ion-permeable electrocatalysts (i.e., oxyhydroxide cocatalysts for water oxidation). Recently, a model was proposed to describe so-called adaptive junctions and non-permeable metallic junctions (e.g., $Ni(OH)_2$/NiOOH vs IrO_x) where the potential drop develops only on the semiconductor for the former case whereas the potential drop occurs both in the electrolyte and in the semiconductor for the latter case [64]. The ion-permeable adaptive junction has the main benefit of redox-active species that charge-up, creating adaptive barriers and increasing the apparent photovoltage generated at the interface. In contrast, an impermeable metallic buried junction creates a constant barrier–height interface [65, 66].

Figure 11 summarizes the transient potential shift caused by a single electron injected into a hemispherical or cubic metal particle. It assumes a constant capacitance of 20 μF cm^{-2} for metals, although the capacitance actually varies slightly with the potential and the facet [27]. As the particle becomes larger, a single electron generates a smaller potential shift. This small potential suggests that multiple electrons have to reach the particle within a reasonable time scale to cause a significant negative shift of the potential. The smaller particles require fewer electrons to shift the potential, but the contact area with the semiconductor also decreases, suggesting that there is an optimum metal particle size for the

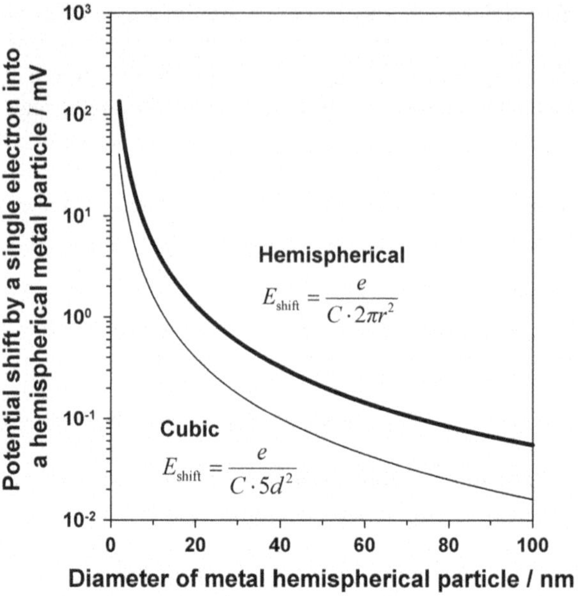

Fig. 11 The transient potential shift caused by a single electron injected into a hemispherical or cubic metal particle assuming a constant capacitance of 20 μF cm^{-2}

charge-up effect derived from excited electrons. This charge-up occurs up to the chemical potential of the electrons injected from the semiconductor (conduction band), and it is equilibrated at the catalyst on the surface.

There are a few examples of direct measurements of the potential shifts on metal particles driven by light absorption by semiconductor particles [67–71]. The semiconductors exhibit shifts in the Fermi level under irradiation, which provides the driving force for redox reactions. The equilibrated potential of the metal particles under irradiation was more negative than the H_2 evolution potential at a given pH, which is consistent with the steady-state H_2 evolution observed. This shift depends on the light intensity, and this charge-up phenomenon plays an essential role in the photocatalysis. It is also possible to measure the potential of semiconductors under various conditions (electrolyte, solution) and applied potentials with and without irradiation when a semiconductor photocatalyst is employed as an electrode [72]. The rest potential measurement of a semiconductor should provide a good estimate of the potentials at equilibrium under dark conditions and steady-state light illumination. A lack of equilibration is often observed between the metal/semiconductor and the redox potential in the solution; this can be attributed to corrosion of the semiconductor, formation of a surface film (e.g., an oxide), or the inherently slow electron transfer across the interface [73], making the situation more complex. It is thus preferable to measure the potentials directly during the photocatalytic process.

The size of the photocatalyst affects a number of parameters. The smaller particles lead to a high specific surface area, a shorter travel distance for the charge carriers to the surface, a lower degree of band bending, and possibly a wider bandgap by quantum size effects [57]. Although the travel distances of the generated electrons and the holes to the surfaces are minimized by decreasing the size of the particles [16], a high surface area does not directly increase, and often even decreases, the photocatalytic activity [74]. The surface is a "defect" possessing dangling bonds and potential determining ions, which have different potential states and may serve as recombination sites for the excited carriers [57]. If the particle radius is smaller than the width of the space charge layer, the degree of depletion does not penetrate into the bulk, as shown in Fig. 9. Because of the small volumes of the particles, the number of carriers in one particle is very limited, leading to limited influence by the semiconductor–electrolyte interface [22]. Under such conditions, the barrier height ΔV at a distance l from the center of the spherical particles is given by the following equation [75]:

$$\Delta V = 1/6\left(l/L_{\mathrm{p}}\right)^{2}, \tag{43}$$

where L_p is the Debye length of the semiconductor. The lower degree of band bending leads to a higher probability that the photogenerated charge carriers transfer simply via diffusion.

The electrolyte is strongly influenced by the surface state and potential-determining ions on the surface [76]. In water, the isoelectric point of the semiconductor provides a useful indication of whether the surface is negatively or positively charged [77]. Therefore, it is an interesting approach to isolate the bare surface from the water electrolyte by using some oxide (e.g., SiO_2, Al_2O_3, or TiO_2) [78–80]. The oxide protective layer isolates the semiconductor surface from the electrolyte, thus avoiding the photocorrosion prevalent for some semiconductor compounds [78]. Knowing the majority carrier density and the size of the cocatalysts, their location (separation distance) needs to be selected appropriately.

2.5 Catalytic Efficiency

On the semiconductor surface, the photogenerated charge carriers need to be successfully consumed by an effective electrocatalytic process. To achieve efficient water splitting under visible light irradiation where there is no significant overpotential for electrocatalysis, the electrocatalysts need to transfer the received electrons and holes to the relevant reactants in the water splitting redox reactions. The efficiencies of these electrocatalytic processes can be measured separately. The water splitting reactions can be described by the following two half reactions:

$$2H^+ + 2e^- \rightarrow H_2\,(\text{acid}) \text{ or } 2H_2O + 2e^- \rightarrow H_2 + 2OH^-\,(\text{base}) \quad E = 0 \text{ V vs. RHE} \tag{44}$$

$$2H_2O \rightarrow O_2 + 4H^+ + 4e^-\,(\text{acid}) \text{ or } 4OH^- \rightarrow O_2 + 2H_2O + 4e^-\,(\text{base}) \quad E = 1.23 \text{ V vs. RHE} \tag{45}$$

In general, hydronium ions (protons) are more easily reduced than water molecules [81], and hydroxyl ions are more easily oxidized than water molecules [82]. Therefore, extreme pH conditions are generally chosen for water electrolysis. The pH effects are further discussed in the following section. An alkaline electrolyzer is a commercialized technology for water splitting. Alkaline electrolytes seem to be chosen because many materials (e.g., nickel- or iron-based) are stable during electrolysis, maintaining the high activity of the hydroxyl ions as the reactants. Typical conditions can be found in the literature: cathode NiZn, anode Pt/IrO_2, at 1.55 V, 2.5 kA m^{-2}, 80°C [82].

Electrochemical reactions occur at the steady-state potential of the electrocatalysts on the surface of photocatalysts. The potential is determined as a consequence of all the photophysical and photochemical events discussed up to now. If there is a way to shift the potentials of the electrocatalysts immobilized on the surface of semiconductor photocatalysts, as discussed previously, then it is possible to obtain the rates from electrocatalytic knowledge, which can be measured separately. Electrocatalytic activity can be described in the form of the Tafel equation by neglecting the reverse reaction. The reaction rate for the photoelectrochemical half reaction, r (mol s^{-1}), can be described by [73]

$$r = \frac{i_0}{nF}\exp\frac{\alpha nF\left(E_D - E^0\right)}{RT}, \tag{46}$$

where i_0 is the exchange current of the given metal (A), α is the transfer coefficient, n is the number of electrons involved in the reaction, F is the Faraday constant, E_D and E^0 are the Fermi level of the metal and the redox potential in solution, respectively, R is the universal gas constant, and T is the absolute temperature. The main problem to be solved is to determine precisely the potential applied to the photocatalyst and cocatalyst. Photoexcited electrons and holes are generally known to undergo relaxation immediately after excitation to the levels at the edges of the CB and VB, respectively. The reduction in the bandgap implies a reduction in the driving force for the reaction, which is associated with the potential difference between the semiconductor band edges and the redox potential in the solution ($E_D - E^0$). Therefore, it is more difficult to achieve photocatalysis at high rates with smaller bandgap materials. Photocatalysts would simply provide an external bias (i.e., apply a potential) to metals when under photoexcitation.

Identifying the *electrocatalytic activity* indicators is very important. For hydrogen evolution, the conventional volcano plot reported by Trasatti [83] or more recently by Nørskov and co-workers [84] for H_2 production in acid solution with different metals shows that there is an optimal value for the free energy of hydrogen

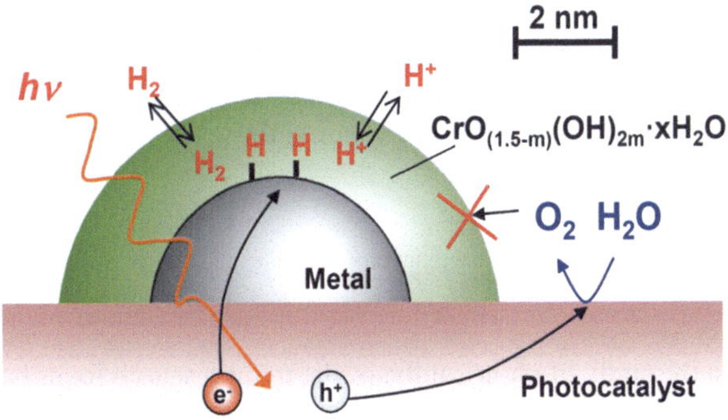

Fig. 12 Metal-Cr core-shell nano-structure for selective hydrogen evolution that does not cause back reaction from H_2 and O_2 to H_2O [94]

adsorption on metals. Similarly, for the oxygen evolution reaction, the volcano trend for metal oxides as a function of several thermodynamic descriptors has been claimed by several authors [85–87]. Markovic and coworkers reported that islands of nickel or cobalt species on noble metal surfaces (such as Pt) further enhance the water redox reactions for both hydrogen and oxygen evolution [88]. Mixed oxyhydroxides, such as iron-nickel, and perovskites have also been reported as low overpotential electrocatalysts that do not use noble metals [89–92]. The metal particle sizes alone significantly affect the overall efficiency of both the electrochemical and photocatalytic reactions [69, 93]. Therefore, the metal particle size should also be rigorously accounted for, but this type of data analysis provides good hints for forming strategies to develop highly efficient photoabsorbers using efficient cocatalysts.

Dark reactions involving the back reaction (H_2 and O_2 going to H_2O), a typical problem for cogeneration of H_2 and O_2 from overall water splitting, must be prevented. Noble metals typically cause the back reaction either catalytically or electrochemically (oxygen reduction reaction). Successful suppression of the oxygen reduction reaction without affecting the hydrogen evolution activity has been achieved using chromium and other materials [94, 95]. For Cr, the experiments suggest that the Cr layer works as a selective membrane for H_2 but is not permeable for O_2, as depicted in Fig. 12 [94]. Metallic nickel does not seem to cause oxygen reduction even under the conditions of hydrogen evolution, and nickel (hydr)oxide works for oxygen evolution and is distributed on the semiconductor surfaces [95].

It is obvious that when either reduction (H_2 formation) or oxidation (O_2 formation) is accelerated, the potential shifts in the direction reflecting the remaining electrons or holes. Enhancement of the rate of reduction or oxidation improves the overall efficiency of water splitting based on this charge-up theory, which is determined by the photon flux and the electron efficiency from the photocatalyst to the metal particles because the accelerated electron or hole process affects the potential, which in turn perturbs the rates of the counter-side process. In this sense, there is no rate-determining step in overall water splitting, where the reactions occur in parallel for reduction (electron path) and oxidation (hole path). In other words, each component in the respective steps during photocatalysis should be improved to achieve overall water splitting. To demonstrate this concept, co-loading of H_2 and O_2 evolution cocatalysts was found to be effective to some extent [96]. Integrated studies are essential to establish highly efficient photocatalysts for overall water splitting.

2.6 Mass Transfer

In the study of photocatalysis, the main research focus has been developing efficient materials, including cocatalysts. The mass transfer of the reactants has often been overlooked during the photocatalytic efficiency determination. Photocatalytic water splitting is nothing but electrocatalysis, which is driven by the photocatalyst-assisted excited carriers. Therefore, rigorous and quantitative arguments from electrocatalysis regarding the thermodynamic and kinetic information should be appropriately put into practice. For example, a two-compartment photoelectrochemical cell is able to separate the oxidation and reduction products (e.g., O_2 and H_2). However, the H_2 evolution reaction causes an increase in pH (according to the reaction at (44)), and the O_2 evolution reaction causes a decreases in the pH (according to the reaction at (45)). The complete isolation of ions leads to a high concentration overpotential (shifting the thermodynamic potential of 59 mV pH^{-1}), which stops the reaction and determines the overall efficiency. Therefore, the use of an ion-exchangeable membrane is mandatory [97]. Nafion or an alkaline membrane typically works in extreme pH media. When acidic conditions are chosen, the development of a non-noble metal electrodes with acid tolerance is required. The recent development of metal phosphide materials is of significant interest because they contain only abundant transition metals, such as Ni, Fe, and Co [98–100]. One of the most significant benefits of co-producing an H_2/O_2 mixture is avoiding this pH gradient, which minimizes the concentration overpotential. The coproduction of H_2/O_2 also avoids the use of (potentially expensive) membranes in the photoreactor and makes it possible to operate at neutral pH, which is otherwise impossible even for highly buffered solutions [97].

An interesting consideration regarding the activity of the reactants is the use of water vapor as a reactant (water liquid vs water vapor). Using vapor phase water has advantages such as the easily controlled supply and simple reactor designs, e.g., a fixed bed, for powder systems [101]. However, it encounters considerable difficulties because of the additional adsorption term for water vapor, which may strongly decrease the overall efficiency. In contrast, liquid phase water or associated ions (H^+ or OH^-) as reactants can achieve high activities (close to unity for water, or increasing or decreasing pH).

One of the benefits and simultaneously demerits of overall water splitting is the cogeneration of H_2 and O_2, which allows for the direct use of pure water or eventually seawater because there is no significant concentration (activity) loss in the reactor [97]. The H_2-O_2 coproduction system allows for nanometer separation of the reduction and oxidation sites, thus minimizing the concentration gradient of the ions. The study of water electrolysis under neutral pH is therefore very important. In water at pH 0 or 14, one H^+ or OH^- ion is present among 55 H_2O molecules. Kinetically speaking, the reactions with hydronium ions (protons) or hydroxyl ions are more facile than those with water molecules for reduction and oxidation, respectively [82]. In neutral conditions, buffering actions are effective based on the reactant switching over varying pH. For electrochemical measurements, the supporting electrolyte is an essential component to avoid solution resistance (iR drop). To date, it seems that there is a lack of information in the literature regarding the supporting electrolyte effects in overall water splitting. The effects of the supporting electrolyte (identity and concentrations) have to be rigorously and quantitatively taken into account because they may cause reactant switching and additional adsorption onto active sites, which may enhance or decrease the overall catalytic efficiency.

2.7 Other Considerations and Standardization of Measurements of Photocatalysis

There cannot be a sole kinetically relevant step in photocatalysis, as discussed previously. The concept of the rate-determining step is only applicable for a series of sequential reaction steps. In photocatalysis, there are always parallel pathways for efficiency loss (either the forward direction for photocatalysis or recombination). Thus, the total efficiency relies on the first-order rate constants of all processes associated with the lifetime (reciprocal of the sum of the first order: $\tau = \frac{1}{\sum_i k_i}$). For better visualization of the photocatalytic efficiencies of the involved processes, multiplication of the efficiencies for the designated processes wherever isolated should be used. After all, the photocatalytic activity for solar energy conversion can be compared based only on the effects of the absorbed (or incident) photons, i.e., quantum yield or quantum efficiency, at a specified wavelength and its integration through varied wavelengths. (The use of terminology such as quantum yield and quantum efficiency in photocatalysis can cause some confusion because they are different in homogeneous and heterogeneous catalysis [30]. In addition, the solar cell or the photoelectrochemical community use quantum efficiency.) It is important to note that the quantum efficiency is not a function that is proportional to the surface area or the mass of the photocatalyst. The photocatalytic activities consider the fate of photons whether reacted or recombined. Therefore, the rates per surface area or per mass of catalyst cannot be used as indicators to compare the intrinsic photocatalytic activities of the materials unless those parameters are of particular interest for comparison [102].

It is important to compare the results obtained using lasers and conventional lamps (including solar radiation and lights with band pass filters) when discussing the behavior of photocatalyst powders because the photon flux (the number of photons per unit area) drastically affects the nature of the chemical process. The photocatalytic process induced by laser radiation readily causes multielectron reactions. It is important to consider carefully the light intensity whenever discussing the photocatalytic activity. It is even recommended to report the photon distribution for all of the illumination used as a function of wavelength.

Many studies use dye degradation to compare the (visible light) photocatalytic activity of different materials. Using photon-responsive dyes as a substrate for a photocatalytic test has long been known to bring ambiguity to the activity results [103–105]. More importantly, promoting this test as a measure of photocatalyst activity must be avoided because there is a tremendous contribution from the dye-sensitized pathway; the excited dye is used in photocatalysis rather than bandgap excitation of the semiconductor [104]. Furthermore, the International Organization for Standardization (ISO) has developed standardized experimental procedures for these tests, which should be followed rigorously [105].

3 Concluding Remarks

Efficient overall water splitting to date is limited to the UV range of light where high STH is not expected according to the solar irradiance [106, 107]. It is essential to develop visible-light-responsive photocatalyst materials. Although cocatalyst design has enabled water splitting using visible light [108, 109], the efficiency has to be greatly improved. This review focuses on the fundamental parameters involved in the photocatalytic processes for overall water splitting. As demonstrated in this chapter, photocatalytic water splitting is a complex process involving photon absorption, exciton separation, carrier diffusion, carrier transport, catalytic efficiency, and mass transfer of the reactants. Isolation of such parameters and their quantitative measurements and descriptions are becoming more and more important for developing novel materials.

Powder semiconductors have tremendous potential for solar fuel generation, partly because of their synthesis scalability. They can also be synthesized by wet chemistry [22], which allows chemists to contribute greatly to the field of solar energy conversion. It is, however, noted that the quality of the semiconductor powders needs to be very high, preferably at the level used in solar cells. Based on the above discussion, the four most critical key parameters on which to focus research investigations to improve photocatalytic performance are:

1. *Electronic structure*: determining most of the semiconductor properties that are strongly supported by advanced DFT calculations
2. *Charge carrier concentrations*: correlating with carrier lifetime and mobility
3. *Electrocatalysts*: enabling the redox reactions at low overpotentials
4. *Interface*: minimizing electronic barriers and protecting unstable components

Integrating the knowledge gained from studying these parameters can enable the concept of "photocatalysts by design," which can lead to improvements in photocatalytic efficiency to allow us to meet our energy demands through solar H_2 production.

Acknowledgement This study presented in this chapter was supported by King Abdullah University of Science and Technology (KAUST). The author thanks Dr. A. Ziani, Mr. A.T. Garcia-Esparza, Mrs. E. Nurlaela, and Mr. T. Shinagawa at KAUST for proofreading the manuscript.

References

1. Lewis NS, Nocera DG (2006) Proc Natl Acad Sci 103:15729
2. International Energy Agency (2010) World Energy Outlook 2010. International Energy Agency, Paris
3. National Renewable Energy Laboratory (NREL) (1999) http://rredc.nrel.gov/solar/spectra/am1.5
4. Pinaud BA, Benck JD, Seitz LC, Forman AJ, Chen Z, Deutsch TG, James BD, Baum KN, Baum GN, Ardo S, Wang H, Miller E, Jaramillo TF (2013) Energy Environ Sci 6:1983
5. Takanabe K, Domen K (2011) Green 1:313
6. Turro NJ, Ramamurthy V, Scaiano JC (eds) (2010) Modern molecular photochemistry of organic molecules. University Science, Sausalito
7. Nozik AJ (1978) Annu Rev Phys Chem 29:189
8. Nosaka Y, Ishizuka Y, Miyama H (1986) Ber Bunsenges Phys Chem 90:1199
9. Memming R (1988) Top Curr Chem 143:79
10. Hagfeldt A, Grätzel M (1995) Chem Rev 95:49
11. Kaneko M, Okura I (eds) (2002) Photocatalysis science and technology. Kodansha/Springer, Tokyo/Berlin
12. Domen K (2003) In: Horvath IT (ed) Encyclopedia of catalysis. Wiley, Hoboken
13. Maeda K, Domen K (2007) J Phys Chem C 111:7851
14. Kamat PV (2007) J Phys Chem C 111:2834
15. Osterloh FE (2008) Chem Mater 20:35
16. Kudo A, Miseki Y (2009) Chem Soc Rev 38:253
17. Inoue Y (2009) Energy Environ 2:364
18. Walter MG, Warren EL, McKone JR, Boettcher SW, Mi Q, Santori EA, Lewis NS (2010) Chem Rev 110:6446
19. Abe R (2010) J Photochem Photobiol C 11:179
20. Maeda K, Domen K (2010) J Phys Chem Lett 1:2655
21. Hisatomi T, Minegishi T, Domen K (2012) Bull Chem Soc Jpn 85:647
22. Takanabe K, Domen K (2012) ChemCatChem 4:1485
23. Tong H, Ouyang S, Bi Y, Umezawa N, Oshikiri M, Ye J (2012) Adv Mater 24:229
24. Tachibana Y, Vayssieres L, Durrant JR (2012) Nat Photonics 6:511
25. Osterloh FE (2013) Chem Soc Rev 42:2294
26. Hisatomi T, Takanabe K, Domen K (2015) Catal Lett 145:95
27. Takanabe K, Domen K (2014) Photocatalysis in generation of hydrogen from water. In: Tao F, Schneider WF, Kamat PV (eds) Heterogeneous catalysis at nanoscale for energy applications. Wiley, Hoboken, pp 239–270
28. Bohren CF, Huffman DR (eds) (2004) Absorption and scattering of light by small particles. Wiley, Weinheim
29. Dahm DJ, Dahm KD (eds) (2007) Interpreting diffuse reflectance and transmittance. NIR, Chichester
30. Braslavsky SE, Braun AM, Cassano AE, Emeline AV, Litter MI, Palmisano L, Parmon VN, Serpone N (2011) Pure Appl Chem 83:931
31. Chen Z, Dinh HN, Miller E (eds) (2013) Photoelectrochemical water splitting, standards, experimental methods, and protocols. Springer, New York
32. Wemple SH, Seman JA (1973) Appl Opt 12:2947

33. Di Giulio M, Micocci G, Rella R, Siciliano P, Tepore A (1993) Phys Status Solidi A 136: K101
34. Lodenquai JF (1994) Sol Energy 53:209
35. Swanepoel R (1983) J Phys E Sci Instrum 16:1214
36. Chen LF, Ong CK, Neo CP, Varadan VV, Varadan VK (2005) Microwave electronics: measurement and materials characterization. Wiley, Chichester
37. Le Bahers T, Rérat M, Sautet P (2014) J Phys Chem C 118:5997
38. Green MA (2008) Sol Energy Mater Sol Cells 92:1305
39. Džimbeg-Malčić V, Barbarić-Mikočević Ž, Itrić K (2011) Technical Gazette 18:117
40. Wood DL, Tauc J (1972) Phys Rev B 5:3144
41. Schubert EF (ed) (2006) Light-emitting diodes, 2nd edn. Cambridge University Press, Cambridge
42. Bae D, Pedersen T, Seger B, Malizia M, Kuznetsov A, Hansen O, Chorkendorff I, Vesborg PCK (2015) Energy Environ Sci 8:650
43. Sze SM, Ng KK (eds) (2006) Physics of semiconductor devices. Wiley, New York
44. Kittel C (2005) Introduction to solid state physics, 8th edn. Wiley, Hoboken
45. Kim DW, Leem YA, Yoo SD, Woo DH, Lee DH, Woo JC (1993) Phys Rev B 47:2042
46. Liang WY (1970) Phys Educ 5:226
47. Bastard G, Mendez EE, Chang LL, Esaki L (1982) Phys Rev B 26:1974
48. Gerischer H (1984) J Phys Chem 88:6096
49. van der Pauw LJ (1958) Philips Res Rep 13:1
50. Heaney MB (2000) Electrical conductivity and resistivity. In: The measurement, instrumentation and sensors handbook. CRC, Boca Raton
51. Nagel H, Berge C, Aberle AG (1999) J Appl Phys 86:6218
52. Law ME, Solley E, Liang M, Burk DE (1991) IEEE Electron Device Lett 12:401
53. Shockley W, Read WT Jr (1952) Phys Rev 87:835
54. Hall RN (1952) Phys Rev 87:387
55. Auger P (1952) C R A S 177:169
56. Zhang Z, Yates JT Jr (2012) Chem Rev 112:5520
57. Yoneyama H (1993) Crit Rev Solid State Mater Sci 18:69
58. Grätzel M (2001) Nature 414:338
59. Gelderman K, Lee L, Donne SW (2007) J Chem Educ 84:685
60. van de Krol R, Grätzel M (2012) Photoelectrochemical hydrogen production. Springer, New York
61. Sato N (1998) Electrochemistry at metal and semiconductor electrodes. Elsevier, Amsterdam
62. Tung RT (2014) Appl Phys Rev 1:011304
63. Cohen ML (1979) J Vac Sci Technol 16:1135
64. Cendula P, Tilley SD, Gimenez S, Bisquert J, Schmid M, Grätzel M, Schumacher JO (2014) J Phys Chem C 118:29599
65. Mills TJ, Lin F, Boettcher SW (2014) Phys Rev Lett 112:148304
66. Lin F, Boettcher SW (2014) Nat Mater 13:81
67. Kamat PV (2002) Pure Appl Chem 74:1693
68. Jakob M, Levanon H, Kamat PV (2003) Nano Lett 3:353
69. Subramanian V, Wolf EE, Kamat PV (2004) J Am Chem Soc 126:4943
70. Yoshida M, Yamakata A, Takanabe K, Kubota J, Osawa M, Domen K (2009) J Am Chem Soc 131:13218
71. Lu X, Bandara A, Katayama M, Yamakata A, Kubota J, Domen K (2011) J Phys Chem C 115:23902
72. Chen Z, Jaramillo TF, Deutsch TG, Kleiman-Shwarsctein A, Forman AJ, Gaillard N, Garland R, Takanabe K, Heske C, Sunkara M, McFarland EW, Domen K, Miller EL, Turner JA, Dinh HN (2010) J Mater Res 25:3
73. Bard AJ, Faulkner LR (2001) Electrochemical methods, 2nd edn. Wiley, New York, pp 736–768

74. Fukasawa Y, Takanabe K, Shimojima A, Antonietti M, Domen K, Okubo T (2011) Chem Asian J 6:103
75. Albery WJ, Bartlett PN (1984) J Electrochem Soc 131:315
76. Chamousis RL, Osterloh FE (2014) Energy Environ Sci 7:736
77. Butler MA, Ginley DS (1978) J Electrochem Soc 125:228–232
78. Paracchino A, Laporte V, Sivula K, Grätzel M, Thimsen E (2011) Nat Mater 10:456
79. Esposito DV, Levin I, Moffat TP, Talin AA (2013) Nat Mater 12:562
80. Hu S, Shaner MR, Beardslee JA, Lichterman M, Brunschwig BS, Lewis NS (2014) Science 344:1005
81. Shinagawa T, Garcia-Esparza AT, Takanabe K (2014) ChemElectroChem 1:1497
82. Hamann CH, Hamnett A, Vielstich W (eds) (2007) Electrochemistry, 2nd edn. Wiley, Weinheim
83. Trasatti S (1972) J Electroanal Chem 32:163
84. Greeley J, Jaramillo TF, Bonde J, Chorkendorff I, Nørskov JK (2006) Nat Mater 5:909
85. Matsumoto Y, Sato E (1986) Mater Chem Phys 14:397
86. Man IC, Su H-Y, Calle-Vallejo F, Hansen HA, Martinez JI, Inoglu NG, Kitchin J, Jaramillo TF, Nørskov JK, Rossmeisl J (2011) ChemCatChem 3:1159
87. Grimaud A, May KJ, Carlton CE, Lee YL, Risch M, Hong WT, Zhou J, Shao-Horn Y (2013) Nat Commun 4:3439
88. Subbaraman R, Tripkovic D, Chang KC, Strmcnik D, Paulikas AP, Hirunsit P, Chan M, Greeley J, Stamenkovic V, Markovic NM (2012) Nat Mater 11:550
89. Suntivich J, May KJ, Gasteiger HA, Goodenough JB, Shao-Horn Y (2011) Science 334:1383
90. Smith RDL, Prévot MS, Fagan RD, Zhang Z, Sedach PA, Siu JMK, Trudel S, Berlinguette CP (2013) Science 340:60
91. Gong M, Li Y, Wang H, Liang Y, Wu JZ, Zhou J, Wang J, Regier T, Wei F, Dai H (2013) J Am Chem Soc 135:8452
92. Gong M, Zhou W, Tsai MC, Zhou J, Guan M, Lin MC, Zhang B, Hu Y, Wang DY, Yang J, Pennycook SJ, Hwang BJ, Dai H (2014) Nat Commun 5:5695
93. Muller BR, Majoni S, Memming R, Meissner D (1997) J Phys Chem B 101:2501
94. Yoshida M, Takanabe K, Maeda K, Ishikawa A, Kubota J, Sakata Y, Ikezawa Y, Domen K (2009) J Phys Chem C 113:10151
95. Yoshida M, Maeda K, Lu D, Kubota J, Domen K (2013) J Phys Chem C 117:14000
96. Townsend TK, Browning ND, Osterloh FE (2012) Environ Sci 5:9543
97. Jin J, Walczak K, Singh MR, Karp C, Lewis NS, Xiang C (2014) Energy Environ Sci 7:3371
98. Popczun EJ, McKone JR, Read CG, Biacchi AJ, Wiltrout AM, Lewis NS, Schaak RE (2013) J Am Chem Soc 135:9267
99. Jiang P, Liu Q, Liang Y, Tian J, Asiri AM, Sun X (2014) Angew Chem Int Ed 53:12855
100. Popczun EJ, Read CG, Roske CW, Lewis NS, Schaak RE (2014) Angew Chem Int Ed 53:5427
101. Dionigi F, Vesborg PCK, Pedersen T, Hansen O, Dahl S, Xiong A, Maeda K, Domen K, Chorkendorff I (2011) Energy Environ Sci 4:2937
102. Kisch H (2010) Angew Chem Int Ed 49:9588
103. Mills A, Wang J (1999) J Photochem Photobiol A 127:123
104. Yang X, Ohno T, Nishijima K, Abe R, Ohtani B (2006) Chem Phys Lett 429:606
105. Mills A (2012) Appl Catal B 128:144
106. Kato H, Asakura K, Kudo A (2003) J Am Chem Soc 125:3082
107. Sakata Y, Matsuda Y, Nakagawa T, Yasunaga R, Imamura H, Teramura K (2011) ChemSusChem 4:181
108. Maeda K, Teramura K, Lu D, Takata T, Saito N, Inoue Y, Domen K (2006) Nature 440:295
109. Maeda K, Xiong A, Yoshinaga T, Ikeda T, Sakamoto N, Hisatomi T, Takashima M, Lu D, Kanehara M, Setoyama T, Teranishi T, Domen K (2010) Angew Chem Int Ed 49:4096

Top Curr Chem (2016) 371: 105–142
DOI: 10.1007/128_2015_633

Published online: 22 April 2015

Nanoscale Effects in Water Splitting Photocatalysis

Frank E. Osterloh

Abstract From a conceptual standpoint, the water photoelectrolysis reaction is the simplest way to convert solar energy into fuel. It is widely believed that nanostructured photocatalysts can improve the efficiency of the process and lower the costs. Indeed, nanostructured light absorbers have several advantages over traditional materials. This includes shorter charge transport pathways and larger redox active surface areas. It is also possible to adjust the energetics of small particles via the quantum size effect or with adsorbed ions. At the same time, nanostructured absorbers have significant disadvantages over conventional ones. The larger surface area promotes defect recombination and reduces the photovoltage that can be drawn from the absorber. The smaller size of the particles also makes electron–hole separation more difficult to achieve. This chapter discusses these issues using selected examples from the literature and from the laboratory of the author.

Keywords Fermi Level • Junction • Metal Oxide • Recombination • Surface Photovoltage

Contents

F.E. Osterloh (✉)
Department of Chemistry, University of California, Davis, CA 95616, USA
e-mail: fosterloh@ucdavis.edu

1 Photoelectrochemical Water Splitting as a Pathway to Sustainable Energy

The goal of keeping global temperature increases below 2°C compared to pre-industrial times requires the rapid deployment of carbon free energy technology that can compete with the price of fossil fuels. The solar energy received on the Earth's surface meets current and future human energy demand [1, 2]. Solar energy can be converted into electricity with up to 46% efficiency using the photovoltaic cells that are available today [3]. However, electricity is difficult to store and distribute over long distances. These problems can be avoided by converting the photochemical energy directly into fuel. The simplest scheme employs the photoelectrolysis reaction shown in (1) to produce hydrogen from water:

$$H_2O \rightarrow \frac{1}{2}\,O_{2(g)} + H_{2\,(g)}\,; \quad \Delta G = +237\ \text{kJ/mol}\,. \tag{1}$$

Hydrogen from this reaction could become the central energy carrier of a hydrogen-based economy [2]. Alternatively, it could be used as an electron source for the exothermic formation of conventional fuels using atmospheric CO_2 as a carbon feedstock. The technology for creating hydrogen by oxidation of water (solar water splitting) already exists in the form of multi-junction photovoltaic systems coupled to water electrolyzers. These can achieve over 18% energy efficiency [4] but they are so expensive that nobody uses them – especially not anyone living on less than \$10 a day, i.e., the majority of people in the world [5, 6]. Photovoltaics based on lead iodide perovskite might be a promising alternative [7] if problems with photo- and thermal stability, scaling, and lead toxicity can be resolved [8]. Potential cost savings are also possible with photoelectrochemical cells (PEC) that combine photovoltaic and electrolytic functions in one unit (Fig. 1) [9]. PECs can reach efficiencies of up to 12.4% [10], about half of the theoretical efficiency limit for these devices (24.4% for a tandem [11, 12] and 30% for a multijunction device [4]). However, issues with stability continue to plague these devices [13, 14].

Of all known solar hydrogen technologies, solar water electrolysis with *suspended photocatalysts* (PCs) has the greatest potential to induce a revolution in fuel production on this planet. Photocatalysts generate hydrogen and oxygen upon exposure of a particle-water mixture to sunlight (Fig. 2a) [15–24]. Because of the total integration of components for light absorption, charge separation, and

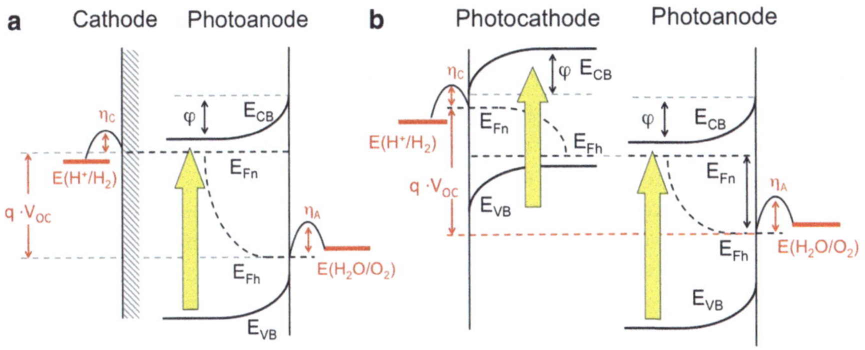

Fig. 1 Energetics of (**a**) type 1 (single absorber) and (**b**) type 2 (dual absorber, tandem) suspended photocatalysts for water photoelectrolysis. For photochemical water splitting, the quasi Fermi levels E_{Fn} and E_{Fh} of the illuminated catalysts need to be above and below the water redox potentials. Band bending φ, maximum possible energy output ($q \times V_{OC}$), and electrochemical overpotentials for anodic η_A and cathodic η_C processes are also shown

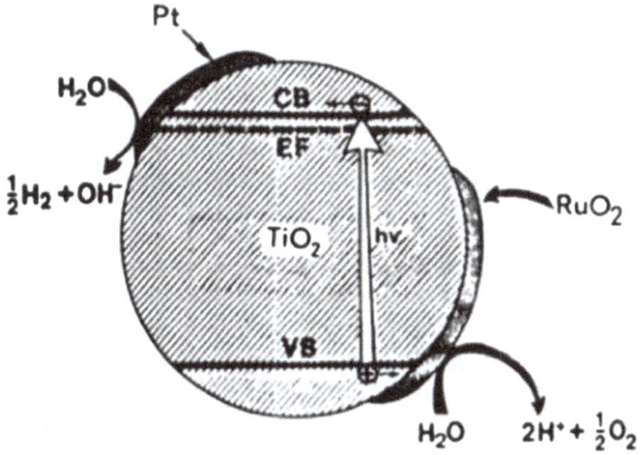

Fig. 2 Photoelectrochemical diode after Duonghong et al. Reproduced with permission from [42]. Copyright 1981, American Chemical Society

water electrolysis, PCs are nearly one order of magnitude cheaper than photoelectrochemical cells at equal efficiency [25, 26]. At 10% energy efficiency, PCs could deliver hydrogen at a cost of \$1.63 per kg, and outcompete oil as an energy carrier.

Photocatalysts operate as either type 1 (Schottky) or type 2 (tandem) devices (Fig. 1) [9, 27–31]. In type 1 PCs, light absorption and charge separation occur at a single absorber particle connected to one or several co-catalysts to complete the circuit for water electrolysis. The ideal limiting Solar-to-Hydrogen (STH) efficiency of this Schottky-type configuration is $\eta = 14.4\%$, based on a light absorber with a 2.0 eV band gap and conversion losses of $E_{Loss} = 0.8$ eV per electron [11, 12]. Actual type 1 photocatalysts such as NiO-modified La:$KTaO_3$ [32] and Cr/Rh-modified GaN:ZnO (QE = 2.5%, pure water, visible light) [33–35] achieve less than 0.1% STH because of low sunlight absorption (E_G of La:$KTaO_3$ is 3.6 eV) and recombination losses at the surfaces of the particles. A number of other catalysts, including $In_{1-x}Ni_xTaO_4$ ($x = 0$–0.2) [36, 37], CoO [38], and Cu_2O [39], have also been reported. However, their performance has not yet been reproduced in the literature and remains in question [40, 41].

The type 2 photocatalyst is based on the tandem (or Z-scheme) concept shown in Fig. 1b. Here, two (or more) separate light absorbers are connected in series. Because the device voltage is divided into several contributions, semiconductors with smaller band gaps can be used which absorb a much greater fraction of the solar flux [43]. The ideal limiting STH efficiency of a dual absorber configuration is 24.4% (for a combination of absorbers with band gaps of 2.25 and 1.77 eV). This is nearly twice that of the Schottky junction catalysts [11, 12]. However, because of the greater complexity (two absorbers instead of one), functional tandem PCs have only been known for about 15 years [16]. Of these, the combination of Rh:$SrTiO_3$ and $BiVO_4$connected with a $Fe^{3+/2+}$ redox shuttle, gives the highest STH efficiency (0.1%) [44].

Currently, there are three main strategies to develop more efficient, more stable, and less expensive PCs. One is to encapsulate traditional absorber materials from photovoltaics (e.g., IV, III/V, II–VI semiconductors) with protecting layers to inhibit photocorrosion. This strategy is has recently been applied to silicon, GaAs, and GaP photoanodes [45], and to cuprous oxide photocathodes [46], with reasonable success. Another approach involves the development of new metal oxide materials that combine suitable properties (visible band gap, high carrier mobilities, long carrier lifetimes) with greater chemical stability for photoelectrochemical water splitting. Such materials can be made by directed synthesis, sometimes guided by theory [47], or they can be made by combinatorial approaches, as described by Bruce Parkinson [48–50], Eric McFarland [51], and Nathan Lewis [52]. The third strategy is to exploit scaling laws and specific effects at the nanoscale [53–56] to overcome the limiting problems of metal oxide absorbers, such as their short electron-hole lifetimes and low mobility. This nanostructuring strategy has gained significant interest in the last 20 years [57–67] and is the focus of this chapter.

2 Brief History of Nanoscale Water Splitting Photocatalysis

The photosynthetic systems of bacteria are the earliest examples of nanostructured solar energy conversion devices; they date back to the beginning of life on this planet [68]. In contrast, artificial nanostructures for solar energy to fuel conversion have emerged only in the last three decades. The first experimental work dates back to 1968, when Boddy established photoelectrochemical water oxidation with an illuminated TiO_2 electrode [69], and Freund and Gomez reported similar reactions on ZnO, TiO_2, and WO_3 [70]. However, the potential of the reaction for artificial photosynthesis was only recognized broadly with Fujishima and Honda's report in Nature [71, 72]. The idea to carry out this process with 'miniaturized photoelectrochemical cells' suspended in water gained traction after Arthur Nozik formulated the concept of 'photochemical diodes' [29]. This was quickly followed by experimental demonstrations of photocatalytic effects in suspended semiconductor particles by Bard [43], and on 'colloidal microelectrodes' by Henglein [54, 73]. In 1979, Michael Grätzel reported photocatalytic water oxidation by a suspended RuO_2 nanoparticle in the presence of a $[Ru(bipy)_3]^{2+}$ complex for visible light absorption and methyl viologen (MV^{2+}) as sacrificial electron acceptor [74]. An improved system consisted of a TiO_2 colloidal light absorber and Pt and RuO_2 co-catalysts for water reduction and oxidation, respectively (Fig. 2) [42]. Even though water splitting was not achieved in this system (O_2 evolution was later attributed to air contamination) [28, 75], the structure exemplifies the design principles of a 'photochemical diode' [76]. Contemporary research on nanoscale water splitting catalysts is motivated by the prospect of extracting charge carriers more efficiently and thereby overcoming the limitations of metal oxide absorbers (short carrier lifetimes and low mobility). The underlying concepts are the topics for the next sections.

3 Size Dependency of Free Energy Creation in Photocatalysts

The effects of nanostructuring the light absorber on free energy conversion can be illustrated well with an equivalent circuit diagram for a photovoltaic cell. In the diagram in Fig. 3, the light-absorbing component corresponds to a photon-driven current source, and the rectifying (charge separating) component is shown as a diode connected in parallel to it. In addition, there are parallel and serial resistances, R_{SH} and R_S. The former is associated with the non-selective charge transfer which leads to leakage or shunting, whereas the latter is produced by the transport of charge carriers from the absorber interior to the interface. For a photovoltaic cell, the load is an electrical device, and for a fuel-producing photocatalyst the load is water electrolysis [77].

In the circuit, I_{SH}, R_S, and R_{SH} are dependent upon the physical size of the junction. For two otherwise identical junctions, the one with twice the surface area of the other is expected to have half the shunt resistance and double the leakage current I_{SH} because it has twice the junction area across which current can leak. It also has half the series resistance R_S because it has twice the cross-sectional area through which current can flow. For a nanostructured photocatalyst, the decrease in R_{SH} corresponds to the leakage currents J_{et} and J_T in Fig. 12. If the leakage current exceeds the generation current I_L, water electrolysis comes to a stop. On the other

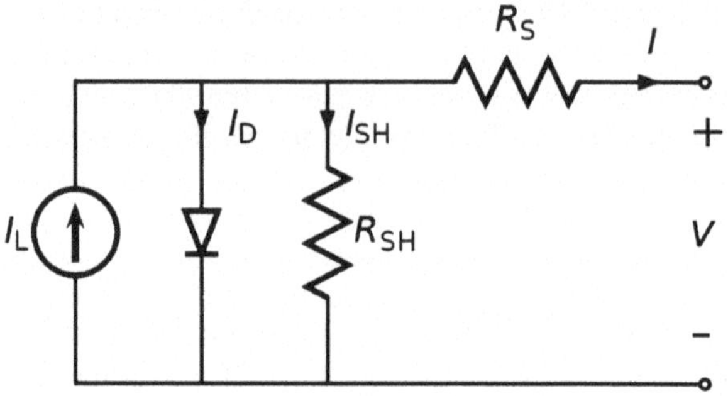

Fig. 3 Equivalent circuit diagram for a photovoltaic cell. R_{SH}: shunt resistance, R_S: serial resistance, I_{SH}: leakage current, I_L: generation current

hand, the decrease of R_S caused by the reduction in electron-hole transport resistances improves the photocurrent. Finally, as the semiconductor particles approach the nanoscale, reduced space charge layer effectiveness diminishes the rectifying properties. This corresponds to reduction of R_{SH}.

4 Light Absorption

The ability of a material to absorb light is determined by the Lambert–Beer law and the wavelength-dependent absorption coefficient α. The light penetration depth α^{-1} refers to the distance after which the light intensity is reduced to $1/e$ of the incident illumination. For example, for Fe_2O_3, $\alpha^{-1} = 118$ nm at $\lambda = 550$ nm [7]; for CdTe, $\alpha^{-1} = 106$ nm (550 nm) [78]; and for Si, $\alpha^{-1} = 680$ nm (510 nm). To ensure >90% absorption of the incident light, the film thickness d must be >2.3 times the value of α^{-1} (Fig. 4). The dimensions of nanostructured photocatalysts are usually smaller than α^{-1}, so each nanoparticle only absorbs a small fraction of the incident light. However, complete light absorption by the suspension can be achieved by adjusting the particle concentration and optical pathlength of the reaction container. Nanostructured surfaces also reduce the reflection losses and increase the light scattering. A solid film with one nanostructured surface (Fig. 4b) increases the horizontal light distribution. A particle suspension (Fig. 4c) increases the light distribution in three dimensions. This occurs by refraction at the surfaces of the particles and by Mie scattering. As a result, the light in a suspension impinges on the particles from all directions. This has an important consequence on the ability of the particles to generate a photovoltage at the solid–liquid interface and to generate the necessary thermodynamic driving force for water electrolysis. According to the Shockley diode equation (2) [77], the open circuit voltage V_{OC} of a solar cell is a logarithmic function of the absorbed flux j_{phot} and of the reverse saturation current j_0 of the diode:

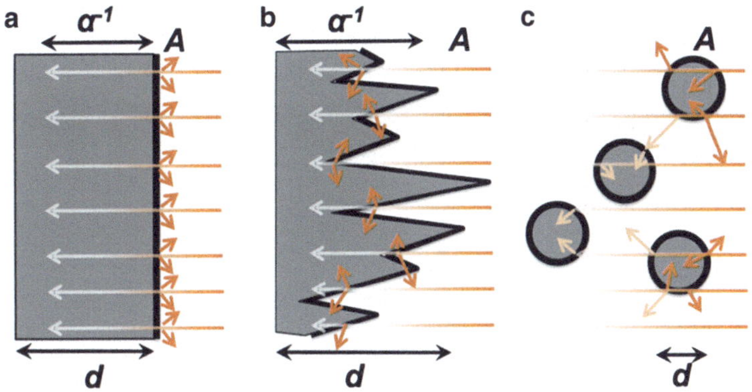

Fig. 4 Light distribution (**a**) in flat films, (**b**) in rough films, and (**c**) in particle suspensions. *A*: surface area; *d*: film or particle thickness; α^{-1}: optical penetration depth. *Short arrows* show scattered or reflected light

$$V_{OC} = \frac{kT}{e} \ln\left(\frac{j_{phot}}{j_0} + 1\right). \tag{2}$$

By increasing the junction area A of the absorber (Fig. 4), the absorbed flux j_{phot}, defined as the photon rate per unit area of the junction, decreases proportionally and the photovoltage drops. For example, if $j_{phot} >>> j_0$, the photovoltage decreases by 0.059 V for every decadic decrease of j_{phot} (i.e., decadic increase of junction area). This has been experimentally observed for silicon microwire arrays [79]. Equation (2) also impacts the construction of photocatalysts. It emphasizes the need to minimize the solid–liquid junction area through inert coatings or by replacing the solid–liquid junction with localized solid-solid junctions on the surface of the light absorber.

5 Majority and Minority Carrier Transport

Photoexcitation produces charge carriers with finite mobility and lifetime, depending on the material, the carrier type, and the light intensity. To drive water redox reactions, these carriers need to reach the material's interfaces at the electrolyte and at the co-catalysts. In the absence of an external field, charge carriers move by diffusion. Their range is defined by the mean free diffusion length L which depends on the carrier diffusion constant D, the carrier lifetime τ, and a dimensionality factor ($q = 2$, 4, or 6 for one-, two-, or three-dimensional diffusion):

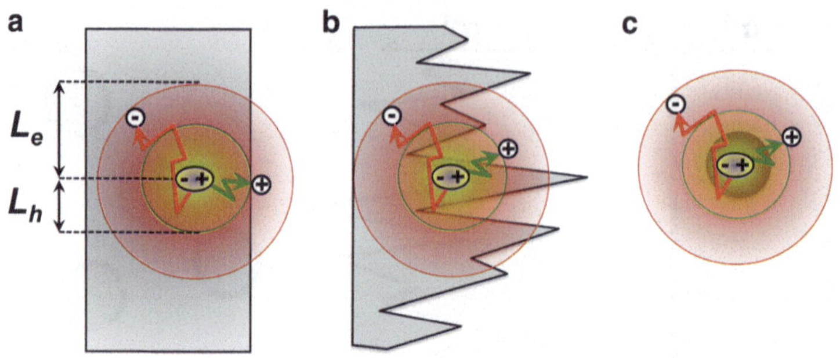

Fig. 5 Charge collection (**a**) in flat films, (**b**) in nanostructured films, and (**c**) in particle suspensions. d: film or particle thickness; L_e: electron diffusion length; L_h: hole diffusion length

$$\overline{L^2} = qD\tau \tag{3}$$

For intrinsic semiconductors, usually $L_e > L_h$ because of the larger diffusion constant D of electrons compared to holes. For example, Si has $D_e = 49\ cm^2\ s^{-1}$ and $D_h = 13\ cm^2\ s^{-1}$ (calculated from mobilities, $\mu_e = 1{,}900\ cm^2\ V^{-1}\ s^{-1}$, and $\mu_h = 500\ cm^2\ V^{-1}\ s^{-1}$ at 298 K using the Einstein–Smoluchowski relation) [80] assuming $\tau_e = \tau_h = 10^{-6}$ s, $L_e = 98\ \mu m$, and $L_h = 51\ \mu m$ for one-dimensional diffusion ($q = 2$). Upon doping, the concentration of the majority carriers increases, and with it their τ and L values. On the other hand, the lifetime and diffusion length of the minority carriers decrease. For optimum collection of both carrier types at the interface, the semiconductor film thickness d has to be in the same range as L_e and L_h (Fig. 5). This can be achieved by increasing the surface roughness of the film, as shown in Fig. 5b. Such a surface nanostructuring approach is particularly useful for first-row transition metal oxides (Fe_2O_3), which suffer from low hole mobility and short hole lifetimes [81, 82]. Ideal electron-hole collection is possible with suspended nanoparticles if their particle size $d < L_e, L_h$.

The impact of the absorber size on charge collection has been experimentally verified with nanoparticles and bulk particles of $KCa_2Nb_3O_{10}$ [83, 84]. Chemical exfoliation of this layered Dion–Jacobson phase affords 1–2 nm thick sheet-like tetrabutylammonium stabilized $HCa_2Nb_3O_{10}$ nanoparticles, whereas sonication leads to 227 ± 202-nm particles referred to as 'bulk' (Fig. 6). Both types of particles photocatalyze hydrogen evolution from aqueous methanol under UV light illumination, but hydrogen evolution rates for the nanosheets are consistently higher than those for the bulk particles, even in the presence of co-catalysts. Rates can be fitted to the kinetic model as in Fig. 6c and as expressed in (4):

$$\mathrm{ER} = \left(\frac{1}{R_G - R_R^L - R_R^S} + \frac{r}{J_{CT}^-} + \frac{r}{J_{CT}^+} + \frac{d_{OX}}{J_{OX}} + \frac{d_{RED}}{J_{RED}} + \frac{1}{R_{OX}} + \frac{1}{R_{RED}}\right)^{-1}. \tag{4}$$

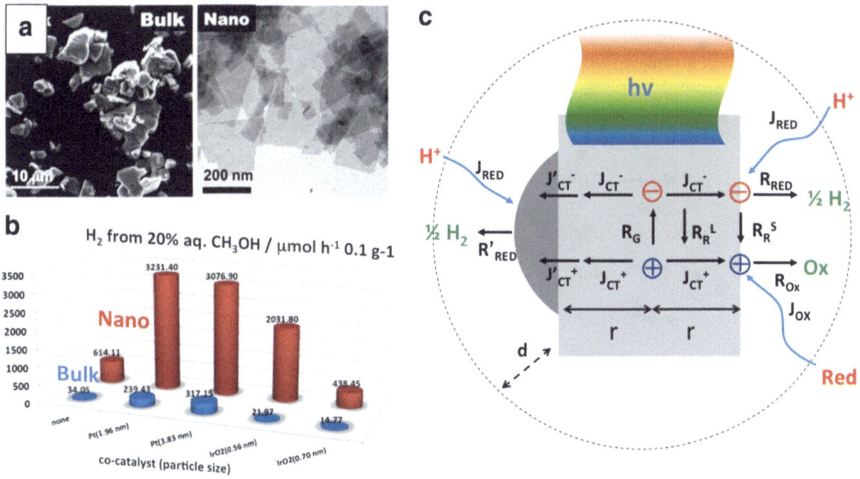

Fig. 6 (**a**) Electron micrographs of bulk and nanoscale niobate particles. (**b**) H_2 evolution rates of bare and co-catalyst modified particles. (**c**) Model to relate the electronic rate of the catalyst (mol s^{-1} cm^{-3}) to the rates for light-induced electron-hole generation (R_G), electron-hole recombination in the lattice ($R_R{}^L$) and on the surface ($R_R{}^S$), rates for charge and mass transfer to the catalyst-water interface (J_{CT}, J_{MT}), and to the rates for the redox reactions with the substrates (R_{RED}, R_{OX}). Reproduced with permission from [84]. Copyright 2012, American Chemical Society

The model calculates the electronic rate of the catalysts as a sum of the inverse rates of charge generation R_G, charge and mass transport J_{CT}, chemical conversion $J_{OX/RED}$, and charge recombination R_R. This one-dimensional continuity analysis [66, 85–88] is equivalent to Kirchhoff's law, according to which the total resistance of a series of resistors equals the sum of the individual resistances. The analysis shows that the activity of the photocatalysts is limited by the slowest kinetics in the series. Under the experimental conditions (sacrificial donor), these are the rates of charge transport to the water–catalyst interface and of proton reduction. Mass transport in the solution phase does not play a significant role, and neither does surface recombination.

The effect of nanoscaling on the ability to extract short-lived minority carriers is also evident in colloidal Fe_2O_3 dispersions, which catalyze the oxygen evolution reaction with aqueous 0.1 M $AgNO_3$ as the sacrificial electron acceptor [89]. Even though the performance of the system is limited by silver deposition on the Fe_2O_3 nanocrystals (Fig. 7), a correlation between O_2 production rates and particle size is seen. This can be attributed to improved hole extraction in the smaller nanocrystals [89]. In principle, by reducing the nanocrystal size further, and by increasing the thermodynamic driving force for charge extraction, most metal oxide materials can be turned into photocatalysts. The photocatalytic properties of 2 nm IrO_2 nanocrystals illustrate this principle. Because IrO_2 is metallic [90], the lifetime of the photogenerated holes is on the picosecond time scale. Nevertheless, visible and UV-light induced photocatalytic oxygen evolution from IrO_2 sols can be observed

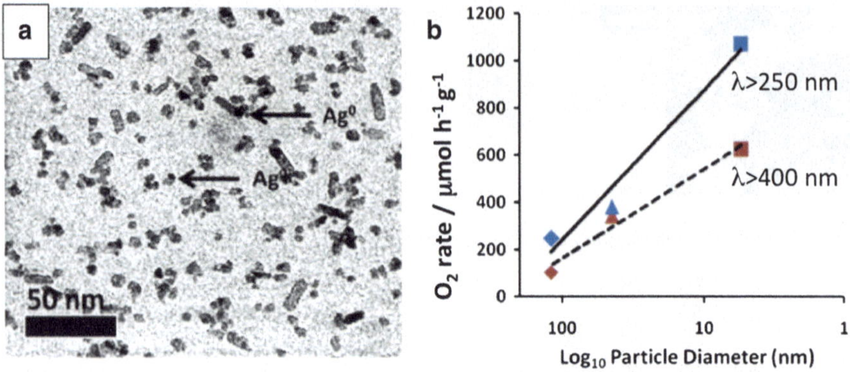

Fig. 7 (**a**) Silver nanoparticles formed on nano-Fe_2O_3 after irradiation in aqueous $AgNO_3$. (**b**) Correlation between Fe_2O_3 particle size and O_2 evolution rate. Reproduced with permission from [89]. Copyright 2011, Royal Society of Chemistry

in the presence of silver nitrate or sodium persulfate sacrificial electron acceptors [91]. However, the quantum efficiency for the process is too low (0.19% at 530 nm) to be useful for solar energy to fuel conversion. This principle also explains photocatalysis with Plasmonic gold nanoparticles.

6 Quantum Size Confinement

The observation of quantum size effects in thin films and quantum dots (QDs) dates back to the work by Dingle [53] and Louis Brus [56]. Quantum size effects not only depend on the material and crystal size, but also the nanocrystal shape [92]. The expanded band gap shifts the conduction and valence band edges to more reducing and more oxidizing potentials, respectively. From Marcus–Gerischer theory it is expected that this increase in thermodynamic driving force raises the rates of interfacial charge transfer and water electrolysis [88, 93, 94]. Increases in solid-solid electron transfer were experimentally confirmed for CdQ(Q=S, Se)-TiO_2 composites [95, 96], and quantitatively described with Marcus theory [97]. In 2013, the author's group established a similar rate dependence for proton reduction with CdSe quantum dots (Fig. 8) [98, 99]. For the study, monodisperse CdSe QDs with diameter 1.8–6.0 nm were synthesized in the presence of 2-mercaptoethanol as a ligand [100], followed by size-selective precipitation with 2-propanol. As can be seen from Fig. 8a, the hydrogen evolution rates from the CdSe QD suspensions in Na_2SO_3 solution show a logarithmic dependence on the crystal diameter.

A logarithmic plot of the rates against the driving force of the reaction is also linear (Fig. 8b). The quasi Fermi energies E_{Fn} were obtained from either the photocurrent onset in photoelectrochemical measurements or extrapolated from the electrochemical reduction peak of the QDs. The trend is expected for free energy-controlled interfacial charge transfer, as described by the Butler–Volmer equation [101]. As

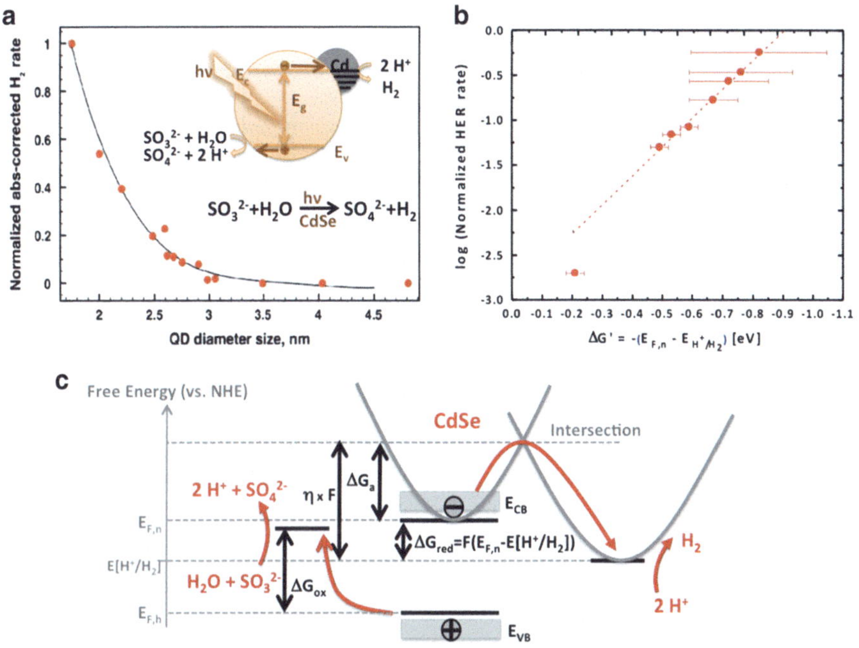

Fig. 8 (**a**) H_2 evolution rates vs QD size. Rates are normalized with regard to catalyst amount and absorbed photons. *Inset*: Schematic diagram of hydrogen evolution from CdSe QDs in the presence of sodium sulfite. (**b**) Weighted least square linear fit (*dashed line*) of experimental H_2 rates vs Fermi energies from electrochemistry. The lowest point is excluded from the fit because the rate is close to experimental error ($\pm 10^{-3}$) of the H_2 rate measurement. (**c**) Reaction energy profile for CdSe QD redox system. The kinetic activation energy ΔG_a for proton reduction is controlled by the Gibbs free energy change ΔG_{red} of the reaction, as given by the quasi-Fermi level of electrons under illumination $E_{F,n}$ and the Nernst potential E [H^+/H_2]. Reproduced with permission from [99]. Copyright 2013, American Chemical Society

shown in Fig. 8c, there are two half reactions at the QD interface, the oxidation of sulfite and the reduction of protons. The driving force ΔG_{ox} for the oxidation reaction is given by the difference between $E_{F,h}$ and E [SO_4^{2-}/SO_3^{2-}] and the driving force for the reduction ΔG_{red} by the difference between $E_{F,n}$ and $E[H^+/H_2]$. Compared to ΔG_{red}, ΔG_{ox} for sulfite oxidation is large and relatively constant across the series of QDs. Thus, the proton reduction kinetics are the rate-limiting factor for the photocatalytic hydrogen evolution over CdSe QDs. Overall, these findings establish a quantitative experimental basis for quantum-confinement-controlled proton reduction with semiconductor nanocrystals.

In metal oxides, quantum confinement effects usually require crystal sizes below 2 nm. That is because electron-hole pairs are less delocalized, and, correspondingly, their Bohr exciton radius is small [82]. Such conditions are fulfilled for 0.71 nm thick WO_3 nanosheets obtained by exfoliation of the layered compound $Bi_2W_2O_9$ (Fig. 9) [102]. Diffuse reflectance optical spectra reveal an absorption edge of 430 nm, consistent with a band gap of 2.88 eV, compared to 2.68 eV for bulk

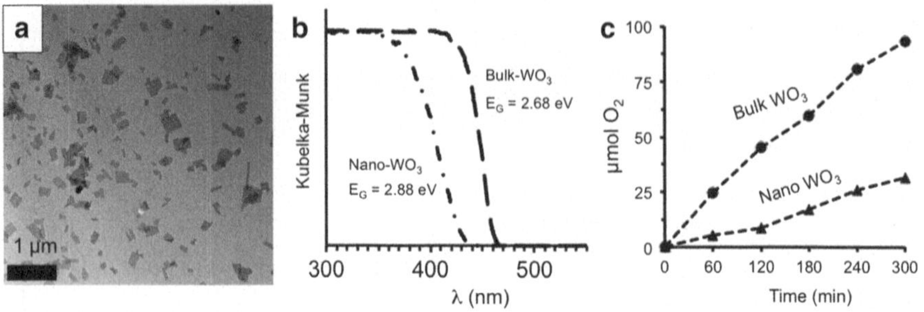

Fig. 9 (**a**) TEM image of nano-WO_3. (**b**) Optical absorption of nano-WO_3 and bulk particles. (**c**) O_2 evolution from aqueous 0.0083 mM $AgNO_3$ with 50 mg WO_3 and under >400 nm light from 300 W Xe lamp (315 mW cm^{-2}). Reproduced with permission from [102]. Copyright 2012, American Chemical Society

WO_3. Under visible light illumination, the photocatalytic O_2 evolution from such nanosheets is lower than the bulk because of reduced light absorption. However, the quantum efficiency for oxygen evolution at 375 nm (1.55%) exceeds that of the bulk (1.43%). This is because of the greater free energy stored in the nanosheets.

7 Multiple Exciton Generation

The altered electronic structure of strongly size-confined nanocrystals gives rise to multiple exciton generation (MEG), i.e., the formation of several (n) electron hole/ pairs after absorption of one photon with an energy n times that of the particle's band gap (Fig. 10). The MEG effect has been made responsible for the abnormally high efficiency of PbSe-sensitized TiO_2 photoelectrochemical cells [103] and PbSe photovoltaic cells [104].

The MEG effect has not yet been demonstrated for water splitting photocatalysts. The problem is that, for MEG, the band gap of the absorber needs to be an integer fraction of the exciting light. For the upper range of solar photons with 3.0 eV, this would correspond to $E_G = 1.50$ or less. Such a low band gap is not sufficient for water electrolysis, considering additional voltage losses from charge transfer and recombination. However, the MEG effect could potentially work for tandem or multi-junction photocatalysts, which utilize several small band gap absorbers in series.

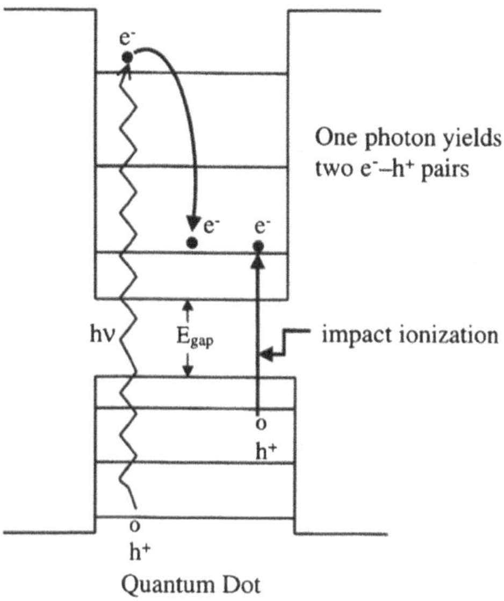

Fig. 10 Enhanced photovoltaic efficiency in QD solar cells by impact ionization (inverse Auger effect). This is promoted in QDs because of the lower rates of carrier relaxation. Reproduced with permission from [105]. Copyright 2002, Elsevier Science Ltd

8 Ion Effects

Because the dimensions of nanomaterials are only a fraction of the space charge layer thickness in solid materials, screening effects from electrolytes and specifically adsorbed ions are dominant. For pure water, the most important ions are hydroxide and hydronium ions, and their effect on metal oxides, including TiO_2 [106] and Fe_2O_3 [107, 108] gives rise to the well-known 59 mV pH^{-1} variation of the flatband potential with the solution pH [109]. In surface water, phosphate, silicate, and fluoride ions are often strongly adsorbing [110], which determines the redox stability of many minerals [109]. In contrast, the flatband potentials of II/VI, III/V, and group IV semiconductors are more susceptible to adsorption of soft ligands, including sulfur [111], HS^- [112, 113], HTe^- [114, 115], and Cl^- [116]. These often bind to specific crystal surfaces [114, 115] and control the open circuit voltage of photoelectrochemical cells [116, 117] and the water redox rate of suspended photocatalysts.

The effect of specifically adsorbed ions follows directly from the definition of the Fermi energy of the electrons E_F in a particle, which is equal to the chemical potential μ minus the electrical potential ϕ of the material (*F*: Faraday constant) [118].

$$E_F = \mu - \phi F. \tag{5}$$

Adsorbed ions can modify either μ or ϕ, or both. In the literature, proton adsorption is usually considered as an effect on the chemical potential μ. This leads to the known Nernstian dependence of the semiconductor flatband potential on solution

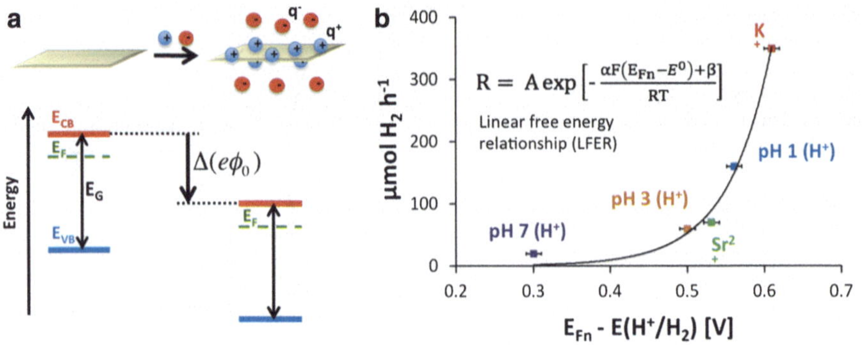

Fig. 11 (**a**) Effect of specifically adsorbed ions on the energetics of a sheet-like nanocrystal. (**b**) Calculated rate of hydrogen evolution vs ΔG for electron transfer with illuminated cation modified $KCa_2Nb_3O_{10}$ nanosheets [123]. *Constants*: Faraday constant $F = 96{,}485$ C mol^{-1}, ideal gas constant $R = 8.314$ J mol^{-1} K^{-1}, $T = 298$ K, free energy parameters $A = 0.0127$, $\alpha = 0.441$, and $\beta = 633$ J mol^{-1}. Additional symbols in (6): relative permittivity ε and dielectric constant ε_0. Reproduced from [123] with permission from The Royal Society of Chemistry

pH [106, 107, 109]. Equation (5) works because redox reactions with protons are fast, allowing the protons to be in electrochemical equilibrium with the electrons in the particle. For ions such as Al^{3+} (−1.66 V vs NHE) and Mg^{2+} (−2.37 V vs NHE) [119], whose negative reduction potentials preclude reduction in water, an electrostatic approach is more suitable (Fig. 11). Here it can be assumed that ion adsorption to the surface A generates the charge density $\sigma = q/A$, which modifies the surface potential ϕ_0. Neglecting space charge layer effects, the potential ϕ felt inside the particle is the same as ϕ_0, causing E_F to change according to (5). The relation between the surface charge density σ and the surface potential ϕ_0 is given by the Grahame equation (for definition of symbols see Fig. 11 caption) [120, 121]:

$$\phi_0 = \frac{2RT}{zF} \sinh^{-1}\left(\frac{\sigma}{\sqrt{8RT\varepsilon\varepsilon_0 c^0}}\right). \qquad (6)$$

In it, z and c^0 describe the charge and molar concentration of the counterions in solution, which surround the particle. Specifically, adsorbed cations produce a positive surface potential, shifting the energy bands down to more oxidizing potentials, and anions move the band edges to more reducing potentials. This shifts the Fermi energy in (5) and with it the driving force for photochemical charge transfer. In general, the variation of the electron transfer rate constant with E_F can be understood using free energy relationships [122], including Butler–Volmer [99, 101] and Marcus theory [94].

These theoretical predictions were recently verified for M-modified $KCa_2Nb_3O_{10}$ nanosheets (M=H^+, K^+, Sr^{2+}) [123]. Hydrogen evolution rates from the illuminated nanosheets are plotted in Fig. 11b against the thermodynamic driving force for proton reduction $F(E_{Fn}-E^0)$, obtained from photoelectrochemistry. The data can be fitted with the linear free energy relationship (LFER) [122] shown in the figure (constants in caption). The numerical values of α and β do not have physical significance, because the fitted rate R (μmol h^{-1}) does not convey any information about the value of the rate constant for the process, which depends on the electro-active area of the nanoparticles, the absorbed photons flux, space charge layer effects, and other unknown parameters [30, 124–128].The model provides a physical explanation for the observed correlation between nanosheet energetics and hydrogen evolution rates, and it confirms the dependence of photocatalytic activity on the presence of specifically adsorbed ions.

9 Interfacial Charge Transfer

The larger specific surface area of nanomaterials promotes charge transfer across the material interfaces (solid–solid and solid–liquid), allowing water redox reactions to occur at lower current densities and, correspondingly, lower overpotentials. This is a direct consequence of the Butler–Volmer equation which relates the current density to the overpotential [101]. This boost is particularly important for the slow, multi-step water oxidation reaction [129, 130], which normally requires highly active and often expensive co-catalysts based on Ir, Rh, or Pt [131]. It is one of the reasons why nanostructured electrocatalysts are so effective [132, 133]. The problem for photocatalysts is that the increase in junction area also increases the rate for reverse charge transfer, as shown by J_T and J_{et} in Fig. 12. These currents oppose the electron drift away from the surface (thick arrow) and reduce the rectifying character of the junction. The effect on the open circuit voltage of the junction is described with the Shockley diode equation (2). Every decadic increase of the reverse saturation current J_0 can be expected to decrease the open circuit voltage by 59 mV. The only way to overcome this fundamental limitation is by making the junction area smaller, as discussed above under Sect. 4.

10 Electron–Hole Recombination

Electron–hole recombination is the major loss mechanism in excitonic solar cells and in photocatalysts [135]. Photogenerated charge carriers recombine through radiative and nonradiative processes in the bulk phase of the semiconductor, in the depletion region, and at surface defects (Fig. 12) [134, 136–138]. These processes diminish the steady state concentrations of usable charge carriers, their charge transfer rates (thick black arrows in Fig. 12), and the driving force for water electrolysis. Furthermore, the larger specific surface area of the particles promotes non-radiative and interfacial recombination rates. The effect of these parameters on photocatalytic activity and electron-hole lifetime are commonly

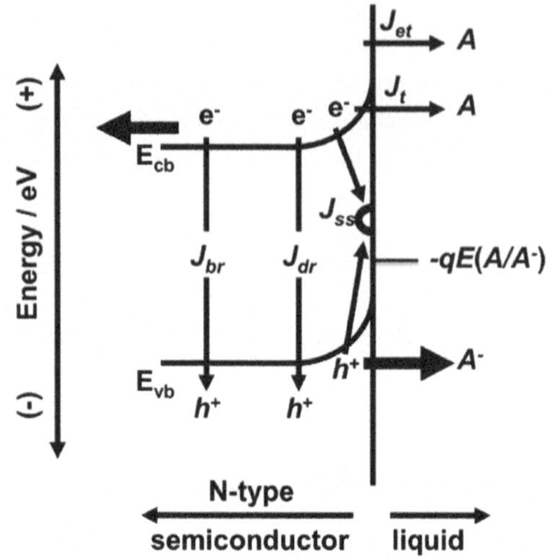

Fig. 12 Recombination pathways for photoexcited carriers in a semiconductor PEC [9, 134]. The *arrows* signify bulk recombination (J_{br}), depletion-region recombination (J_{dr}), and surface recombination (J_{ss}). Additional loss mechanisms caused by undesired charge transfer are also shown. Electron tunneling through and over the barrier produces the current densities (J_t) and (J_{et}). Electron collection by the back contact and hole collection by the redox couple (e.g., oxidation of water to O_2) are desired processes shown by thick *black arrows*. Reproduced with permission from [134]. Copyright 2005, American Chemical Society

acknowledged in the literature [89, 139–143], but quantitative studies on this topic are rare [65, 79, 144–148].

We recently employed graphitic carbon nitride g-C_3N_4 to observe the effect of structure defects on the ability of the material to reduce protons photocatalytically and to generate a photovoltage [149]. Specifically, we observe an inverse relation between the photocatalytic hydrogen production rate from aqueous methanol and the calcination temperature (Fig. 13). Higher temperatures also decrease the photoluminescence (PL) of the material and the photovoltage. Based on the PL and surface photovoltage (SPV) data, there are two types of defects present near the conduction and valence band edges of the material. These defects promote electron–hole recombination and reduce the ability of the material to generate a photovoltage.

The reduction of surface defects must be a major goal for the future if nanostructured photocatalysts are to be used for unbiased solar water splitting [150]. The literature suggests that performance enhancements can be achieved with chemical treatments. For example, reaction of silicon with HF can suppress surface recombination and improve the performance of photovoltaic devices [138]. For Fe_2O_3 photoanodes it has been shown that application of Al_2O_3 [151] or SnO_2 [152] overlayers also improves the performance. For $BiVO_4$ photoanodes, both application of an SnO_2 underlayer [153], and incorporation of W dopants boost the electrical power output, which is attributed to reduction of surface recombination

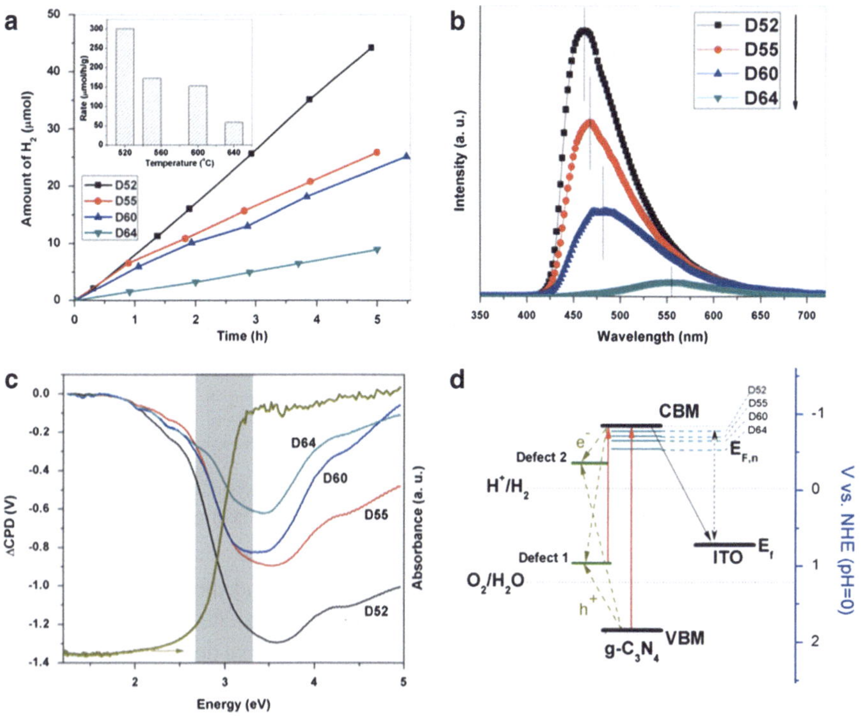

Fig. 13 Properties of g-C_3N_4 prepared at 520, 550, 600, and 640°C. (**a**) H_2 evolution from platinated samples (30 mg) in methanol (20 vol.%) aqueous solution at pH 4.5 under visible light (>400 nm). (**b**) PL spectra at 350 nm excitation. (**c**) SPV spectra of g-C_3N_4 films on ITO substrate with UV–vis spectrum (*dark yellow line*) of D52. (**d**) Energy diagram of g-C_3N_4 with defect levels at +0.97 V and −0.38 V. Reproduced from [149] with permission from The Royal Society of Chemistry

[154]. Further work is needed to determine the general merit of these and related surface passivation approaches for suspended photocatalysts.

11 Excited State Entropy

Theoretically, the degree of quantum confinement of a light absorber determines the excited state entropy and its free energy [155]. According to (7) [156], the free energy of a semiconductor absorber $\mu_{e,h}$ is determined by the band gap E_G, the temperature, the electron and hole concentrations n_e and n_h, and the effective density of states $N_{CB/VB}$ near the band edges. The greater the $N_{CB/VB}$, the more diluted the charge carriers and the higher their entropy loss. In quantum dots, a reduction of $N_{CB/VB}$ can readily be achieved through quantum size effects (Fig. 14). Here, for $N_{CB} = 18$, 12, 6, and $n = 6$ electrons, the number of microstates

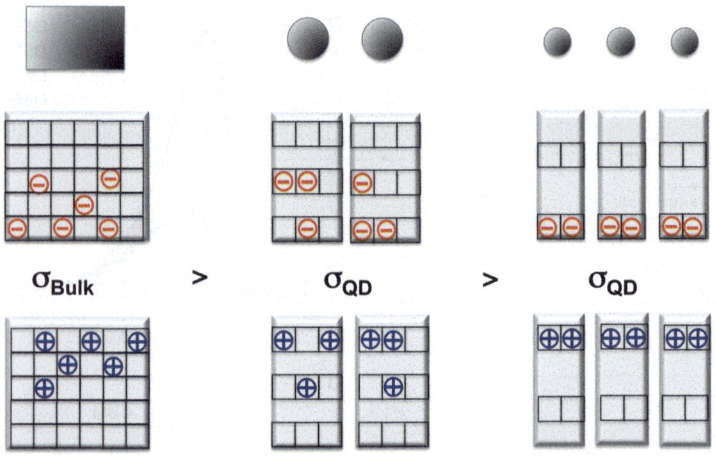

Fig. 14 Effect of quantum sizing on $N_{CV/VB}$ and entropy for three systems of six electron-hole pairs. Reproduced from [155] with permission from The American Chemical Society

$W_{Tot} = 18!/[(18-6)!6!]$, $12!/[(12-6)!6!]$, $6!/[(6-6)!6!]$, and the entropy per electron $\sigma = 1.5 \times 10^{-23}$, 1.1×10^{-23}, 0.0 J/K. Thus, the free energy increases with quantum confinement because of the reduction of the effective density of states near the band edges. Physically, this represents a concentration of charge carriers near the band edges. This is in addition to the increase of the potential energy of the charge carriers which results from the widening of the band gap.

$$\mu_{e,h} = E_G - kT\ln\frac{N_{CB}N_{VB}}{n_e n_h}. \tag{7}$$

The entropy effect is difficult to observe in actual photocatalysts because of the presence of other loss mechanisms (e.g., non-radiative recombination). Furthermore, the energetics of nanoscale systems is very sensitive to variations in size, shape, and molecular environment. The corresponding increase in the ground state entropy of polydisperse and randomly packed quantum dots can reverse the entropy reduction of the individual subsystems. Additional entropy losses in particulate absorbers can occur as a result of the redistribution of light. Potentially, these losses can decrease the photovoltage by as much as 315 mV [157].

12 Electron–Hole Separation

In nanomaterials, carrier separation (arrows in Fig. 12) is more difficult to achieve than in the bulk because at average doping concentrations ($n_0 = 10^{17}$ cm^{-3}) space charge layers are not effective on the nanoscale [30, 59, 158, 159]. Also, for

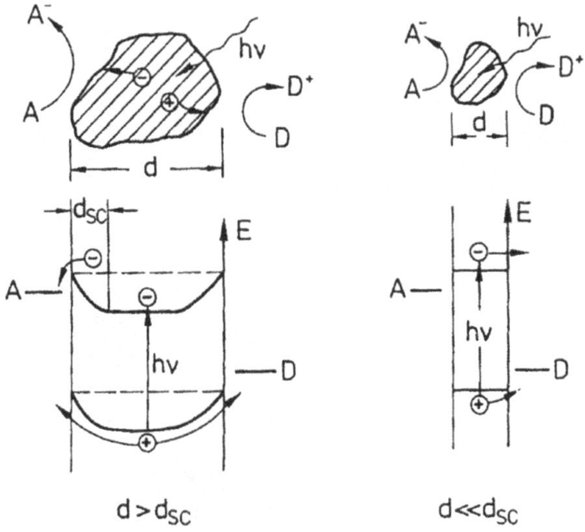

Fig. 15 Space charge layers in large and small particles. Reproduced with permission from [30]. Copyright 1994, Springer Berlin/Heidelberg

spherical nanoparticles, the space charge layer thickness d_{sc} cannot exceed the radius $d/2$ of the particle (Fig. 15), which restrains the possible barrier height at the interface. For example, for 16-nm TiO_2 nanocrystals ($\varepsilon = 160$) with a charge carrier concentration of $n_0 = 10^{17}$ cm^{-3}, O'Regan calculated a barrier height of 0.3 meV under maximum depletion [160]. This means that, in the absence of a strong applied bias, the bands in a nanoparticle are essentially flat, as shown in Fig. 15.

In the absence of an interior electrical field, the relative rates of electron-hole injection into the electrolyte are governed by the kinetics of interfacial charge transfer alone [158]. This leads to an increase of undesirable processes (current densities (J_t) and (J_{et}) in Fig. 12) which short-circuit the device. The absence of an internal mechanism for charge separation also increases electron-hole recombination, as discussed above. This is the reason why efficient collection of majority carriers in dye-sensitized cells with nanocrystalline TiO_2 substrates depends on the presence of iodide as an easily oxidizable electron donor [160, 161]. Without iodide, only 4% of electrons can be harvested at the back contact because of recombination with holes on the Ru dye (this corresponds to J_{et} and J_t in Fig. 12). Furthermore, in nanostructures with quantum size effects (e.g., in many metal chalcogenides), the electron-hole pairs are confined to a space smaller than the Bohr exciton radius. This means that additional energy is necessary to separate the charges. This is analogous to organic photovoltaic cells, where a higher exciton binding energy results from the lower dielectric constant of organic polymers [162].

Photochemical labeling can provide information about charge separation in suspended light absorbers [32, 163, 164]. Typically, a photocatalyst is irradiated in the presence of metal salt that forms insoluble deposits (e.g., Pt, Au, MnO_x, IrO_x) after accepting photogenerated electrons or holes. This helps to pinpoint the locations of the photocharges on the absorber surface. For nanosheets derived

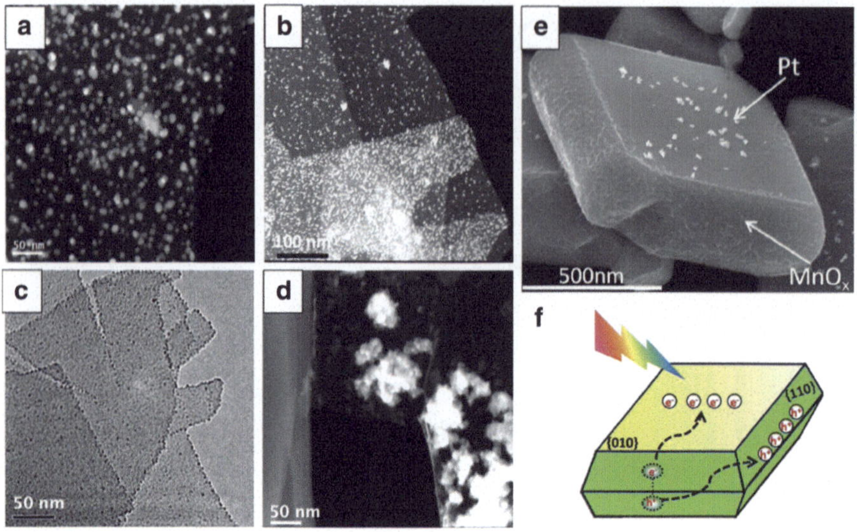

Fig. 16 $HCa_2Nb_3O_{10}$ nanosheets after labeling with (**a**) Ag, (**b**) Pt, (**c**) IrOx, and (**d**) MnOx [165]. (**e**) $BiVO_4$ microcrystals after labeling with Pt and MnOx (**f**) [166]. Reproduced from [166] with permission from The Royal Society of Chemistry

from the layered perovskite $KCa_2Nb_3O_{10}$, non-selective labeling results support the lack of intra-sheet charge separation (Fig. 16) [165]. Better charge separation is generally found in microscale particles, especially those where facets with different surface potentials are present. This is shown for $BiVO_4$ microcrystals in Fig. 16e, f. The selective deposition of Pt and MnOx suggests that charge carriers accumulate at (010) and (111) facets of the crystal [166].

In ferroelectric materials, local dipoles (Fig. 17) can aid photochemical charge separation [167, 168]. Materials such as $BaTiO_3$ support a spontaneous electrical polarization that stems from the displacement of mobile cations in the unit cell. Photochemical labeling experiments confirm that these dipoles can guide the accumulation of photochemical charge carriers [169]. The ferroelectric polarization appears to promote photocatalytic reactions [169], but its use for photocatalytic water splitting has not been tested.

Gradient doping provides an alternative way to generate a potential energy gradient inside of small crystals. This approach was recently demonstrated by van de Krol's group for tungsten-doped $BiVO_4$ films. [170] A tungsten concentration gradient inside the films moves photoelectrons towards the electrode and holes towards the solid–liquid interface. This is an elegant way to control charge separation, but it is probably only usable in defect-tolerant semiconductors, such as $BiVO_4$.

For the majority of other metal oxide particles, the use of electron or hole selective acceptors may be a more suitable approach to promote charge separation. Many transparent n- or p-type metal oxides are known to accept electrons or holes

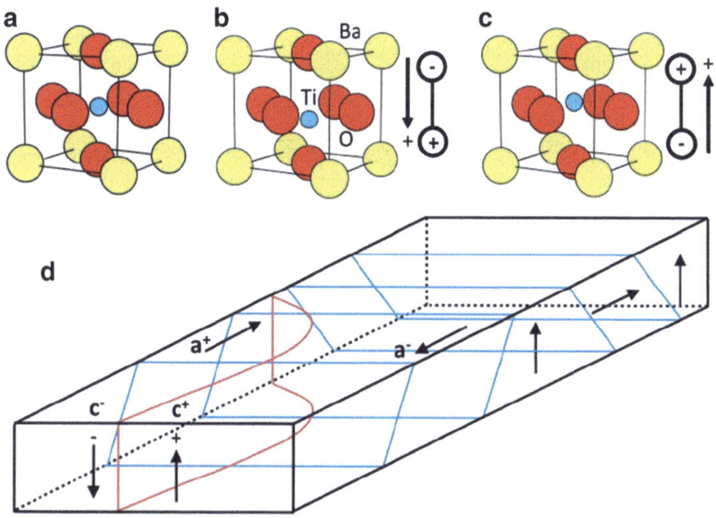

Fig. 17 Cubic unit cell of $BaTiO_3$ above (**a**) and below (**b**, **c**) the Curie temperature. Note displacement of the central Ti(IV) ion and corresponding dipoles. (**d**) Crystal with domains of constant polarization in the *arrow* direction. Reproduced from [169]

selectively [171]. For example, n-SnO_2 underlayers on $BiVO_4$ photoanodes have been found to boost the photocurrent by 300% [153] and p-type NiO layers on Fe_2O_3 improved the photovoltage by nearly 0.4 V [172]. Many of these metal oxides can be deposited from solution and could potentially be applied to photocatalysts [173].

13 Interparticle Charge Transport

In nanocrystalline films, charge carriers move by diffusion instead of drift [59, 174–176]. As a result, charge transport is much slower than it is in the bulk phase, increasing the chances for recombination and back reactions [177]. If the nanoparticles are not fused together, additional barriers arise from interparticle charge transport, which occurs by thermally activated hopping and by electron tunneling (Fig. 18). Charge transport depends on the interparticle distance and the electrostatic charging energy of the donor acceptor nanocrystal couple [178, 179].

The problem of charge transport over macroscopic distance is less important with suspended catalysts. However, it does play a role for charge transport in photocatalyst films, as used for photoelectrochemical measurements, or with immobilized photocatalysts.

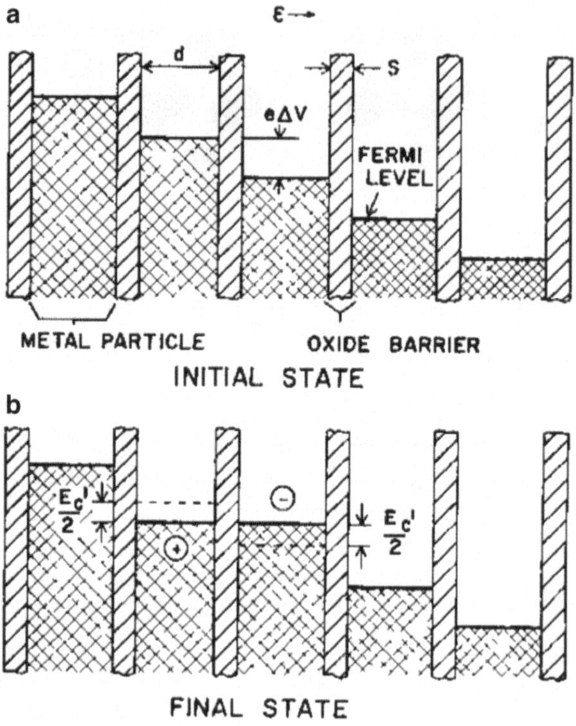

Fig. 18 Electron hopping in nanostructured films. Reproduced from [178] by permission of the publisher (Taylor & Francis Ltd., http://www.tandf.co.uk/journals)

14 Examples of Nanoscale Photocatalysts for Overall Water Splitting

Because of the problems above, the number of known nanoscale photocatalysts for overall water splitting is limited [57–66]. Only five nanoscale photocatalysts have been reported in the literature. Four of them are Schottky type devices and require ultraviolet light for operation. The most active catalyst was reported by Kondo's group in 2011. It consists of aggregates of NiO_x-loaded $NaTaO_3$ nanocrystals (Fig. 19) [180]. The catalytic rate of this system was very high (2.0 mmol h^{-1} of H_2 with stoichiometric O_2) but required photons with >4.0 eV because of the band gap of bulk $NaTaO_3$ [32]. The activity of the 20 nm nanoparticles was three times higher than that of the 50, 100, and 200 nm particles tested for comparison. This was attributed to the higher surface area of the smallest absorber size and higher crystallinity.

Another system was developed by Akihiko Kudo's group and consists of $LiNbO_3$ nanowires (70 nm × 10 μm) with a band gap of 4.0 eV (Fig. 20) [142]. After modification with 1 mass % RuO_2 co-catalyst, the nanocomposite can split water with 0.7% quantum efficiency at 254 nm (O_2 was evolved in slight excess). The lower activity of the bulk compared to the nanowires was attributed to surface defects resulting from ball-milling.

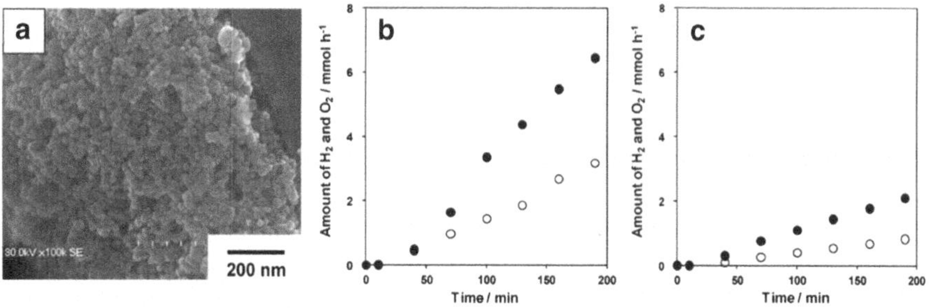

Fig. 19 NiO-NaTaO$_3$ nanoscale photo catalyst. (**a**) SEM of NaTaO$_3$, (**b**, **c**) H_2/O_2 evolution from water with 20 nm and 50 nm particles, respectively, under UV irradiation ($\lambda > 200$ nm) from a 450-W high-pressure mercury lamp (UM-452, Ushio) using 0.3 g of catalyst. Reproduced from [180] with permission from The Royal Society of Chemistry

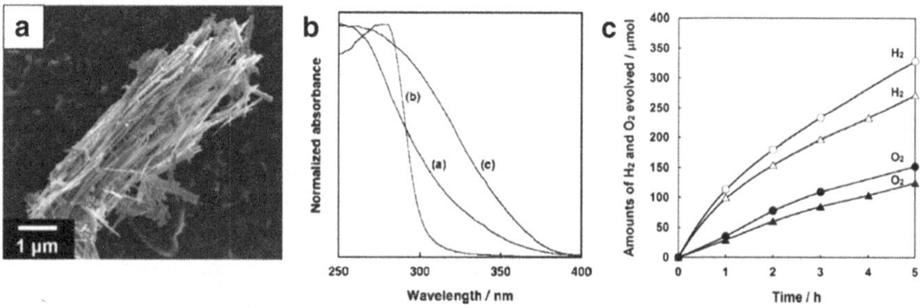

Fig. 20 RuO$_2$-LiNbO$_3$ nanoscale photo catalyst. (**a**) SEM of LiNbO$_3$. (**b**) Optical absorption (*a*: LiNbO$_3$-Nanowire, *b*: LiNbO$_3$–Bulk, *c*: Nb$_2$O$_5$-Nanowire). (**c**) H_2/O_2 evolution from water with bulk (*triangles*) and nano RuO$_2$-LiNbO$_3$ (*squares*). Conditions: 0.3 g catalyst, pure water, 370 mL, 400-W high pressure Hg lamp; inner-irradiation cell made of quartz. Reproduced from [142] with permission from The Royal Society of Chemistry

In a third example (Fig. 21), Yan et al. reported overall water splitting with 3 mass % RuO$_2$-modified Zn$_2$GeO$_4$ nanorods (100 × 150 nm) under UV light from a 400 W Hg UV lamp to excite the large band gap (>4.5 eV) of the material. Here, the H_2 evolution rate was 17.4 μmol h^{-1} (stoichiometric O_2) with 100 mg of the catalyst [143]. The lower activity of the bulk material is attributed to surface defects and lower surface area.

The last example consists of 6–30 nm SrTiO$_3$ nanocrystals with a 3.3 eV indirect band gap (Fig. 22) [181]. After modification with an NiO$_x$ co-catalyst, this system produces 19.4 μmol H_2 g^{-1} h^{-1} (with stoichiometric O_2) from pure water under >3.2 eV illumination (26.3 mW cm^{-2}). The activity of the 30 nm particles was three times higher than the 6 nm particles, probably as a result of surface defects in the latter. These defects are visible as the tail in the absorption spectrum. They resulted from the lower preparation temperature of the 6 nm particles. The 30 nm particles gave 35% lower rates for H_2/O_2 evolution than the bulk particles (synthesized by solid state reaction). This is attributed to a quantum confinement effect,

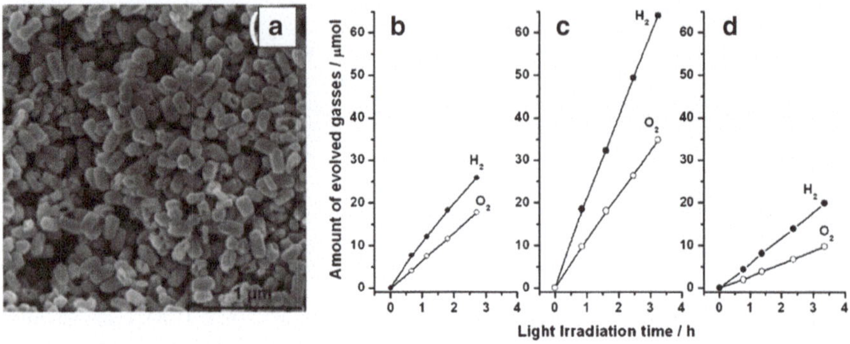

Fig. 21 RuO_2-Zn_2GeO_4 nanoscale photo catalyst. (**a**) SEM of Zn_2GeO_4. (**b**) H_2/O_2 evolution from water with particles prepared at 40°C. (**c**) Same for particles made at 100°C. (**d**) Same for bulk particles from a solid state reaction at 1,300°C. Conditions: UV irradiation ($\lambda > 200$ nm) of 0.1 catalyst in pure water with a 400-W high-pressure mercury lamp. Reproduced from [143] with permission from The Royal Society of Chemistry

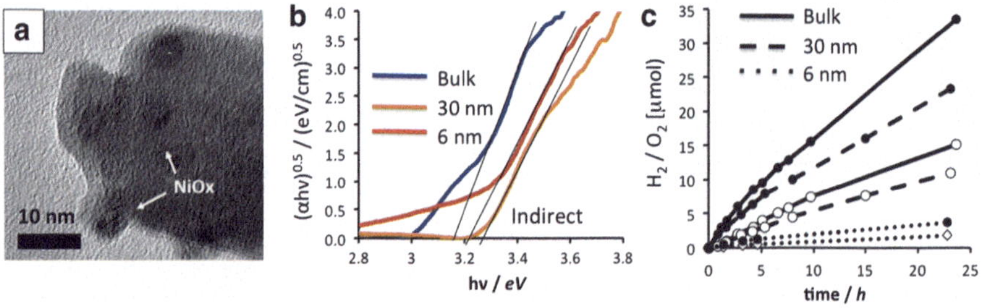

Fig. 22 NiO-$SrTiO_3$nanoscalephotocatalyst. (**a**) SEM of NiO-$SrTiO_3$. (**b**) Tauc plots for bulk (>100 nm), 30 nm, and 6 nm $SrTiO_3$ particles. (**c**) H_2/O_2 evolution from water with 100 mg of catalyst under UV irradiation (250–380 nm at ~26 mW cm^{-2} from a 300 W Xe lamp). Reproduced from [181]with permission from The American Chemical Society

which increases the band gap of the smaller particles and reduces the absorbed photons.

In 2014, Kudo's group reported the first nanostructured tandem catalyst for overall water splitting [44]. The system employs physical mixtures of Ru-modified Rh:$SrTiO_3$ and $BiVO_4$ powders in the presence of $FeCl_3$ as a redox shuttle. Whereas 50–70 nm Rh:$SrTiO_3$ particles gave 16/7.5 μmol h^{-1} H_2/O_2, 300 nm particles gave 128/64 μmol h^{-1}.This inverse particle size dependency of the efficiency suggests sub-optimal charge separation and recombination in the smaller particles. The best system achieved a quantum efficiency of 3.9–4.2% (at 420 nm) and a solar to hydrogen efficiency of 0.1%.

Finally, in 2014, Peidong Yang's group reported direct water splitting from a mesh consisting of Rh:$SrTiO_3$ and $BiVO_4$ nanowires of 100–200 nm diameter

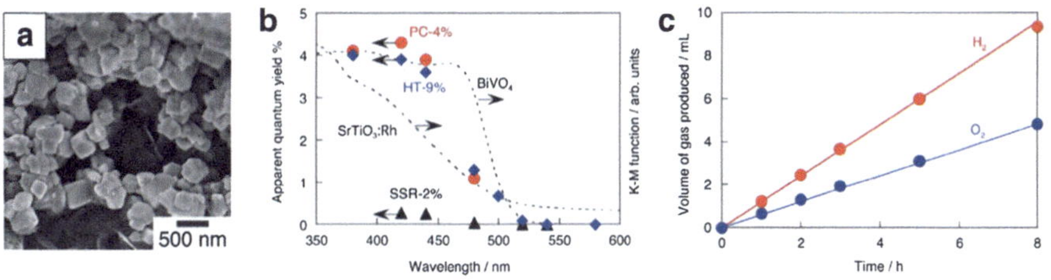

Fig. 23 Sr:$SrTiO_3$-Pt/$BiVO_4$ Tandem Photocatalyst system. (**a**) SEM of Rh:$SrTiO_3$. (**b**) Photoaction spectrum for Tandem system. (**c**) H_2/O_2 evolution under simulated sunlight (AM 1.5). Conditions: 50 mg of catalyst, 2 mM $FeCl_3$ solution. Reproduced from [44] with permission from The Royal Society of Chemistry

(Fig. 23). However, the H_2/O_2 evolution rate (<0.2 μmol h^{-1}) and the turnover number were too low to designate this process as catalytic [182].

15 Measuring Photovoltage in Nanoscale Photocatalysts

The low performance of the examples above emphasizes the need for a better understanding of charge transfer in nanoscale absorbers. Photoelectrochemistry is the most obvious way to probe photochemical charge separation in photocatalyst particles. However, because electrical contact to a working electrode is necessary, measurements need to be conducted on particle films. As noted above, particle mediated charge transport is generally slow, and reduces photocurrents to the microscale (Fig. 24). There is also a substantial potential drop across the film, which obscures the photovoltage of the individual particle junctions. Recently, Domen demonstrated that these charge transfer problems can be alleviated somewhat by evaporating metal electrode layers directly onto particle films [183–185]. The photocurrents from photoelectrodes made from Sc:$La_5Ti_2CuS_5O_7$ and $LaTiO_2N$ electrodes are strongly enhanced and reach over -0.5 mA cm^{-2}/ +3.0 mA cm^{-2} for water reduction/oxidation respectively (A.M. 1.5 illumination, at +0.3 V/+1.2 V vs RHE). This allows for more accurate studies of junction potentials in small particles.

Alternatively, surface photovoltage spectroscopy (SPS) can provide a direct assessment of the photovoltage of particle junctions. In SPS, a semi-transparent Kelvin electrode measures the contact potential difference (CPD) of an illuminated photocatalyst film (Fig. 25) as a function of the excitation energy [187, 188]. The measured photovoltage (ΔCPD) is produced by the transfer of free charge carriers or polarization of bound charge carrier pairs (polarons) in the sample [189–191]. Because the technique relies on the detection of voltage and not current, even small concentrations ($<10^{10}$ cm^{-2}) of charge carriers can be observed [189, 192–197]. This sensitivity exceeds that of photoelectrochemistry by at least 1,000 times. Another advantage of SPS is that it can be performed selectively on solid–

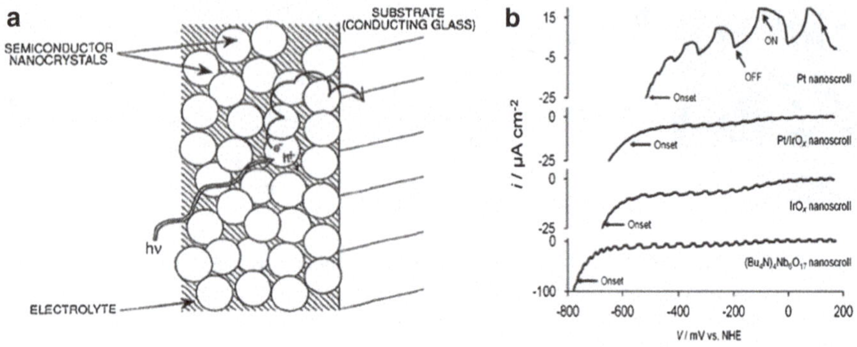

Fig. 24 (**a**) Charge transport limitation in particulate electrodes. Reproduced from [158] by permission of The Electrochemical Society. (**b**) Photocurrent scans for niobate nanocrystal films in methanol solution. Reproduced from [186] with permission (2011 Wiley-VCH Verlag GmbH & Co. KGaA, Weinheim)

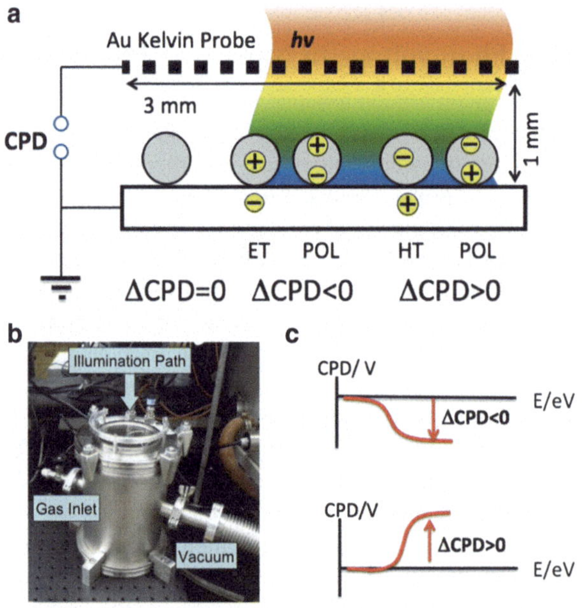

Fig. 25 (**a**) Geometry of SPV measurement. Contact potential changes are caused by polarization (POL) of electron-hole pairs or by charge transfer (ET/HT) into the substrate. The ΔCPD value can be interpreted as the open circuit voltage of the sample-substrate junction. (**b**) SPV Instrument. (**c**) Example spectra. A negative/positive ΔCPD signal corresponds to electron-hole movement towards the substrate. The photo-onset energy provides information about the effective band gap. Reprinted with permission from [190]. Copyright 2014, American Chemical Society

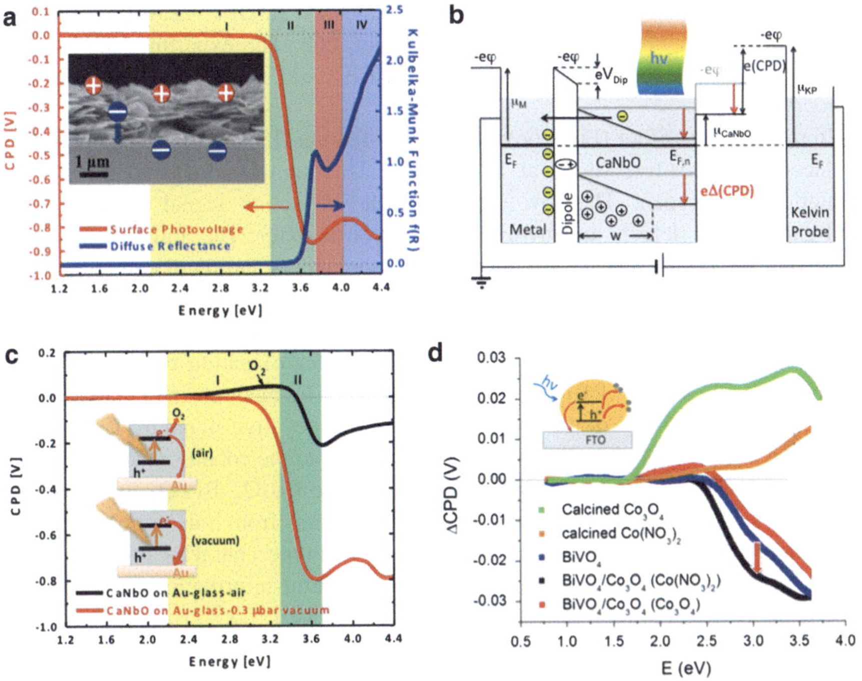

Fig. 26 (**a**) Surface photovoltage and diffuse reflectance spectra of $HCa_2Nb_3O_{10}$ nanosheet film on Au (*inset*: SEM of film). (**b**) Energy diagram for the metal-nanosheet-Kelvin probe configuration under band gap illumination. *Symbols*: e: electron charge, φ: electric potential; ΔCPD: light-induced contact potential difference change; *w*: space charge layer width; μ_i: chemical potential (work function) of $HCa_2Nb_3O_{10}$, metal substrate, and of Kelvin probe; V_{Dip}: potential drop from interfacial dipole; E_F: Fermi level (electrochemical potential) in the dark; E_{Fn}: quasi-Fermi level of electrons under illumination. (**c**) Photovoltage spectra of nanosheet film in vacuum and air. Reprinted with permission from [189] Copyright 2014, American Chemical Society. (**d**) SPV spectra of Co_3O_4, $BiVO_4$, and Co_3O_4–$BiVO_4$ on FTO substrate in vacuum (*insert*: charge transfer direction). Reproduced from [208] with permission from The Royal Society of Chemistry

solid interfaces without the need for a liquid phase or a redox couple. Because the photovoltage is entirely de-coupled from solid-liquid charge transfer, it allows direct measurement of solid-solid junction potentials.

Thomas Dittrich's group has used SPS extensively over the past decade on nanocrystal [198–200], molecular [201–203], and thin film [204] photovoltaics. However, as shown in the following example, SPS is also well suited for the characterization of nanostructured photocatalysts, where it can provide a quantitative understanding of charge transfer at solid-solid and solid-molecule contacts [189]. Figure 26 illustrates this for $HCa_2Nb_3O_{10}$ nanocrystals, a known large band gap (3.5 eV) water splitting photocatalyst. [205, 206]. The main signal in the SPV spectrum of this compound is negative and corresponds to photochemical electron transfer from $HCa_2Nb_3O_{10}$ to the gold substrate. The observed photovoltage (−1.025 V) is controlled by the built-in potential, i.e., E_F(Au)−

$E_{F,n}(HCa_2Nb_3O_{10} = -5.3$ eV [207] – (-3.5 eV) [205] $= -1.8$ eV, of the semiconductor–metal junction, as shown in Fig. 26b. The experimental value is lower because of the low illumination intensity and screening effects in the nanocrystal film.

In the presence of air (Fig. 26c) the negative feature II is decreased and partially replaced by a positive signal at 2.4–3.2 eV. This signal belongs to the photo-reduction of oxygen. The low energy of the reduction signal suggests that mid-gap surface defects are involved in electron transfer. The ability to observe these states in the SPV spectrum may promote the understanding of photochemical reactions between photocatalyst and molecules. In Fig. 26d the method is applied to the characterization of a contact between $BiVO_4$, a known water oxidation photocatalyst [209], and a Co_3O_4 nanoparticle water oxidation co-catalyst. The negative ΔCPD signal of $BiVO_4$ alone is produced by electron injection into the FTO substrate. Addition of a Co_3O_4 layer boosts the negative signal by 12 mV (red arrow) because of hole (minority carrier) injection into the cobalt oxide [208]. The boost can be interpreted as the photovoltage of the Co_3O_4–$BiVO_4$ junction. By itself, p-Co_3O_4 has a positive ΔCPD signal which stems from hole majority carrier transfer to the FTO substrate. This confirms the p-type character of this oxide. These examples illustrate how SPS can provide useful information about carrier type and charge separation in nanoscale contacts.

16 Conclusion

Research activity on nanostructured photocatalysts for solar water splitting has increased significantly over the last 30 years. Nanoscaling has been shown to improve charge extraction from absorbers with low carrier mobility and short carrier lifetimes. For selected metal chalcogenides the quantum size effect has been useful for increasing the stored free energy and for enhancing the rate of charge transfer and photocatalytic proton reduction. It was also shown that the energetics of nanoscale absorbers can be controlled with specifically adsorbed ions. At the same time, it has been found that nanoscaling decreases the ability of an absorber to generate free chemical energy from solar energy. The smaller size of the particles makes electron–hole separation more difficult to achieve because of non-selective interfacial charge transfer and reduced electric fields. Additionally, the larger specific area of nanostructures promotes defect recombination, which reduces the concentration of free charge carriers. These issues limit the photo-current and the photovoltage that can be generated by the absorber, and with it the solar to hydrogen conversion efficiency. Contemporary research should aim to overcome these problems by developing new chemical methods for surface passiv-ation and selective charge transfer and by more quantitatively assessing photovoltage and surface recombination in nanoparticles. The results from such studies not only benefit solar energy conversion but also promote our understanding of interfaces in general, as relevant to electronics, corrosion science, catalysis, materials science, and geology. We already know that useful solar energy conversion is possible at the nanoscale – the photosynthetic systems of bacteria are the earliest examples of nanostructured solar energy conversion devices [68]. Effective artificial devices are sure to follow once we improve our understanding of nanoscale interfaces.

Acknowledgement This material is based upon work supported by the National Science Foundation under CHE – 1152250 and CBET – 1133099. Any opinions, findings, conclusions, or recommendations expressed in this material are those of the author and do not necessarily reflect the views of the National Science Foundation. The author thanks the Research Corporation for Science Advancement for a Scialog award, and Kathryn A. Newton for help with proofreading the manuscript and with obtaining copyrights.

References

1. Lewis NS, Nocera DG (2006) Powering the planet: chemical challenges in solar energy utilization. Proc Natl Acad Sci U S A 103(43):15729–15735
2. Lewis NS, Crabtree G, Nozik AJ, Wasielewski MR, Alivisatos AP (2005) Basic research needs for solar energy utilization. Department of Energy http://science.energy.gov/bes/news-and-resources/reports/
3. Green MA, Emery K, Hishikawa Y, Warta W, Dunlop ED (2015) Solar cell efficiency tables (Version 45). Prog Photovolt Res Appl 23(1):1–9
4. Peharz G, Dimroth F, Wittstadt U (2007) Solar hydrogen production by water splitting with a conversion efficiency of 18%. Int J Hydrogen Energy 32(15):3248–3252
5. Nocera DG (2012) The Artificial Leaf. Acc Chem Res 45(5):767–776
6. Reece SY, Hamel JA, Sung K, Jarvi TD, Esswein AJ, Pijpers JJH, Nocera DG (2011) Wireless solar water splitting using silicon-based semiconductors and Earth-abundant catalysts. Science 334(6056):645–648
7. Luo J, Im J-H, Mayer MT, Schreier M, Nazeeruddin MK, Park N-G, Tilley SD, Fan HJ, Grätzel M (2014) Water photolysis at 12.3% efficiency via perovskite photovoltaics and Earth-abundant catalysts. Science 345(6204):1593–1596
8. Gratzel M, Park NG (2014) Organometal halide perovskite photovoltaics: a diamond in the rough. Nano 9(5):1440002–1440009
9. Walter MG, Warren EL, McKone JR, Boettcher SW, Mi QX, Santori EA, Lewis NS (2010) Solar water splitting cells. Chem Rev 110(11):6446–6473
10. Khaselev O, Turner JA (1998) A monolithic photovoltaic-photoelectrochemical device for hydrogen production via water splitting. Science 280(5362):425–427
11. Bolton JR, Strickler SJ, Connolly JS (1985) Limiting and realizable efficiencies of solar photolysis of water. Nature 316(6028):495–500
12. Varghese OK, Grimes CA (2008) Appropriate strategies for determining the photoconversion efficiency of water photo electrolysis cells: a review with examples using titania nanotube array photoanodes. Sol Energy Mat Sol C 92(4):374–384
13. Sathre R, Scown CD, Morrow WR, Stevens JC, Sharp ID, Ager JW, Walczak K, Houle FA, Greenblatt JB (2014) Life-cycle net energy assessment of large-scale hydrogen production via photoelectrochemical water splitting. Energy Environ Sci 7:3264–3278
14. Ronge J, Bosserez T, Martel D, Nervi C, Boarino L, Taulelle F, Decher G, Bordiga S, Martens JA (2014) Monolithic cells for solar fuels. Chem Soc Rev 43:7963–7981
15. Osterloh FE (2008) Inorganic materials as catalysts for photochemical splitting of water. Chem Mater 20(1):35–54

16. Maeda K (2013) Z-Scheme water splitting using two different semiconductor photocatalysts. ACS Catal 3(7):1486–1503
17. Kudo A (2011) Z-Scheme photocatalyst systems for water splitting under visible light irradiation. MRS Bull 36(1):32–38
18. Maeda K, Teramura K, Saito N, Inoue Y, Kobayashi H, Domen K (2006) Overall water splitting using (oxy)nitride photocatalysts. Pure Appl Chem 78(12):2267–2276
19. Rajeshwar K (2007) Hydrogen generation at irradiated oxide semiconductor-solution interfaces. J Appl Electrochem 37(7):765–787
20. Maeda K (2011) Photocatalytic water splitting using semiconductor particles: history and recent developments. J Photoch Photobio C 12(4):237–268
21. Maeda K, Domen K (2007) New non-oxide photocatalysts designed for overall water splitting under visible light. J Phys Chem C 111(22):7851–7861
22. Kudo A, Miseki Y (2009) Heterogeneous photocatalyst materials for water splitting. Chem Soc Rev 38(1):253–278
23. Abe R (2010) Recent progress on photocatalytic and photoelectrochemical water splitting under visible light irradiation. J Photoch Photobio C 11(4):179–209
24. Osterloh FE, Parkinson BA (2011) Recent developments in solar water splitting photocatalysis. MRS Bull 36(1):17–22
25. Pinaud BA, Benck JD, Seitz LC, Forman AJ, Chen ZB, Deutsch TG, James BD, Baum KN, Baum GN, Ardo S, Wang HL, Miller E, Jaramillo TF (2013) Technical and economic feasibility of centralized facilities for solar hydrogen production via photocatalysis and photoelectrochemistry. Energy Environ Sci 6(7):1983–2002
26. James BD, Baum GN, Perez J, Baum KN http://www1.eere.energy.gov/hydrogenandfuelcells/pdfs/pec_technoeconomic_analysis.pdf
27. Bard AJ, Fox MA (1995) Artificial photosynthesis – solar splitting of water to hydrogen and oxygen. Acc Chem Res 28(3):141–145
28. Mills A, LeHunte S (1997) An overview of semiconductor photocatalysis. J Photoch Photobio A 108(1):1–35
29. Nozik AJ (1978) Photoelectrochemistry – applications to solar-energy conversion. Ann Rev Phys Chem 29:189–222
30. Memming R (1994) Photoinduced charge-transfer processes at semiconductor electrodes and particles. Electron Transfer I 169:105–181
31. Krol R (2012) Principles of photoelectrochemical cells. In: van de Krol R, Grätzel M (eds) Photoelectrochemical hydrogen production, vol 102. Springer, USA, pp 13–67
32. Kato H, Asakura K, Kudo A (2003) Highly efficient water splitting into H-2 and O-2 over lanthanum-doped $NaTaO_3$ photocatalysts with high crystallinity and surface nanostructure. J Am Chem Soc 125(10):3082–3089
33. Ohno T, Bai L, Hisatomi T, Maeda K, Domen K (2012) Photocatalytic water splitting using modified GaN:ZnO solid solution under visible light: long-time operation and regeneration of activity. J Am Chem Soc 134(19):8254–8259
34. Maeda K, Teramura K, Lu DL, Takata T, Saito N, Inoue Y, Domen K (2006) Characterization of Rh-Cr mixed-oxide nanoparticles dispersed on (Ga1-xZnx)(N1-xOx) as a cocatalyst for visible-light-driven overall water splitting. J Phys Chem B 110(28):13753–13758
35. Maeda K, Teramura K, Lu DL, Takata T, Saito N, Inoue Y, Domen K (2006) Photocatalyst releasing hydrogen from water – enhancing catalytic performance holds promise for hydrogen production by water splitting in sunlight. Nature 440(7082):295–295
36. Zou ZG, Ye JH, Sayama K, Arakawa H (2001) Direct splitting of water under visible light irradiation with an oxide semiconductor photocatalyst. Nature 414(6864):625–627
37. Zou ZG, Arakawa H (2003) Direct water splitting into H-2 and O-2 under visible light irradiation with a new series of mixed oxide semiconductor photocatalysts. J Photoch Photobio A 158(2–3):145–162

38. Liao L, Zhang Q, Su Z, Zhao Z, Wang Y, Li Y, Lu X, Wei D, Feng G, Yu Q, Cai X, Zhao J, Ren Z, Fang H, Robles-Hernandez F, Baldelli S, Bao J (2014) Efficient solar water-splitting using a nanocrystalline CoO photocatalyst. Nat Nano 9(1):69–73
39. Hara M, Kondo T, Komoda M, Ikeda S, Shinohara K, Tanaka A, Kondo JN, Domen K (1998) Cu_2O as a photocatalyst for overall water splitting under visible light irradiation. Chem Commun 3:357–358
40. de Jongh PE, Vanmaekelbergh D, Kelly JJ (1999) Cu_2O: a catalyst for the photochemical decomposition of water? Chem Commun 12:1069–1070
41. Malingowski AC, Stephens PW, Huq A, Huang QZ, Khalid S, Khalifah PG (2012) Substitutional mechanism of Ni into the wide-band-gap semiconductor in TaO_4 and its implications for water splitting activity in the wolframite structure type. Inorg Chem 51(11):6096–6103
42. Duonghong D, Borgarello E, Gratzel M (1981) Dynamics of light-induced water cleavage in colloidal systems. J Am Chem Soc 103(16):4685–4690
43. Bard AJ (1979) Photoelectrochemistry and heterogeneous photocatalysis at semiconductors. J Photochem 10(1):59–75
44. Kato H, Sasaki Y, Shirakura N, Kudo A (2013) Synthesis of highly active rhodium-soped $SrTiO_3$ powders in Z-scheme systems for visible-light-driven photocatalytic overall water splitting. J Mater Chem A 1(39):12327–12333
45. Hu S, Shaner MR, Beardslee JA, Lichterman M, Brunschwig BS, Lewis NS (2014) Amorphous TiO_2 coatings stabilize Si, GaAs, and GaP photoanodes for efficient water oxidation. Science 344(6187):1005–1009
46. Paracchino A, Laporte V, Sivula K, Graetzel M, Thimsen E (2011) Highly active oxide photocathode for photoelectrochemical water reduction. Nat Mater 10(6):456–461
47. Joshi UA, Palasyuk A, Arney D, Maggard PA (2010) Semiconducting oxides to facilitate the conversion of solar energy to chemical fuels. J Phys Chem Lett 1(18):2719–2726
48. Woodhouse M, Parkinson BA (2008) Combinatorial discovery and optimization of a complex oxide with water photoelectrolysis activity. Chem Mater 20(7):2495–2502
49. Woodhouse M, Herman GS, Parkinson BA (2005) Combinatorial approach to identification of catalysts for the photoelectrolysis of water. Chem Mater 17(17):4318–4324
50. Woodhouse M, Parkinson BA (2009) Combinatorial approaches for the identification and optimization of oxide semiconductors for efficient solar photoelectrolysis. Chem Soc Rev 38(1):197–210
51. Jaramillo TF, Baeck SH, Kleiman-Shwarsctein A, Choi KS, Stucky GD, McFarland EW (2005) Automated electrochemical synthesis and photoelectrochemical characterization of Zn1-xCoxO thin films for solar hydrogen production. J Comb Chem 7(2):264–271
52. Katz JE, Gingrich TR, Santori EA, Lewis NS (2009) Combinatorial synthesis and high-throughput photopotential and photocurrent screening of mixed-metal oxides for photoelectrochemical water splitting. Energy Environ Sci 2(1):103–112
53. Dingle R, Wiegmann W, Henry CH (1974) Quantum states of confined carriers in very thin AlxGa1-xAs-GaAs-AlxGa1-xAs heterostructures. Phys Rev Lett 33(14):827–830
54. Henglein A (1982) Photo-degradation and fluorescence of colloidal-cadmium sulfide in aqueous-solution. Phys Chem Chem Phys 86(4):301–305
55. Fojtik A, Weller H, Koch U, Henglein A (1984) Photo-chemistry of colloidal metal sulfides. 8. Photo-physics of extremely small CDS particle S – Q-state CDS and magic agglomeration numbers. Phys Chem Chem Phys 88(10):969–977
56. Brus LE (1983) A simple-model for the ionization-potential, electron-affinity, and aqueous redox potentials of small semiconductor crystallites. J Chem Phys 79(11):5566–5571
57. Vayssieres L (2009) On solar hydrogen & nanotechnology. Wiley, Singapore/Hoboken, p xxi, 680 pp, 16 p
58. Hoertz PG, Mallouk TE (2005) Light-to-chemical energy conversion in lamellar solids and thin films. Inorg Chem 44(20):6828–6840
59. Hagfeldt A, Gratzel M (1995) Light-induced redox reactions in nanocrystalline systems. Chem Rev 95(1):49–68

60. Zhu JF, Zach M (2009) Nanostructured materials for photocatalytic hydrogen production. Curr Opin Colloid Interface Sci 14(4):260–269
61. Kamat PV (2007) Meeting the clean energy demand: nanostructure architectures for solar energy conversion. J Phys Chem C 111(7):2834–2860
62. Kamat PV, Dimitrijevic NM (1990) Colloidal semiconductors as photocatalysts for solar-energy conversion. Sol Energy 44(2):83–98
63. Kamat PV, Tvrdy K, Baker DR, Radich JG (2010) Beyond photovoltaics: semiconductor nanoarchitectures for liquid-junction solar cells. Chem Rev 110(11):6664–6688
64. Zhang JZ (2011) Metal oxide nanomaterials for solar hydrogen generation from photoelectrochemical water splitting. MRS Bull 36(1):48–55
65. Foley JM, Price MJ, Feldblyum JI, Maldonado S (2012) Analysis of the operation of thin nanowire photoelectrodes for solar energy conversion. Energy Environ Sci 5(1):5203–5220
66. Jaegermann W, Tributsch H (1988) Interfacial properties of semiconducting transition-metal chalcogenides. Prog Surf Sci 29(1–2):1–167
67. van de Krol R, Liang YQ, Schoonman J (2008) Solar hydrogen production with nanostructured metal oxides. J Mater Chem 18(20):2311–2320
68. Umena Y, Kawakami K, Shen JR, Kamiya N (2011) Crystal structure of oxygen-evolving photosystem II at a resolution of 1.9 Å. Nature 473(7345):55–U65
69. Boddy PJ (1968) Oxygen evolution on semiconducting TIO_2. J Electrochem Soc 115(2):199
70. Freund T, Gomes WP (1970) Electrochemical methods for investigating catalysis by semiconductors. Catal Rev 3(1):1–36
71. Fujishima A, Honda K (1971) Studies on photosensitive electrode reactions. 3. Electrochemical evidence for mechanism of primary stage of photosynthesis. B Chem Soc Jpn 44(4):1148–1150
72. Fujishima A, Honda K (1972) Electrochemical photolysis of water at a semiconductor electrode. Nature 238(5358):37–38
73. Henglein A (1989) Small-particle research – physicochemical properties of extremely small colloidal metal and semiconductor particles. Chem Rev 89(8):1861–1873
74. Kalyanasundaram K, Gratzel M (1979) Cyclic cleavage of water into H-2 and O-2 by visible-light with coupled redox catalysts. Angew Chem Int Ed Engl 18(9):701–702
75. Mills A, Porter G (1982) Photosensitized dissociation of water using dispersed suspensions of N-type semiconductors. J Chem Soc Faraday T I 78:3659–3669
76. Nozik AJ (1977) Photochemical diodes. Appl Phys Lett 30(11):567–569
77. Würfel P (2005) Physics of solar cells. Wiley-VCH, Weinheim, p 244
78. Berger LI (2008) Optical properties of selected inorganic and organic solids. In: Lide DR (ed) CRC handbook of chemistry and physics, vol 88. CRC/Taylor and Francis, Boca Raton
79. Maiolo JR, Atwater HA, Lewis NS (2008) Macroporous silicon as a model for silicon wire array solar cells. J Phys Chem C 112(15):6194–6201
80. Berger LI (2008) Properties of semiconductors. In: Lide DR (ed) CRC handbook of chemistry and physics, vol 88. CRC/Taylor and Francis, Boca Raton
81. Huda MN, Al-Jassim MM, Turner JA (2011) Mott insulators: an early selection criterion for materials for photoelectrochemical H(2) production. J Renew Sustain Energy 3(5):053101-1–053101-10
82. Cox PA (2010) Transition metal oxides: an introduction to their electronic structure and properties. Clarendon/Oxford University Press, Oxford/New York
83. Sabio EM, Chamousis RL, Browning ND, Osterloh FE (2012) Correction: photocatalytic water splitting with suspended calcium niobium oxides: why nanoscale is better than bulk – a kinetic analysis. J Phys Chem C 116(35):19051–19051
84. Sabio EM, Chamousis RL, Browning ND, Osterloh FE (2012) Photocatalytic water splitting with suspended calcium niobium oxides: why nanoscale is better than bulk – a kinetic analysis. J Phys Chem C 116(4):3161–3170
85. Laser D, Bard AJ (1976) Semiconductor electrodes. 9. digital-simulation of relaxation of photogenerated free carriers and photocurrents. J Electrochem Soc 123(12):1837–1842

86. Morrison SR (1980) Electrochemistry at semiconductor and oxidized metal electrodes. Plenum, New York, p xiv, 401
87. Pleskov YV, Gurevich YY (1986) Semiconductor photoelectrochemistry. Consultants Bureau, New York, p xxv, 422
88. Salvador P (2001) Semiconductors' photoelectrochemistry: a kinetic and thermodynamic analysis in the light of equilibrium and nonequilibrium models. J Phys Chem B 105(26): 6128–6141
89. Townsend TK, Sabio EM, Browning ND, Osterloh FE (2011) Photocatalytic water oxidation with suspended alpha-Fe_2O_3 particles – effects of nanoscaling. Energy Env Sci 4:4270–4275
90. de Almeida JS, Ahuja R (2006) Electronic and optical properties of RuO_2 and IrO_2. Phys Rev B 3(16)
91. Frame FA, Townsend TK, Chamousis RL, Sabio EM, Dittrich T, Browning ND, Osterloh FE (2011) Photocatalytic water oxidation with non-sensitized IrO_2 nanocrystals under visible and UV light. J Am Chem Soc 133(19):7264–7267
92. Yoffe AD (2001) Semiconductor quantum dots and related systems: electronic, optical, luminescence and related properties of low dimensional systems. Adv Phys 50(1):1–208
93. Gerischer H (1990) The impact of semiconductors on the concepts of electrochemistry. Electrochim Acta 35(11–12):1677–1699
94. Marcus RA (1964) Chemical + electrochemical electron-transfer theory. Ann Rev Phys Chem 15:155–196
95. Sant PA, Kamat PV (2002) Interparticle electron transfer between size-quantized CdS and TiO_2 semiconductor nanoclusters. Phys Chem Chem Phys 4(2):198–203
96. Robel I, Kuno M, Kamat PV (2007) Size-dependent electron injection from excited CdSe quantum dots into TiO_2 nanoparticles. J Am Chem Soc 129(14):4136–4137
97. Tvrdy K, Frantsuzov PA, Kamat PV (2011) Photoinduced electron transfer from semiconductor quantum dots to metal oxide nanoparticles. PNAS 108(1):29–34
98. Holmes MA, Townsend TK, Osterloh FE (2012) Quantum confinement controlled photocatalytic water splitting by suspended CdSe nanocrystals. Chem Commun 48(3): 371–373
99. Zhao J, Holmes MA, Osterloh FE (2013) Quantum confinement controls photocatalysis – a free energy analysis for photocatalytic proton reduction at CdSe nanocrystals. ACS Nano 7(5):4316–4325
100. Rogach AL, Kornowski A, Gao MY, Eychmuller A, Weller H (1999) Synthesis and characterization of a size series of extremely small thiol-stabilized CdSe nanocrystals. J Phys Chem B 103(16):3065–3069
101. Bard AJ, Faulkner LR (2001) Electrochemical methods: fundamentals and applications. 2nd edn. Wiley, New York, p xxi, 833
102. Waller M, Townsend TK, Zhao J, Sabio EM, Chamousis RL, Browning ND, Osterloh FE (2012) Single-crystal tungsten oxide nanosheets: photochemical water oxidation in the quantum confinement regime. Chem Mater 24(4):698–704
103. Sambur JB, Novet T, Parkinson BA (2010) Multiple exciton collection in a sensitized photovoltaic system. Science 330(6000):63–66
104. Semonin OE, Luther JM, Choi S, Chen HY, Gao JB, Nozik AJ, Beard MC (2011) Peak external photocurrent quantum efficiency exceeding 100% via MEG in a quantum dot solar cell. Science 334(6062):1530–1533
105. Nozik AJ (2002) Quantum dot solar cells. Phys E Low Dimens Syst Nanostruct 14(1–2): 115–120
106. Kavan L, Gratzel M, Gilbert SE, Klemenz C, Scheel HJ (1996) Electrochemical and photoelectrochemical investigation of single-crystal anatase. J Am Chem Soc 118(28): 6716–6723
107. Khan SUM, Akikusa J (1999) Photoelectrochemical splitting of water at nanocrystalline n-Fe_2O_3 thin-film electrodes. J Phys Chem B 103(34):7184–7189

108. Atkinson RJ, Posner AM, Quirk JP (1967) Adsorption of potential-determining ions at ferric oxide-aqueous electrolyte interface. J Phys Chem 71(3):550–558
109. Brown GE, Henrich VE, Casey WH, Clark DL, Eggleston C, Felmy A, Goodman DW, Gratzel M, Maciel G, McCarthy MI, Nealson KH, Sverjensky DA, Toney MF, Zachara JM (1999) Metal oxide surfaces and their interactions with aqueous solutions and microbial organisms. Chem Rev 99(1):77–174
110. Hingston FJ, Atkinson RJ, Posner AM, Quirk JP (1967) Specific adsorption of anions. Nature 215(5109):1459–1461
111. Meissner D, Memming R, Kastening B (1988) Photoelectrochemistry of cadmium-sulfide. 1. Reanalysis of photocorrosion and flat-band potential. J Phys Chem 92(12):3476–3483
112. Ginley DS, Butler MA (1978) Flatband potential of cadmium-sulfide (Cds) photoanodes and its dependence on surface ion effects. J Electrochem Soc 125(12):1968–1974
113. Frese KW, Canfield DG (1984) Adsorption of hydroxide and sulfide ions on single-crystal n-CdSe electrodes. J Electrochem Soc 131(11):2614–2618
114. Lincot D, Vedel J (1988) Adsorption of telluride ions on cadmium telluride – consequences for photoelectrochemical cells. J Phys Chem 92(14):4103–4110
115. Minoura H, Watanabe T, Oki T, Tsuiki M (1977) Effects of dissolved Cd2+ and S2- ions on flatband potential of CdS electrode in aqueous-solution. Jpn J Appl Phys 16(5):865–866
116. Singh P, Singh R, Gale R, Rajeshwar K, Dubow J (1980) Surface-charge and specific ion adsorption effects in photoelectrochemical devices. J Appl Phys 51(12):6286–6291
117. Butler MA, Ginley GS (1978) Prediction of flatband potentials at semiconductor-electrolyte interfaces from atomic electronegativies. J Electrochem Soc 125(2):228–232
118. Bard AJ, Faulkner LR (2001) Electrochemical methods: fundamentals and applications, 2nd edn. Wiley, New York, p 60
119. Vanysek P (2008) Electrochemical series. In: CRC handbook of chemistry and physics, vol 88. CRC/Taylor and Francis, Boca Raton
120. Bard AJ, Faulkner LR (2001) Electrochemical methods: fundamentals and applications, 2nd edn, Wiley, New York, p 550
121. Grahame DC (1947) The electrical double layer and the theory of electrocapillarity. Chem Rev 41(3):441–501
122. Mayer JM (2004) Proton-coupled electron transfer: a reaction chemist's view. Ann Rev Phys Chem 55:363–390
123. Chamousis RL, Osterloh FE (2014) Use of potential determining ions to control energetics and photochemical charge transfer of a nanoscale water splitting photocatalyst. Energy Envi Sci 7(2):736–743
124. Nozik AJ, Memming R (1996) Physical chemistry of semiconductor-liquid interfaces. J Phys Chem 100(31):13061–13078
125. Miller RJD, Memming R (2008) Fundamentals in photoelectrochemistry. In: Archer MD, Nozik AJ (eds) Nanostructured and photoelectrochemical systems for solar photon conversion, vol 3. Imperial College Press, London
126. Chmiel G, Gerischer H (1990) Photoluminescence at a semiconductor electrolyte contact around and beyond the flat-band potential. J Phys Chem 94(4):1612–1619
127. Klahr BM, Hamann TW (2011) Current and voltage limiting processes in thin film hematite electrodes. J Phys Chem C 115(16):8393–8399
128. Tan MX, Laibinis PE, Nguyen ST, Kesselman JM, Stanton CE, Lewis NS (1994) Principles and applications of semiconductor photoelectrochemistry. In: Progress in inorganic chemistry, Wiley, New York, vol 41. pp 21–144
129. Cowan AJ, Tang JW, Leng WH, Durrant JR, Klug DR (2010) Water splitting by nanocrystalline TiO_2 in a complete photoelectrochemical cell exhibits efficiencies limited by charge recombination. J Phys Chem C 114(9):4208–4214
130. Tang JW, Durrant JR, Klug DR (2008) Mechanism of photocatalytic water splitting in TiO (2). Reaction of water with photoholes, importance of charge carrier dynamics, and evidence for four-hole chemistry. J Am Chem Soc 130(42):13885–13891

131. McCrory CCL, Jung SH, Peters JC, Jaramillo TF (2013) Benchmarking heterogeneous electrocatalysts for the oxygen evolution reaction. J Am Chem Soc 135(45):16977–16987
132. Popczun EJ, McKone JR, Read CG, Biacchi AJ, Wiltrout AM, Lewis NS, Schaak RE (2013) Nanostructured nickel phosphide as an electrocatalyst for the hydrogen evolution reaction. J Am Chem Soc 135(25):9267–9270
133. Popczun EJ, Read CG, Roske CW, Lewis NS, Schaak RE (2014) Highly active electrocatalysis of the hydrogen evolution reaction by cobalt phosphide nanoparticles. Angew Chem 126(21):5531–5534
134. Lewis NS (2005) Chemical control of charge transfer and recombination at semiconductor photoelectrode surfaces. Inorg Chem 44(20):6900–6911
135. Shockley W, Queisser HJ (1961) Detailed balance limit of efficiency of p–n junction solar cells. J Appl Phys 32(3):510–519
136. Lewis NS (1990) Mechanistic studies of light-induced charge separation at semiconductor liquid interfaces. Acc Chem Res 23(6):176–183
137. Lewis NS (2001) Frontiers of research in photoelectrochemical solar energy conversion. J Electroanal Chem 508(1–2):1–10
138. Yablonovitch E, Allara DL, Chang CC, Gmitter T, Bright TB (1986) Unusually low surface recombination velocity on silicon and germanium surfaces. Phys Rev Lett 57(2):249–252
139. Diebold U (2003) The surface science of titanium dioxide. Surf Sci Rep 48(5–8):53–229
140. Cummings CY, Marken F, Peter LM, Tahir AA, Wijayantha KGU (2012) Kinetics and mechanism of light-driven oxygen evolution at thin film alpha-Fe_2O_3 electrodes. Chem Commun 48(14):2027–2029
141. Arakawa H (2002) Water photolysis by TiO_2 particles-significant effect of Na_2CO_3 addition on water splitting. In: Kaneko M, Okura I (eds) Photocatalysis science and technology. Springer, New York, pp 235–248
142. Saito K, Koga K, Kudo A (2011) Lithium niobate nanowires for photocatalytic water splitting. Dalton Trans 40(15):3909–3913
143. Yan SC, Wan LJ, Li ZS, Zou ZG (2011) Facile temperature-controlled synthesis of hexagonal Zn(2)GeO(4) nanorods with different aspect ratios toward improved photocatalytic activity for overall water splitting and photoreduction of CO(2). Chem Commun 47(19):5632–5634
144. Pala RA, Leenheer AJ, Lichterman M, Atwater HA, Lewis NS (2014) Measurement of minority-carrier diffusion lengths using wedge-shaped semiconductor photoelectrodes. Energy Environ Sci 7(10):3424–3430
145. Pendlebury SR, Cowan AJ, Barroso M, Sivula K, Ye JH, Gratzel M, Klug DR, Tang JW, Durrant JR (2012) Correlating long-lived photogenerated hole populations with photocurrent densities in hematite water oxidation photoanodes. Energy Environ Sci 5(4):6304–6312
146. Hagedorn K, Forgacs C, Collins S, Maldonado S (2010) Design considerations for nanowire heterojunctions in solar energy conversion/storage applications. J Phys Chem C 114(27): 12010–12017
147. Maruyama M, Iwase A, Kato H, Kudo A, Onishi H (2009) Time-resolved infrared absorption study of $NaTaO_3$ photocatalysts doped with alkali earth metals. J Phys Chem C 113(31): 13918–13923
148. Garnett EC, Yang PD (2008) Silicon nanowire radial p-n junction solar cells. J Am Chem Soc 130(29):9224
149. Wu P, Wang J, Zhao J, Guo L, Osterloh FE (2014) Structure defects in g-C3N4 limit visible light driven hydrogen evolution and photovoltage. J Mater Chem A 2(47):20338–20344
150. Osterloh FE (2014) Boosting the efficiency of suspended photocatalysts for overall water splitting. J Phys Chem Lett 5(15):2510–2511
151. Le Formal F, Tetreault N, Cornuz M, Moehl T, Gratzel M, Sivula K (2011) Passivating surface states on water splitting hematite photoanodes with alumina overlayers. Chem Sci 2(4):737–743
152. Spray RL, McDonald KJ, Choi K-S (2011) Enhancing photoresponse of nanoparticulate alpha-Fe_2O_3 electrodes by surface composition tuning. J Phys Chem C 115(8):3497–3506

153. Liang YQ, Tsubota T, Mooij LPA, van de Krol R (2011) Highly improved quantum efficiencies for thin film $BiVO_4$ photoanodes. J Phys Chem C 115(35):17594–17598
154. Zhong DK, Choi S, Gamelin DR (2011) Near-complete suppression of surface recombination in solar photoelectrolysis by "Co-Pi" catalyst-modified W:BiVO4. J Am Chem Soc 133(45): 18370–18377
155. Osterloh FE (2014) Maximum theoretical efficiency limit of photovoltaic devices: effect of band structure on excited state entropy. J Phys Chem Lett 2014:3354–3359
156. Gerischer H (1966) Electrochemical behavior of semiconductors under illumination. J Electrochem Soc 113(11):1174–1182
157. Polman A, Atwater HA (2012) Photonic design principles for ultrahigh-efficiency photovoltaics. Nat Mater 11(3):174–177
158. Hodes G, Howell IDJ, Peter LM (1992) Nanocrystalline photoelectrochemical cells – a new concept in photovoltaic cells. J Electrochem Soc 139(11):3136–3140
159. Cesar I, Sivula K, Kay A, Zboril R, Graetzel M (2009) Influence of feature size, film thickness, and silicon doping on the performance of nanostructured hematite photoanodes for solar water splitting. J Phys Chem C 113(2):772–782
160. Oregan B, Moser J, Anderson M, Gratzel M (1990) Vectorial electron injection into transparent semiconductor membranes and electric-field effects on the dynamics of light-induced charge separation. J Phys Chem 94(24):8720–8726
161. Dloczik L, Ileperuma O, Lauermann I, Peter LM, Ponomarev EA, Redmond G, Shaw NJ, Uhlendorf I (1997) Dynamic response of dye-sensitized nanocrystalline solar cells: characterization by intensity-modulated photocurrent spectroscopy. J Phys Chem B 101(49): 10281–10289
162. Giebink NC, Wiederrecht GP, Wasielewski MR, Forrest SR (2011) Thermodynamic efficiency limit of excitonic solar cells. Phys Rev B 83(19):195326-1–195326-6
163. Miseki Y, Kato H, Kudo A (2009) Water splitting into H_2 and O_2 over niobate and titanate photocatalysts with (111) plane-type layered perovskite structure. Energy Environ Sci 2(3): 306–314
164. Matsumoto Y, Ida S, Inoue T (2008) Photodeposition of metal and metal oxide at the TiOx nanosheet to observe the photocatalytic active site. J Phys Chem C 112(31):11614–11616
165. Sabio EM, Chi M, Browning ND, Osterloh FE (2010) Charge separation in a niobate nanosheet photocatalyst studied with photochemical labeling. Langmuir 26(10):7254–7261
166. Li RG, Han HX, Zhang FX, Wang DG, Li C (2014) Highly efficient photocatalysts constructed by rational assembly of dual-cocatalysts separately on different facets of $BiVO_4$. Energy Environ Sci 7(4):1369–1376
167. Giocondi JL, Rohrer GS (2001) Spatially selective photochemical reduction of silver on the surface of ferroelectric barium titanate. Chem Mater 13(2):241–242
168. Yang SY, Seidel J, Byrnes SJ, Shafer P, Yang CH, Rossell MD, Yu P, Chu YH, Scott JF, Ager JW, Martin LW, Ramesh R (2010) Above-bandgap voltages from ferroelectric photovoltaic devices. Nat Nanotechnol 5(2):143–147
169. Li L, Salvador PA, Rohrer GS (2014) Photocatalysts with internal electric fields. Nanoscale 6(1):24–42
170. Abdi FF, Han LH, Smets AHM, Zeman M, Dam B, van de Krol R (2013) Efficient solar water splitting by enhanced charge separation in a bismuth vanadate-silicon tandem photoelectrode. Nat Commun 2013:4
171. Dittrich T, Belaidi A, Ennaoui A (2011) Concepts of inorganic solid-state nanostructured solar cells. Sol Energy Mater 95(6):1527–1536
172. Du C, Yang XG, Mayer MT, Hoyt H, Xie J, McMahon G, Bischoping G, Wang DW (2013) Hematite-based water splitting with low turn-on voltages. Angew Chem Int Ed Engl 52(48): 12692–12695
173. Pasquarelli RM, Ginley DS, O'Hayre R (2011) Solution processing of transparent conductors: from flask to film. Chem Soc Rev 40(11):5406–5441

174. Sodergren S, Hagfeldt A, Olsson J, Lindquist SE (1994) Theoretical-models for the action spectrum and the current–voltage characteristics of microporous semiconductor-films in photoelectrochemical cells. J Phys Chem 98(21):5552–5556
175. Hagfeldt A, Bjorksten U, Lindquist SE (1992) Photoelectrochemical studies of colloidal TIO_2-films – the charge separation process studied by means of action spectra in the UV region. Sol Energy Mat Sol C 27(4):293–304
176. Bisquert J, Vikhrenko VS (2004) Interpretation of the time constants measured by kinetic techniques in nanostructured semiconductor electrodes and dye-sensitized solar cells. J Phys Chem B 108(7):2313–2322
177. Hagfeldt A, Lindstrom H, Sodergren S, Lindquist SE (1995) Photoelectrochemical studies of colloidal TiO_2 films – the effect of oxygen studied by photocurrent transients. J Electroanal Chem 381(1–2):39–46
178. Abeles B, Sheng P, Coutts MD, Arie Y (1975) Structural and electrical properties of granular metal-films. Adv Phys 24(3):407–461
179. Terrill RH, Postlethwaite TA, Chen CH, Poon CD, Terzis A, Chen AD, Hutchison JE, Clark MR, Wignall G, Londono JD, Superfine R, Falvo M, Johnson CS, Samulski ET, Murray RW (1995) Monolayers in three dimensions: NMR, SAXS, thermal, and electron hopping studies of alkanethiol stabilized gold clusters. J Am Chem Soc 117(50):12537–12548
180. Yokoi T, Sakuma J, Maeda K, Domen K, Tatsumi T, Kondo JN (2011) Preparation of a colloidal array of NaTaO(3) nanoparticles via a confined space synthesis route and its photocatalytic application. Phys Chem Chem Phys 13(7):2563–2570
181. Townsend TK, Browning ND, Osterloh FE (2012) Nanoscale strontium titanate photocatalysts for overall water splitting. ACS Nano 6(8):7420–7426
182. Liu B, Wu C-H, Miao J, Yang P (2014) All inorganic semiconductor nanowire mesh for direct solar water splitting. ACS Nano 8(11):11739–11744
183. Liu J, Hisatomi T, Ma G, Iwanaga A, Minegishi T, Moriya Y, Katayama M, Kubota J, Domen K (2014) Improving the photoelectrochemical activity of $La_5Ti_2CuS_5O_7$ for hydrogen evolution by particle transfer and doping. Energy Environ Sci 7(7):2239–2242
184. Minegishi T, Nishimura N, Kubota J, Domen K (2013) Photoelectrochemical properties of $LaTiO_2N$ electrodes prepared by particle transfer for sunlight-driven water splitting. Chem Sci 4(3):1120–1124
185. Urabe H, Hisatomi T, Minegishi T, Kubota J, Domen K (2014) Photoelectrochemical properties of SrNbO2N photoanodes for water oxidation fabricated by the particle transfer method. Faraday Discuss 176:213–223
186. Townsend TK, Sabio EM, Browning ND, Osterloh FE (2011) Improved niobate nanoscroll photocatalysts for partial water splitting. ChemSusChem 4(2):185–190
187. Kronik L, Shapira Y (1999) Surface photovoltage phenomena: theory, experiment, and applications. Surf Sci Rep 37:1–206
188. Kronik L, Shapira Y (2001) SPV review-short version. Surf Interface Anal 31:954–965
189. Zhao J, Osterloh FE (2014) Photochemical charge separation in nanocrystal photocatalyst films – insights from surface photovoltage spectroscopy. J Phys Chem Lett 5:782–786
190. Osterloh FE, Holmes MA, Zhao J, Chang L, Kawula S, Roehling JD, Moulé AJ (2014) P3HT: PCBM bulk-heterojunctions: observing interfacial and charge transfer states with surface photovoltage spectroscopy. J Phys Chem C 118(27):14723–14731
191. Osterloh FE, Holmes MA, Chang L, Moule AJ, Zhao J (2013) Photochemical charge separation in poly(3-hexylthiophene) (P3HT) films observed with surface photovoltage spectroscopy. J Phys Chem C 117(51):26905–26913
192. Lagowski J (1994) Semiconductor surface spectroscopies – the early years. Surf Sci 299 (1–3):92–101
193. Luria JL, Hoepker N, Bruce R, Jacobs AR, Groves C, Marohn JA (2012) Spectroscopic imaging of photopotentials and photoinduced potential fluctuations in a bulk heterojunction solar cell film. ACS Nano 6(11):9392–9401

194. Burstein L, Bregman J, Shapira Y (1991) Characterization of interface states at III-V compound semiconductor–metal interfaces. J Appl Phys 69(4):2312–2316
195. Lagowski J, Jastrzebski L, Cullen GW (1981) Electronic characterization of hetero-epitaxial silicon-on-sapphire by surface photo-voltage spectroscopy. J Electrochem Soc 128(12): 2665–2670
196. Musser ME, Dahlberg SC (1980) The surface photo-voltage of polymethine semiconducting-films. J Chem Phys 72(7):4084–4088
197. Moons E, Eschle M, Gratzel M (1997) Determination of the energy diagram of the dithioketopyrrolopyrrole/SnO2:F heterojunction by surface photovoltage spectroscopy. Appl Phys Lett 71(22):3305–3307
198. Gross D, Mora-Sero I, Dittrich T, Belaidi A, Mauser C, Houtepen AJ, Da Como E, Rogach AL, Feldmann J (2010) Charge separation in type II tunneling multi layered structures of CdTe and CdSe nanocrystals directly proven by surface photovoltage spectroscopy. J Am Chem Soc 132(17):5981
199. Dittrich T, Fiechter S, Thomas A (2011) Surface photovoltage spectroscopy of carbon nitride powder. Appl Phys Lett 99(8):084105-1–084105-3
200. Nowotny MK, Bogdanoff P, Dittrich T, Fiechter S, Fujishima A, Tributsch H (2010) Observations of p-type semiconductivity in titanium dioxide at room temperature. Mater Lett 64(8):928–930
201. Zidon Y, Shapira Y, Dittrich T, Otero L (2007) Light-induced charge separation in thin tetraphenyl-porphyrin layers deposited on Au. Phys Rev B 75(19)
202. Mandujano-Ramirez HJ, Gonzalez-Vazquez JP, Oskam G, Dittrich T, Garcia-Belmonte G, Mora-Sero I, Bisquert J, Anta JA (2014) Charge separation at disordered semiconductor heterojunctions from random walk numerical simulations. Phys Chem Chem Phys 16(9): 4082–4091
203. Fungo F, Milanesio ME, Durantini EN, Otero L, Dittrich T (2007) Optically induced switch of the surface work function in TiO2/porphyrin-C-60 dyad system. J Mater Chem 17(20): 2107–2112
204. Herzog C, Belaidi A, Ogacho A, Dittrich T (2009) Inorganic solid state solar cell with ultra-thin nanocomposite absorber based on nanoporous TiO(2) and In(2)S(3). Energy Environ Sci 2(9):962–964
205. Maeda K, Mallouk TE (2009) Comparison of two- and three-layer restacked Dion–Jacobson phase niobate nanosheets as catalysts for photochemical hydrogen evolution. J Mater Chem 19(27):4813–4818
206. Compton OC, Osterloh FE (2009) Niobate nanosheets as catalysts for photochemical water splitting into hydrogen and hydrogen peroxide. J Phys Chem C 113(1):479–485
207. Lide DR (2008) Electron work function of the elements. In: CRC handbook of chemistry and physics, vol 88. CRC/Taylor and Francis, Boca Raton
208. Wang J, Osterloh FE (2014) Limiting factors for photochemical charge separation in $BiVO_4$/Co_3O_4, a highly active photocatalyst for water oxidation in sunlight. J Mater Chem A 2: 9405–9411
209. Kudo A, Ueda K, Kato H, Mikami I (1998) Photocatalytic O-2 evolution under visible light irradiation on $BiVO_4$ in aqueous $AgNO_3$ solution. Catal Lett 53(3–4):229–230

Top Curr Chem (2016) 371: 143–172
DOI: 10.1007/128_2015_636

Published online: 24 May 2015

Heterojunctions in Composite Photocatalysts

Roland Marschall

Abstract Combining different light-absorbing materials for the formation of semiconductor heterojunctions is a very effective strategy for preparing highly active photocatalyst and photoelectrochemical systems. Moreover, the combination of solid state semiconductors with polymers or molecular absorbers expands the possible combinations of materials to alter light absorption and optimize charge carrier separation. In this chapter, different strategies to prepare such composites are presented, highlighting the necessity of intimate interfacial contact for optimum charge carrier transfer. Moreover, the most recent developments and improvements in the formation of heterojunctions and composite photocatalyst systems based on semiconductor solids are presented.

Keywords Charge separation · Composites · Heterojunctions · Photocatalysis · Semiconductors

Contents

R. Marschall (✉)
Institute of Physical Chemistry, Justus-Liebig-University Giessen, Heinrich-Buff-Ring 58, 35392 Giessen, Germany
e-mail: roland.marschall@phys.chemie.uni-giessen.de

1 Introduction

One of the major strategies to enhance photocatalytic efficiency is to optimize the lifetime of photoexcited charge carriers. Increasing the charge carrier lifetime by reducing the recombination probability of photogenerated electron–hole pairs ultimately results in increased probability of the photogenerated charges to be available for oxidation or reduction reactions. The result is an enhancement in photocatalytic activity.

Charge carrier recombination occurs during the diffusion of charge carriers to the surface of the semiconductor, or at the surface itself. Thus, the photocatalytic activity can be, for example, enhanced by optimizing the diffusion of photoexcited charges to the semiconductor surface by reducing lattice defects or reducing the particle size.

The most common strategy to withdraw charge carriers from semiconductors is surface modification with co-catalysts for photocatalytic water splitting, as presented by Nocera and co-workers in this book. Historically, noble metal nanoparticles, e.g., platinum, were especially used for hydrogen generation as co-catalysts through formation of a Schottky contact with most semiconductors for electron extraction from the semiconductor. Moreover, such modification also facilitates the surface reduction and oxidation reactions. Nowadays, more Earth-abundant alternatives are used, either for hydrogen or oxygen evolution.

One of the frequently applied strategies for charge carrier separation is the formation of semiconductor composites, either with carbon scaffolds, polymers, or other semiconductors [1]. The latter case can be differentiated between multiphase and multicomponent heterojunctions, meaning that two or more different materials, or two or more different crystal structures of one material, are combined. The formation of composites and heterojunctions is used for spatial separation of charges, making the recombination of electrons and holes less probable. The absorption threshold can also be increased by combining two semiconductors with different band gaps. Moreover, vectorial charge transfer can be even optimized so that the formation of multiphase heterojunctions leads to even better charge carrier separation than achieved by co-catalyst decoration [2].

This chapter presents the common strategies for preparing semiconductor composites, emphasizing the necessity of in situ formation of composites in contrast to post-synthetic combination. Z-scheme photocatalyst systems as a special type of composite are introduced, together with semiconductor-polymer, semiconductor-carbon, and semiconductor-sensitizer composites.

2 Semiconductor Heterojunctions and Their Preparation

The simplest method to prepare a semiconductor composite is to synthesize the individual components and mix them thoroughly afterwards. However, it is very difficult to achieve good interfacial contact between the two semiconductors for optimum charge transfer. Simple mixing and grinding, often also denoted as a "physical mixture," is regularly shown to be inefficient for charge carrier separation [3]. A number of strategies have therefore been developed to mix different components efficiently.

A viable strategy is to use ultrasonication to mix particle suspensions [4]. Preformed particles are mixed in aqueous solutions, and treatment with an ultrasonic tip sonicator or in an ultrasonic bath is started. However, similar particles tend to form clusters. By carefully adjusting the surface charges of the different semiconductors with opposite charges, a more efficient mixing of multiple particle types can be achieved. For example, efficient interfacial contact between TiO_2 and SnO_2 was achieved in 10^{-6} mol L^{-1} formic acid, in which TiO_2 exhibits a positive surface charge and SnO_2 a negative surface charge. The result after ultrasonication of this suspension is a very homogeneous semiconductor composite (Fig. 1).

As an alternative mixing method with high energy input, ball milling is often used to prepare semiconductor composites. Two materials are prepared separately, and filled into, e.g., an agate crucible with several agate balls. The tribochemical treatment, which can be performed for several hours, applies immense forces to the materials for strong mixing, resulting in high interfacial contact. A secondary effect is often particle size reduction during ball milling, resulting in a larger surface area,

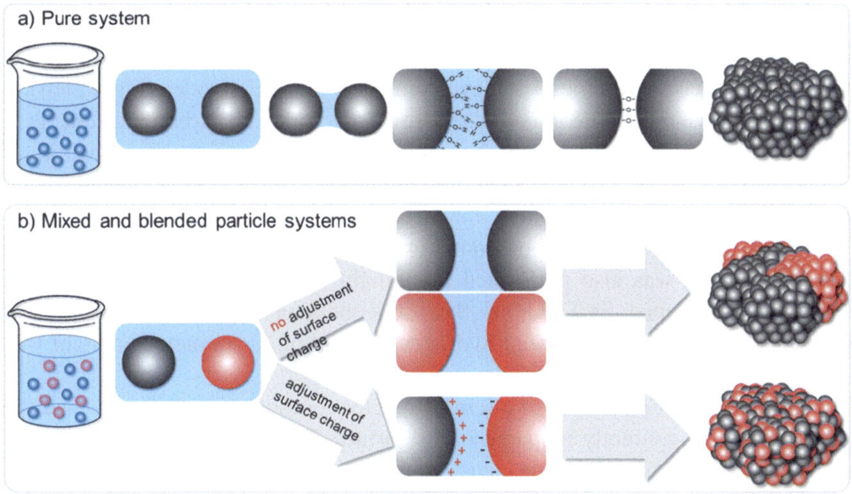

Fig. 1 Interfacial contact approach for (**a**) pure systems and (**b**) mixed particle systems to get efficiently mixed semiconductor composites. Reproduced from Siedl et al. [4] with permission, ©2012 American Chemical Society

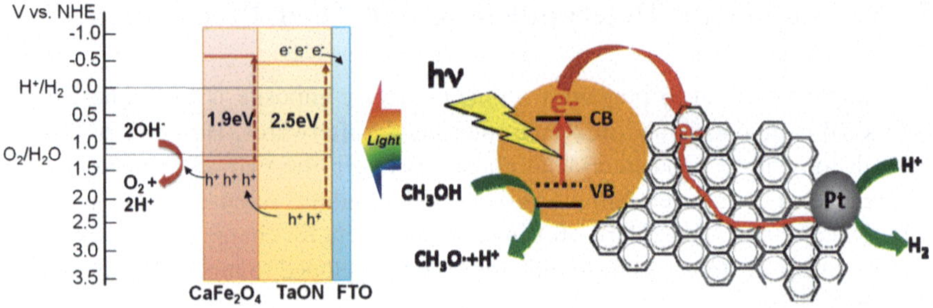

Fig. 2 *Left*: Proposed charge transfer in $TaON/CaFe_2O_4$ composite photoelectrodes prepared via electrophoretic deposition. Reprinted with permission from Kim et al. [6], ©2013 American Chemical Society. *Right*: semiconductor/Reduced Graphene Oxide (rGO) composite for solar hydrogen production. Reproduced with permission from Mukherji et al. [7], ©2011 American Chemical Society

which is advantageous for photocatalytic application [5]. However, whether ball milling leads to a more efficient intermixing compared to the ultrasonic treatment depends on the materials used. Moreover, impurities from the balls or the crucible have to be avoided and checked carefully.

For the preparation of composite photoelectrodes from preformed particles, electrophoretic deposition has been used recently to achieve good back electrode and interfacial contact. In this case, the surface charge also plays an important role for the deposition procedure, as for the ultrasonic treatment shown above. For example, for $TaON/CaFe_2O_4$ composite photoelectrodes, electrophoretic deposition was performed in acetone with addition of iodine, the latter reacting with acetone when under bias to provide protons and make the TaON particles positively charged [6]. By applying 10–20 V of bias, the electrode could be coated with the first layer of particles. The procedure could then be repeated in a second particle suspension in acetone/iodine to achieve composite photoelectrodes (Fig. 2, left).

A well-known procedure to combine semiconductors with carbon scaffolds is to use the photogenerated electrons from the semiconductor for the reduction of graphene oxide (GO) [8]. The resulting semiconductor/reduced graphene oxide (rGO) composite suspension show enhanced activity produced by electron separation onto the conductive carbon compound. Shown for the first time with TiO_2, this combination was also performed in sunlight/visible light using visible-light active nitrogen-doped $Sr_2Ta_2O_7$ as the semiconductor [7]. After deposition of platinum (Pt) onto rGO, enhanced hydrogen production in simulated sunlight was shown from a methanol/water mixture (Fig. 2, right).

Often, only one component of a semiconductor composite is pre-formed and the second material is directly prepared on its surface. This procedure leads to better interfacial contact for optimum charge transfer. The need for efficient material contact when preparing multi-component semiconductor heterojunctions subsequently was already shown by Hoffmann et al. during their investigation of different $Pt\text{-}CdS\text{-}TiO_2$ heterojunctions for hydrogen evolution [9]. They convincingly showed that vectorial charge transfer of electrons under visible light

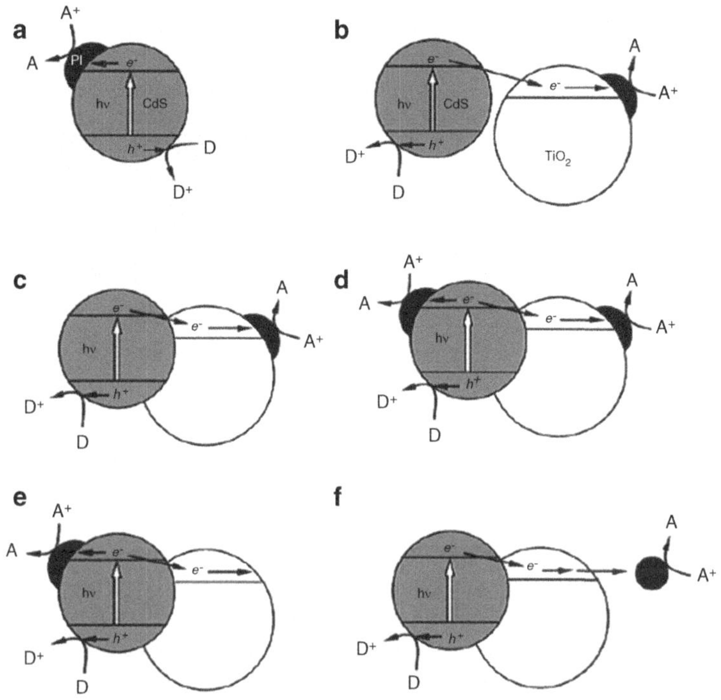

Fig. 3 Schematic illustration of the electron transfers in the photocatalyst composites: (**a**) Pt-CdS; (**b**) CdS + Pt-TiO_2 (physically mixed); (**c**) CdS/(Pt-TiO_2); (**d**) Pt-(CdS/TiO_2); (**e**) sol–gel-TiO_2/(Pt-CdS); (**f**) CdS/TiO_2 + Pt sol; (**c**) shows optimum hydrogen evolution rates. Reproduced with permission from Park et al. [9], ©2008 Royal Society of Chemistry

irradiation can be achieved only when Pt is deposited solely onto TiO_2, before CdS is subsequently deposited onto Pt-TiO_2 (Fig. 3c). Deposition of Pt onto CdS/TiO_2 composites (Fig. 3d) is not as efficient, because visible-light-excited electrons could either transfer to TiO_2 or to the Pt co-catalyst on CdS. Deposition of Pt solely on CdS (Fig. 3e) also showed less activity, because of visible light excited electrons being partly transferred to non-decorated TiO_2, resulting in no hydrogen formation.

Several techniques, such as precipitation, photodeposition, impregnation, solvothermal treatment, and chemical vapor deposition or sol–gel coatings, are known for preparing semiconductor composites in this manner. For example, cobalt-doped hematite nanorods prepared hydrothermally on a titanium mesh were recently impregnated with $Mg(NO_3)_2$ solution followed by thermal treatment at 550 °C, resulting in the Co-Fe_2O_3/$MgFe_2O_4$ composite shown in Fig. 4 [10].

As a result, the photocurrent efficiencies of the composite photoelectrodes were strongly enhanced compared to pristine $MgFe_2O_4$ or Fe_2O_3 photoelectrodes by improved charge carrier separation. This improvement was even more pronounced because of the nanostructuring of the electrode resulting in optimized hole diffusion to the nanorod surface. Nanostructuring is another prominent strategy to improve

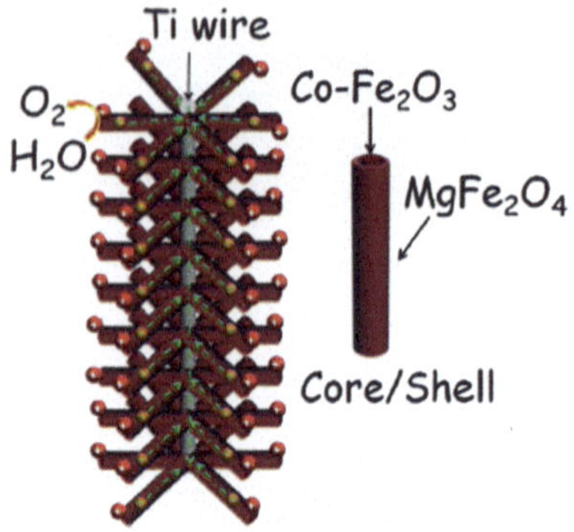

Fig. 4 Core-shell $Co\text{-}Fe_2O_3/MgFe_2O_4$ composite on Ti mesh. Reproduced with permission from Hou et al. [10], ©2013 Wiley-VCH

charge carrier diffusion, as presented by F. Osterloh in an earlier chapter. Other strategies to prepare composite photoelectrodes include successive dip-coating from sol–gel solutions [11], sputter coating [12], and consecutive spray deposition from precursor solutions [13].

Impregnation can also be used to functionalize GO with semiconductors such as TiO_2 from molecular precursors. Ismail et al. dispersed GO in a water/ethanol mixture and added titanium isopropoxide [14]. The resulting TiO_2-GO composites were further treated and reduced to obtain TiO_2-rGO composites, which were finally used in methylene blue (MB) degradation experiments, and were far more active than commercial multiphase TiO_2 nanoparticles (Evonik-Degussa Aeroxide TiO_2 P25) because of electron withdrawal to the carbon scaffold.

For the same reason, and to improve the electron transfer, photocatalyst-carbon junctions can also be prepared via hydrothermal or solvothermal procedures. Multi-walled carbon nanotubes (CNTs) were recently covered with Ta_2O_5 from tantalum ethoxide precursors via a hydrothermal procedure [15]. The resulting materials exhibited very high activities and stabilities for sacrificial hydrogen generation because of the optimized interfacial contact for electron transfer onto the highly electron-conducting carbon scaffold. Whether solvothermal treatment or impregnation led to better interfacial contact could not be clarified. However, the connection has been carefully investigated. The same is valid for composites prepared via chemical vapor deposition (CVD), as was recently shown for Au-V_2O_5@ZnO nanorods [16].

Intimate interfacial contact can even open up new pathways for photoexcitation, as was shown when polymeric carbon nitride (also called polyheptazine) was deposited from the gas phase onto TiO_2 [17]. The surface OH-groups of TiO_2 acted as catalysts for the polymerization of evaporated urea, and “hybrid” photoactive core-shell materials consisting of a polyheptazine-TiO_2 heterojunction were prepared (Fig. 5). Because of the nature of the interfacial contact, charge excitation under visible light from the valence band of polyheptazine was even possible in the conduction band of TiO_2, leaving the hole on the polyheptazine shell for water oxidation. Photoelectrochemical water oxidation was shown after IrO_2 deposition as co-catalyst onto polyheptazine-TiO_2 heterojunction photoanodes

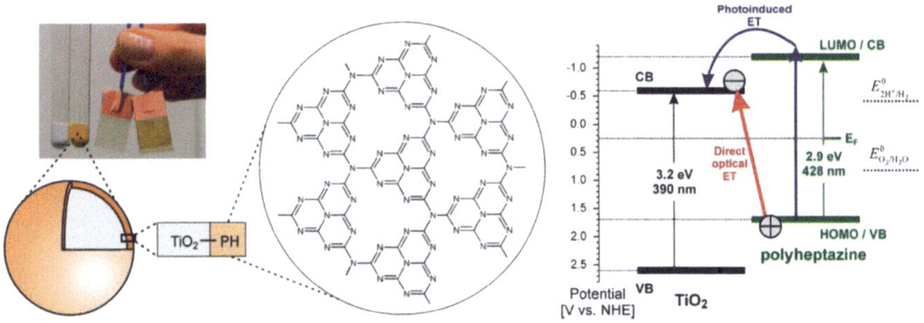

Fig. 5 *Left*: Core-shell particles made of TiO_2 and polyheptazine, assumed structure shown in the *circle*. *Right*: estimated electron transfer process under visible-light illumination. Reproduced with permission from Bledowski et al. [17], ©2011 Royal Society of Chemistry

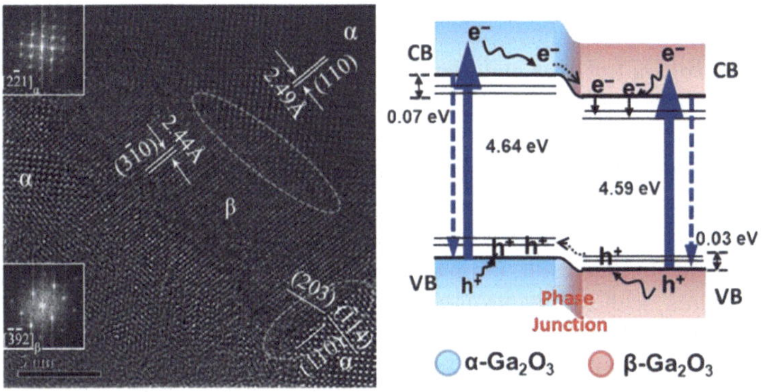

Fig. 6 *Left*: High-resolution TEM image showing the interfacial contact between α-Ga_2O_3 and β-Ga_2O_3.*Right*: assumed charge carrier transfer under illumination. Reproduced with permission from Wang et al. [18] ©2012 Wiley-VCH

under visible light irradiation ($\lambda > 420$ nm, phosphate buffer 0.1 M, pH 7 at 0.5 V vs Ag/AgCl).

When pure α-Ga_2O_3 was treated at elevated temperatures by Li et al., different amounts of β-Ga_2O_3 were evolved, resulting in an intimate mixture of the multiphase semiconductor heterojunction of both phases [18]. The different gallium oxide phases moreover exhibit slightly different conduction and valence band positions, resulting in a type-II multiphase heterojunction. Charge carriers generated upon illumination are therefore separated on the different phases, electrons being transferred onto β-Ga_2O_3 whereas the photogenerated holes accumulate on α-Ga_2O_3 because of the more negative valence band (Fig. 6). As a result, the multiphase heterojunction was shown to be much more active in overall water splitting than pure phase gallium oxides. Especially because the second phase was formed inside the first phase, the contact is estimated to be very homogeneous and intimate, as shown by TEM investigations. New examples of multiphase heterojunctions for photocatalysis are the combination of hexagonal $BiPO_4$ with

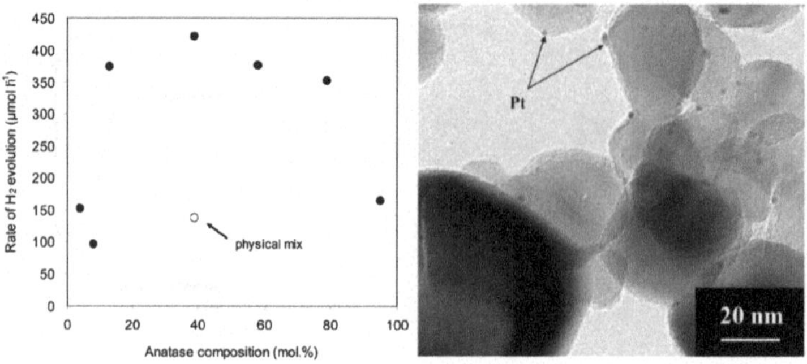

Fig. 7 *Left*: H_2 evolution rates over flame-sprayed TiO_2 samples as a function of anatase content, compared to physical mixture (*open circle*). *Right*: Pt nanoparticles on anatase-rutile mixture after photodeposition and hydrogen evolution. Reproduced with permission of Kho et al. [3], ©2010 American Chemical Society

monoclinic $BiPO_4$ for dyes and phenol degradation [19], and the combination of cubic and orthorhombic $NaNbO_3$ for CO_2 photoreduction [20].

Compared to the strategies presented above, the most intimate contact between two semiconductors can be achieved when both are prepared simultaneously, assuring no phase separation. A very effective preparation technique is flame spray pyrolysis, by combustion of the appropriate precursor solutions in one or more flames [21]. As a result, intimately mixed functional nanoparticle composites can be prepared, with well controlled particle sizes, homogeneity, and morphology. Amal et al. used this preparation technique to prepare the most famous multiphase semiconductor heterojunction system, namely anatase-rutile mixed phase nanoparticles, to investigate synergistic effects in the presence of both phases [3]. Optimum hydrogen evolution after Pt deposition was found for nanoparticles containing 39% anatase and 61% rutile (Fig. 7). Comparable physical mixtures showed drastically lower gas evolution. Many other preparation techniques have been used to prepare anatase-rutile multiphase heterojunctions to investigate synergistic effects between both phases. These details have been recently reviewed [1].

Another multiphase heterojunction of TiO_2 gained considerable attention recently, namely the anatase-brookite heterojunction. Brookite, being a metastable TiO_2 polymorph, is the least investigated phase among the naturally occurring TiO_2 phases. Bahnemann et al. prepared anatase-brookite multiphase heterojunctions and investigated their formation by thermal hydrolysis of titanium bis(ammonium lactate) dihydroxide(TALH) in the presence of urea as an in situ OH^- source, controlling the phase evolution in hydrothermal treatment [22]. The phase composition was tuned by adjusting the urea concentration, and even pure brookite nanorods were achieved in high concentrations of urea (Fig. 8). Brookite-anatase multiphase heterojunctions at a ratio of 25:75 showed higher photonic efficiency for hydrogen evolution from water/methanol mixtures, despite exhibiting an even lower surface area than pure anatase. This was because of enhanced charge carrier separation, as the conduction band potential of brookite is positioned at a 140 mV

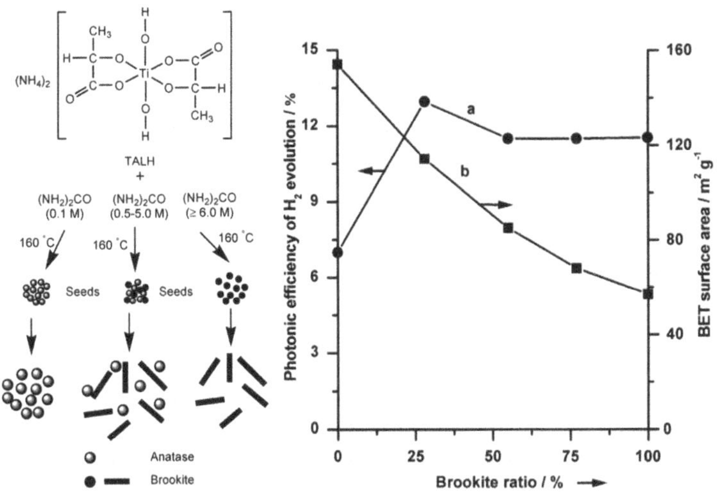

Fig. 8 *Left*: Phase formation via thermal hydrolysis of TALH depending on urea content. *Right*: photonic efficiencies for H_2 evolution of anatase-brookite multiphase heterojunctions. Reproduced with permission of Kandiel et al. [22], ©2010 American Chemical Society

more negative potential than that of anatase. Recently, similar anatase-brookite mixtures prepared with the same ratio were used for photocatalytic CO_2 reduction, confirming the results found earlier [23]. However, for the photocatalytic oxidation of dichloroacetic acid, anatase showed better activities (photonic efficiency $\xi \sim 5.5\%$ decreasing with brookite content and decreasing surface area) because of its larger surface area.

Further calcination of anatase-brookite multiphase heterojunctions leads to three-phase heterojunctions of anatase-brookite-rutile [24]. The synergistic effect between anatase and brookite was confirmed by photocatalytic methanol oxidation experiments, in that anatase-brookite nanoparticles showed higher activities ($\xi \sim 13.5\%$, 65 mg catalyst in 65 mL methanol/H_2O [30 mmol/L], 450-W Xe lamp, $\lambda > 320$ nm) than pure anatase nanoparticles ($\xi \sim 7\%$). Very small additional amounts of rutile (6 wt%) have no further effect, whereas larger amounts of rutile (49 wt%) reduce the activity drastically ($\xi \sim 6\%$), possibly because of hindered charge transfer. However, pure brookite nanoparticles exhibited higher photocatalytic activities ($\xi \sim 16\%$) compared to anatase or anatase-rich composites. Brookite-rutile composites showed no enhanced methanol oxidation ($\xi \sim 3$–7.5%).

As shown by these multiphase TiO_2 heterojunction examples, controlling the evolution of different components or phases can lead to optimum charge transfer properties. This strategy was recently used by Tüysüz et al. in the hydrothermal preparation of sodium tantalates [25]. By adjusting the amount of sodium hydroxide (the sodium precursor) added to tantalum ethoxide solutions or by investigating the hydrothermal reaction time of the resulting suspension, $NaTaO_3$-$Na_2Ta_2O_6$ heterojunctions were prepared for highly efficient H_2 production from water/methanol mixtures (up to 6 mmol/h, 100 mg catalyst in 550 mL H_2O/50 mL methanol,

350-W Hg mid-pressure immersion lamp). The composite with the highest surface area showed comparable photocatalytic performance to pure phase $Na_2Ta_2O_6$ with the same surface area, but superior performance over pure $NaTaO_3$. Although the surface area of the prepared $NaTaO_3$ is quite low, the higher activity of pure $Na_2Ta_2O_6$ and the composite over $NaTaO_3$ can be assumed to arise from both the improved charge carrier separation and the larger band gap going in line with a presumably more negative conduction band of $Na_2Ta_2O_6$, resulting in a higher driving force for photoexcited electrons to reduce protons.

Another example of the in situ formation of photocatalyst heterojunctions using synthetic control of composition and keeping the surface area constant was recently shown in the sol–gel-induced formation of barium tantalate multicomponent heterojunctions [2]. By adjusting the barium precursor in sol–gel complexation synthesis, two-component and three-component semiconductor heterojunctions were prepared and investigated in hydrogen evolution and overall water splitting. With reducing barium precursor, $Ba_3Ta_5O_{15}$ and $BaTaO_6$ evolved as additional "impurities" to the initially pure $Ba_5Ta_4O_{15}$. In the optimum three-component material, strongly enhanced photocatalytic activities for hydrogen evolution (Fig. 9) and overall water splitting (after photodeposition of Rh-Cr_2O_3) were achieved, being superior to pure $Ba_5Ta_4O_{15}$. Moreover, the optimum composition was even more active than rhodium (Rh) decorated pure $Ba_5Ta_4O_{15}$, indicating that charge separation on different components by in situ heterojunction formation via vectorial charge transfer (Fig. 9, right) can be more effective than charge extraction with noble metal co-catalysts.

Sol–gel syntheses can also be used to achieve intimate contact between different oxides derived from precursor mixtures, as shown for WO_3-TiO_2 nanoparticle heterojunctions [26]. Different tungsten precursors ($Na_2WO_4 \cdot 2H_2O$ and $W(OC_2H_5)_6$) and varied ratios were investigated for the preparation of heterojunction photoanodes. Incident photon-to-current efficiencies (IPCE) strongly

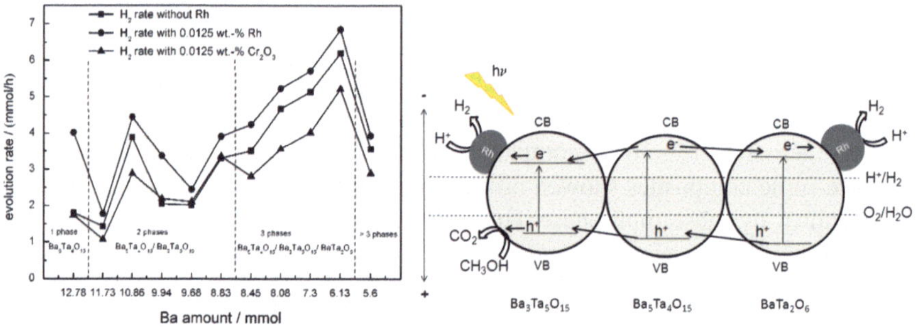

Fig. 9 *Left*: H_2 evolution rates depending on barium tantalate multicomponent heterojunctions prepared by varying the barium amount in the synthesis. *Right*: vectorial charge transfer over the three-component heterojunction after charge carrier excitation. Reproduced with permission from Soldat et al. [2], ©2014 Royal Society of Chemistry

depended on the amount of WO_3 mixed into the TiO_2 matrix; an optimum amount of 3% nominal W/Ti ratio resulted in a value of 16%, higher than measured photoanodes of Evonik-Degussa Aeroxide TiO_2 P25.

Sol–gel solutions of precursor mixtures can also be used to prepare nanostructured multicomponent heterojunctions, as was shown for TiO_2-CuO composite nanofibers via electrospinning [27]. The Cu content had a strong influence on the fiber surface area, and fibers containing 6 mol% Cu and calcined at 450 °C exhibited the highest surface area with 156 m^2/g and thus the highest activity for photocatalytic hydrogen generation (400-W high pressure Hg lamp, 10 vol.% methanol/water mixture, amounts not given). Whether charge carrier transfer or the surface area has the stronger influence could not be unambiguously concluded.

3 Recent Developments in Heterojunction Design

Section 2 has already covered several types of multiphase and multicomponent heterojunctions for photocatalytic reactions when introducing the strategies for their preparation. The following section summarizes some recent developments and new results in the field of semiconductor-containing composites in addition to the already highlighted examples. For even more examples, and also some historical background, the reader is referred to a recently published review article [1].

3.1 Oxide and Sulfide Semiconductor Composites

3.1.1 $BiVO_4$

$BiVO_4$ with a band gap of 2.4 eV has recently gained tremendous attention as a photoanode material for photoelectrochemical water splitting, especially because of the recent work by van de Krol et al. By gradient doping with tungsten (W), the charge carrier separation was strongly increased [28]. Moreover, after modification with CoPi co-catalyst, this photoanode was combined with a double-junction amorphous silicon (2-jn a-Si) photovoltaic (PV) cell to construct a hybrid triple-junction photoelectrode (Fig. 10, left). As an advantage, photons with energy lower than the band gap of $BiVO_4$ are absorbed by the a-Si, forming additional charge carriers for a two-step recombination process (Fig. 10, right), which, together with the W gradient doping, led to a photoelectrochemical solar-to-hydrogen (STH) efficiency of 4.9%.

$BiVO_4$ has also been combined with other oxide semiconductors to form multicomponent heterojunctions, including TiO_2 [29], WO_3 [30], and ZnO [31]. A p–n heterojunction was also recently prepared combining p-type BiOCl with n-type $BiVO_4$ [32]. In a novel type of Z-scheme composite photocatalyst,

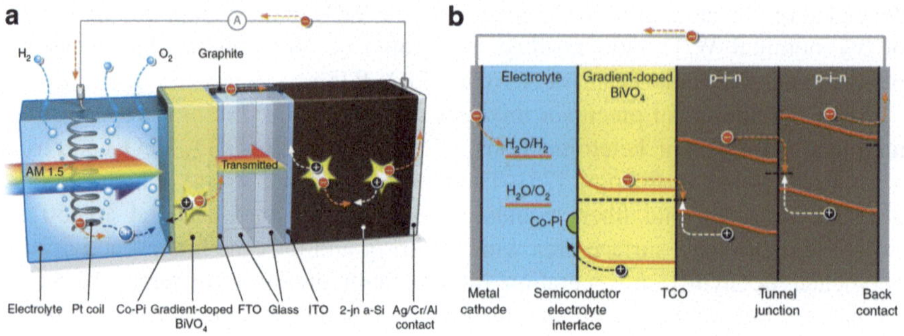

Fig. 10 *Left*: Schematic diagram of the triple-junction CoPi-W:$BiVO_4$ 2-jn a-Si photoanode, hydrogen evolution occurred at a Pt coil. *Right*: band diagram of the triple-junction photoanode, TCO (transparent-conductive oxide): tin-doped indium oxide. Reproduced with permission of Abdi et al. [28], ©2013 Nature Publishing Group

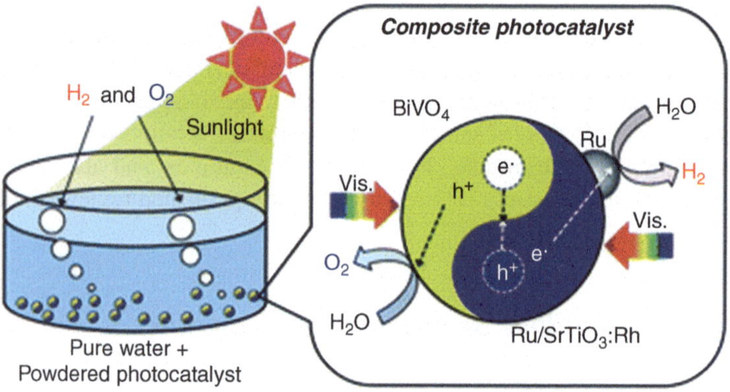

Fig. 11 Composite photocatalyst made of two visible-light absorbing components, Rh-doped Ru-$SrTiO_3$ and $BiVO_4$, for overall water splitting. Reproduced with permission of Jia et al. [33], ©2014 Royal Society of Chemistry

$BiVO_4$ was combined with Rh-doped Ru-$SrTiO_3$, another visible-light absorbing oxide photocatalyst (Fig. 11) [33]. A high apparent quantum yield (AQY) of 1.4% at 420 nm was reported, which is remarkable because, in contrast to earlier reports, no electron mediator was used. Overall water splitting was achieved in stoichiometric ratio. The composite prepared by impregnating Rh-doped Ru-$SrTiO_3$ with $Bi(NO_3)_3$ and NH_4VO_3 followed by calcination showed superior activity compared to Rh-doped Ru-$SrTiO_3$ and $BiVO_4$ powders mixed in suspension, indicating the importance of intimate contact between the components for optimum charge transfer (as described in Sect. 2).

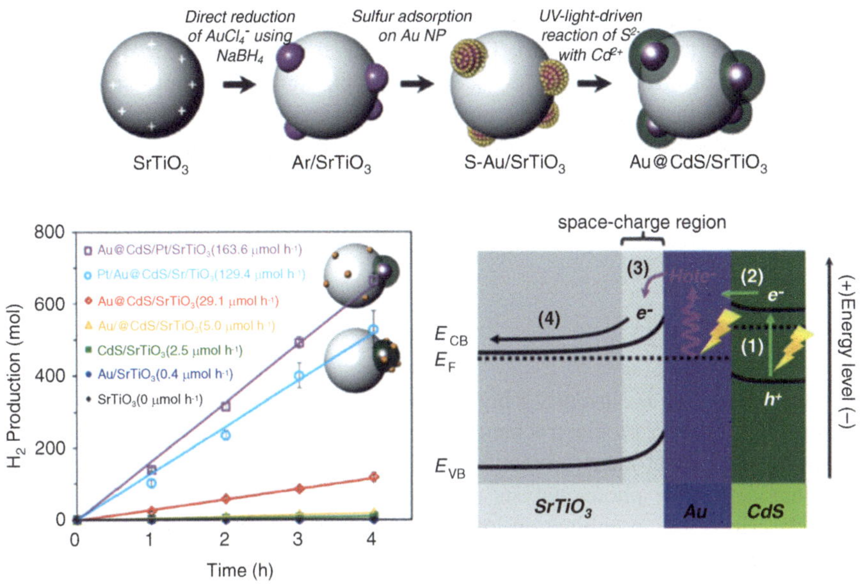

Fig. 12 *Top*: Preparation of Au@CdS/$SrTiO_3$ composites. *Lower left*: H_2 evolution rates depending on the composition and sequence of preparation. *Lower right*: charge transfer mechanism upon visible light irradiation. Reproduced with permission of Yu et al. [48], ©2014 Wiley-VCH

3.1.2 CdS

The combination of CdS with TiO_2 is one of the most investigated multicomponent heterojunctions in heterogeneous photocatalysis. Because CdS suffers from photocorrosion, it is very often used to sensitize wide band gap semiconductors, transferring electrons under visible light illumination to their conduction band (compare Fig. 3), and their combination is still a very active topic in photocatalysis and photoelectrochemistry, especially in combination with TiO_2 nanowires and nanotubes [34, 35]. For better charge extraction, CdS has also been combined with graphene or rGO [36–39], ZnO [40, 41], and recently with numerous other semiconductors such as Nb_2O_5 [42], Cu_2S [43], Bi_2S_3 (forming a type I heterojunction) [44], γ-TaON [45], and other II–VI semiconductors such as ZnSe [46] or CdTe [47].

Yu et al. recently combined CdS with plasmonic gold (Au) on $SrTiO_3$ to reduce the fast decay of hot electrons across Au nanoparticles after photoexcitation [48]. Au nanoparticles deposited on $SrTiO_3$ result in the typical Schottky contact between semiconductor and metal. Two-step deposition of CdS onto Au (Fig. 12) resulted in a Au@CdS core-shell structure. The most effective composite, however, for hydrogen evolution was achieved when the $SrTiO_3$ was decorated with Pt as co-catalyst before Au/CdS deposition. Thus, the resulting mechanism (Fig. 12) included both charge carrier excitation in CdS *and* transfer of hot electrons from plasmonic Au into $SrTiO_3$ by visible light, followed by electron transfer to Pt deposited on $SrTiO_3$ (not shown). Thus, Pt acts as the active site for H_2 evolution and also as the electron sink for the hot electrons coming from Au onto $SrTiO_3$. This example shows that very effective charge transfer can be achieved by combining plasmonic and composite effects for photocatalytic H_2 generation.

Other quantum-dot materials that have been used to sensitize semiconductors by forming composites include CdSe (e.g., with TiO_2 [49], Fe_2O_3 [50], Si [51]) and CdTe [47].

3.1.3 α-Fe_2O_3

Hematite (α-Fe_2O_3) is currently a highly investigated absorber material, especially for photoanodes in photoelectrochemical cells [52]. Moreover, it is also often used in combination with other semiconductors such as TiO_2 [53, 54] or WO_3 [55] to construct composite photocatalysts or photoanodes. Recently, numerous reports have described the combination of low-cost hematite with other iron-based, low-cost ferritic semiconductors such as $MgFe_2O_4$ [10] or $ZnFe_2O_4$, not only for photoelectrochemical work [56, 57] but also for battery applications [58]. Because of the short diffusion length for minority charge carriers ($L_D = 2$–4 nm) and poor majority carrier conductivity, those reports often involve nanostructuring (as shown in Fig. 4). Combinations with graphene usually improve the performance of hematite for photoelectrochemistry and photocatalysis in visible light [59] or for Li-ion batteries [60].

3.1.4 TiO_2

TiO_2 is still the most investigated photocatalyst today, and newly prepared composites containing TiO_2 with other semiconductor materials are reported regularly, including well known combinations with Ag_3PO_4 [61], Bi_2O_3 [62], Bi_2WO_6 [63], BiOCl [64], Cu_2O [12, 65], CuO [66], $CuInS_2$ [67], In_2S_3 [68], V_2O_5 [69], WO_3 [70], $ZnFe_2O_4$ [71], ZnO [72, 73], N-ZrO_2 [74], CNTs [75], and rGO [76]. Details about photocatalytic reactions in visible light for those composites are given in Table 1.

Recent reports on new composites with TiO_2 have covered, for example, combinations with $BiFeO_3$ [77] and Cr-$SrTiO_3$ [78]. In the latter case, Cr-$SrTiO_3$ nanocubes were grown onto TiO_2 nanotube arrays prepared by anodization (Fig. 13). Cr doping strongly enhanced the visible light response of $SrTiO_3$ (band gap 2.3 eV of the Cr-doped material). The resulting heterojunction showed increased photocurrents and photoelectrochemical H_2 generation (9.2 μmol h^{-1}, 0.1 M KOH solution, 0.6 V vs standard calomel electrode (SCE), $\lambda > 420$ nm, 300-W Xe lamp) compared to undoped composites (1.3 μmol h^{-1}) when visible-light excited electrons are transferred from Cr-$SrTiO_3$ over TiO_2 to the back electrode for charge separation.

Table 1 Recent reports on composites of TiO_2 with other semiconductors

Semiconductor 1	Semiconductor 2	Photocatalysis in visible light (>420 nm)	References
TiO_2	Ag_3PO_4	AO7 degradation, *E. coli* treatment	[61]
	Bi_2O_3	MO degradation	[62]
	Bi_2WO_6	MO degradation	[63]
	BiOCl	RhB degradation	[64]
	Cu_2O	H_2 generation, water splitting	[12, 65]
	CuO	H_2 generation	[66]
	$CuInS_2$	2-CP degradation, enhanced photocurrents	[67]
	In_2S_3	Enhanced photocurrents, *p*-nitrophenol degradation	[68]
	V_2O_5	Enhanced photocurrents, RhB degradation	[69]
	WO_3	MB degradation	[70]
	$ZnFe_2O_4$	MB degradation	[71]
	ZnO	–	[72, 73]
	N-ZrO_2	Formaldehyde degradation	[74]

AO7 acid orange 7, *MO* methyl orange, *RhB* rhodamine B, *2-CP* 2-chlorophenol, *MB* methylene blue

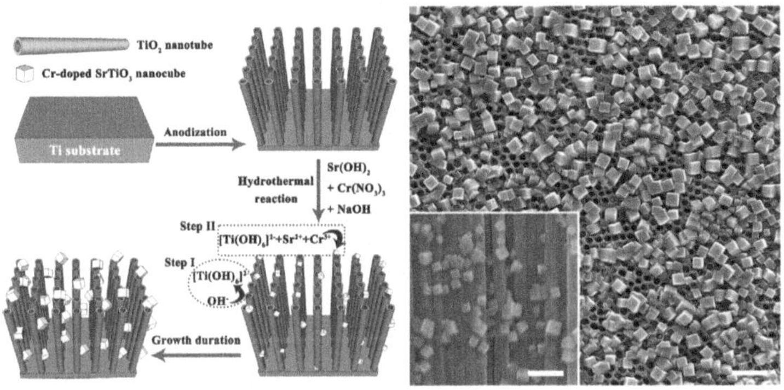

Fig. 13 *Left*: Preparation scheme for Cr-$SrTiO_3$ nanocubes onto TiO_2 nanotube arrays. *Right*: nanocubes on *top* and on the *sides* of TiO_2 nanotube arrays, *scale bars* 200 nm; Reproduced with permission from Jiao et al. [78], ©2014 Wiley-VCH

3.1.5 WO_3

Several examples of recently reported heterojunctions formed with WO_3 have been mentioned in earlier sections. WO_3 is an n-type semiconductor and exhibits a conduction band minimum below the proton reduction potential. WO_3 is often used as a visible-light active photoanode for water oxidation and also in heterojunction photoanodes. For example, Shaner et al. recently prepared an integrated Si/WO_3 tandem photoanode by depositing WO_3 electrochemically onto

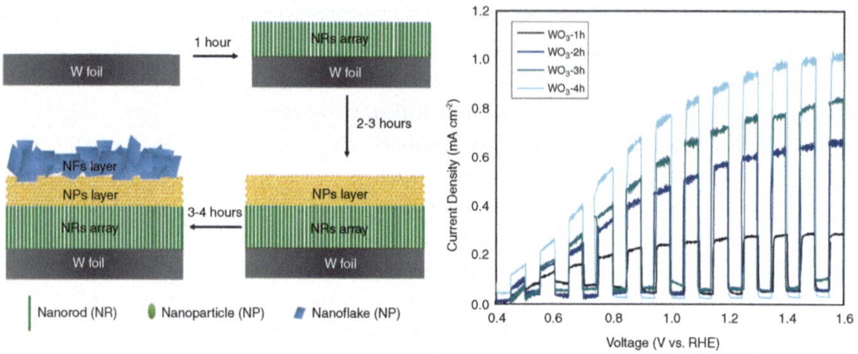

Fig. 14 *Left*: Formation process of triple-layered WO_3 photoanodes. *Right*: chopped light voltammetry of WO_3 photoanodes depending on anodization time. Reproduced with permission from Qi et al. [80], ©2014 Royal Society of Chemistry

n–p$^+$-Si microwire arrays, and combined it with a Pt cathode for unassisted hydrogen evolution [79].

To improve the surface area and optimize photocurrents, a special triple-layered "multiphase" WO_3 photoanode was recently reported, which was prepared by anodization of tungsten [80]. WO_3 nanoparticle and nanoflake layers were synthesized on previously grown WO_3 nanorods, leading to improved photocurrents upon visible light illumination (Fig. 14).

Because the different layers grew by subsequent anodization and etching/anodization processes, the contact between the layers is very intimate, with reduced interface and internal resistances explaining the improved photoelectrochemical performance (Fig. 14, right).

3.1.6 ZnO

ZnO is an n-type semiconductor with a band gap of 3.2 eV. Recently, many reports on ZnO nanostructures combined with other semiconductors showed enhanced light absorption and improved charge carrier transfer.

For example, Kargar et al. grew ZnO nanowires onto CuO nanowires (grown onto Cu foil or Cu mesh) via sputtering ZnO seeds followed by hydrothermal growth (Fig. 15, left) [81]. The resulting branched CuO-ZnO nanowire p–n heterojunctions showed broad light absorption and strongly enhanced cathodic photocurrents produced by the high surface area of the nanostructures, especially when deposited on Cu mesh.

Zhong et al. coated ZnGaON layers on ZnO nanowire arrays via chemical vapor deposition (CVD; Fig. 15, right) [82]. The coating not only enhanced the anodic photocurrent under simulated sunlight (1.75 mA cm^{-2} compared to 1.25 mA cm^{-2} for ZnO), but also showed improved stability of the ZnO nanorods against photocorrosion.

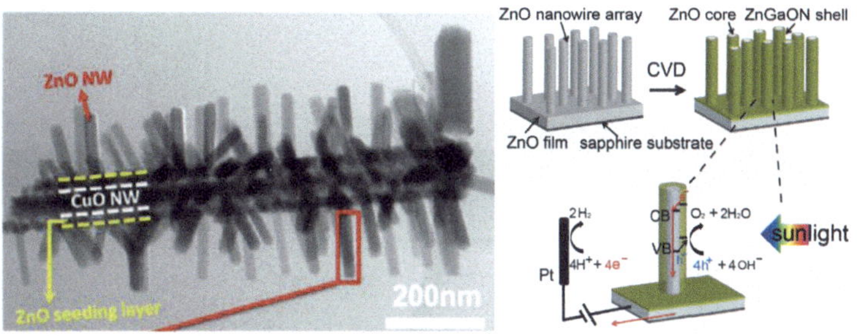

Fig. 15 *Left*: Branched CuO-ZnO nanowire arrays, here peeled-off from Cu foil. Reproduced with permission from Kargar et al. [81], ©2013 American Chemical Society. *Right*: preparation of ZnO nanowires coated with ZnGaON shell as heterojunction photoanodes. Reproduced with permission from Zhong et al. [82], ©2014 Royal Society of Chemistry

Qin et al. deposited $ZnFe_2O_4$ nanoparticles on ZnO nanorods for enhanced photoanodic performance in visible light, because of electron transfer from $ZnFe_2O_4$ to ZnO [83]. In a comparable approach, Guo et al. coated ZnO nanowire arrays completely with $ZnFe_2O_4$ and investigated the performance for RhB degradation (Fig. 16, left) [84]. The degradation mechanism on ZnO was found to be related to dye sensitization, whereas, when using $ZnFe_2O_4$/ZnO or pure $ZnFe_2O_4$ nanorods, reactive radicals formed upon visible light illumination were found to be responsible for RhB degradation. Moreover, the $ZnFe_2O_4$/ZnO composite showed higher activities (89% decolorization, 5 mL 10 ppm RhB solution, 150-W Xe lamp, $\lambda > 400$ nm) compared to pure $ZnFe_2O_4$ (43%).

ZnO nanorods have also been decorated with ZnSe on a Zn foil [85]. The resulting nanostructured heterojunctions (Fig. 16, right) showed enhanced light absorption and improved activity for the degradation of RhB (up to 96% degradation after 120 min, 5 mL of 10^{-5} MRhB aqueous solution, 300-W Xe lamp, $\lambda > 420$ nm) compared to pure ZnO nanorods (5.6%). Different nanostructures of ZnSe (nanoparticles, nanorods, nanosheets) were decorated onto ZnO, giving increased surface areas. Repeating the degradation experiments with the different nano-heterojunctions and adjusting the amount of powder using the exact same surface area, ZnSe-nanosheets onto ZnO nanorods turned out to be the most active composite, indicating an influence of the exposed crystal facets on the activity.

Using ZnO with a plate-morphology, Zamiri et al. prepared Ag_2S-ZnS-ZnO composite photocatalysts with a wet-chemical approach and evaluated their activity in NO_x degradation [86]. Under UV-visible irradiation (300-W Xe lamp), all three systems (ZnO, ZnO-ZnS, and ZnO-ZnS-Ag_2S) showed activity for NO_x removal, but the activity over ZnO and ZnO-ZnS strongly deteriorated (initial rates dropped from 2.40 min^{-1}/1.46 min^{-1} to 0.84 min^{-1}/0.73 min^{-1}) after three runs, whereas the deactivation for the three-component system was less pronounced (from 1.74 to 1.32 min^{-1}). The reason was to be found in electron transfer from ZnO and ZnS to Ag_2S, the latter acting as active site for the photocatalytic reaction.

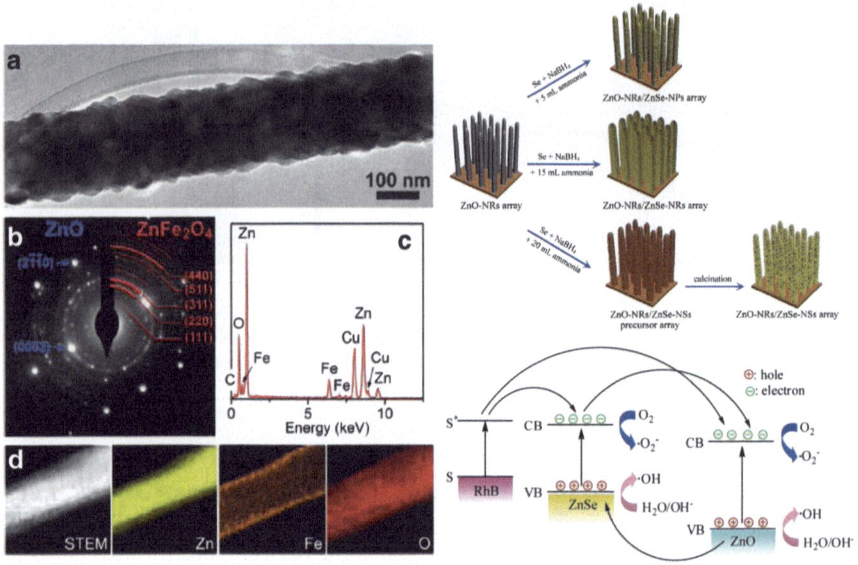

Fig. 16 *Left*: (**a**) $ZnFe_2O_4$/ZnO composite nanorod with corresponding SAED (**b**), EDXS (**c**), and elemental mapping (**d**). Reproduced with permission of Guo et al. [84], ©2014 Elsevier. *Right*: preparation process of ZnSe-decorated ZnO nanorods, and charge carrier transfer processes upon visible light irradiation in the presence of RhB. Reproduced with permission of Wu et al. [85], ©2014 Royal Society of Chemistry

3.1.7 Other Semiconductors

As shown in the sections above, the variety of metal oxides used to construct heterojunctions in composite photocatalysts is vast. Thus, this section can only give a small indication about the possibilities combing semiconductors not including one of the above-mentioned oxides and sulfides. Mostly, at least one of the components is a visible light absorbing material to extend the overall light absorption, and some examples are given in Table 2.

For example, Guo et al. doped the layered perovskite $Sr_2Nb_2O_7$ with nitrogen, and combined it with the visible-light absorbing Ag_3PO_4 [87]. Anion doping with, e.g., nitrogen is a well-known strategy, especially for layered oxides, to shift the valence band maximum upwards [99]. Thus, N-$Sr_2Nb_2O_7$ and Ag_3PO_4 both absorb visible light. Upon light irradiation, electrons are transferred from N-$Sr_2Nb_2O_7$ to Ag_3PO_4 for oxygen reduction, and holes transfer to N-$Sr_2Nb_2O_7$ for improved IPA oxidation (10.3 ppm h^{-1} CO_2, 1,250–1,500 ppm initial gaseous IPA, 0.2 g photocatalyst on 8.5 cm^2 area, 300-W Xe lamp, $\lambda > 420$ nm) compared to N-$Sr_2Nb_2O_7$ (~2 ppm h^{-1}) and Ag_3PO_4 (~0 ppm h^{-1}).

Grinberg et al. prepared ferroelectric solid oxide solutions of $KNbO_3$ with $BaNi_{0.5}Nb_{0.5}O_{3-\delta}$ [95] of different compositions. With 10 mol% $KNbO_3$, the resulting solid solution exhibited a direct band gap of 1.39 eV. The resulting open-circuit voltage of 0.7 mV and short-circuit current of 0.1 μA cm^{-1} of 20 μm

Table 2 Recent combinations of mixed oxides and sulfides for composite photocatalysts

Semiconductor 1	Semiconductor 2	Photocatalysis in visible light (>420 nm)	References
Ag_3PO_4	N-$Sr_2Nb_2O_7$	IPA oxidation	[87]
	$SrTiO_3$	O_2 generation	[88]
$Bi_2O_2CO_3$	β-Bi_2O_3	*o*-Phenylphenol degradation	[89]
	Bi_2S_3	RhB degradation	[90]
		Enhanced photocurrents	
Bi_2WO_6	$Mg_{1-x}Cu_xWO_4$	RhB degradation	[91]
		Enhanced photocurrents	
Cu_2O	NiO/Cu_2MoS_4	Enhanced photocurrents	[92]
	$SrTiO_3$	Enhanced photocurrents	[93]
	TaON	Photoelectrochemical water splitting	[94]
$KNbO_3$	$BaNi_{0.5}Nb_{0.5}O_{3-\delta}$	Enhanced photocurrents	[95]
$NaNbO_3$	$SnNb_2O_6$	O_2 generation, H_2 generation	[96]
Ta_3N_5	La,Rh:$SrTiO_3$	Z-scheme water splitting	[97]
	Ta_2O_5	O_2 generation, H_2 generation	[98]

IPA isopropyl alcohol

thick films at 300 K were about 50 times higher than for comparable ferroelectric film absorbers such as 50-μm (Pb,La)(Zr,Ti)O_3 samples or 0.84-μm (Na,K)NbO_3 samples in UV light.

Saito and Kudo combined perovskite $NaNbO_3$ nanowires with the visible-light absorbing material $SnNb_2O_6$ as a shell [96]. The resulting composite showed higher photocatalytic performance (10 μmol h^{-1} H_2 production, 1 wt% Pt, 0.1 g catalyst, 150 mL solution containing MeOH, 300-W Xe lamp, $\lambda > 420$ nm) than pure $NaNbO_3$ (1.1 μmol h^{-1}), but comparable values with bulk $SnNb_2O_6$ (14 μmol h^{-1} with 0.3 wt% Pt).

Wang et al. combined iridium/cobalt oxide-decorated Ta_3N_5 with Ru-decorated La,Rh:$SrTiO_3$ for redox-free Z-scheme overall water splitting [97]. The optimum dopant amount for La and Rh were found to be 4 mol%, and apparent quantum yields of up to 1.1% at 420 nm were reached (0.05 g catalyst each, aqueous H_2SO_4 solution (pH 3.9, 200 mL), 300-W Xe lamp, different cut-off filters). During the reaction, photoexcited electrons from Ta_3N_5 recombine via iridium (Ir) with photogenerated holes on La,Rh:$SrTiO_3$, whereas holes on Ta_3N_5 oxidize water at cobalt oxide co-catalyst and electrons from La,Rh:$SrTiO_3$ reduce protons at Ru co-catalyst.

3.2 Polymer-Based Semiconductor Composites

As already covered in Sect. 2, polymeric carbon nitride is nowadays often used as a material to construct semiconductor composites. The band positions of carbon nitride are situated at more negative potentials compared to most oxide or sulfide

Table 3 Composites of carbon nitride with other visible light active semiconductors

Semiconductor 1	Semiconductor 2	Photocatalysis in visible light (>420 nm)	References
C_3N_4	N-TiO_2	–	[104]
C_3N_4	CdS	H_2 production, MO degradation, oxidation of aromatic alcohols, nitrobenzene reduction	[105–107]
C_3N_4	$BiIO_4$	RhB degradation, enhanced photocurrents	[108]
C_3N_4	Fe_2O_3	DR81 degradation, enhanced photocurrent	[109]
C_3N_4	Zn_2GeO_4	MB degradation	[110]
C_3N_4	Si	Enhanced photocurrents	[111]
C_3N_4	Bi_2MoO_6	MB degradation, RhB degradation	[112]
C_3N_4	WO_3	Enhanced photocurrents	[113]
C_3N_4	$ZnFe_2O_4$	H_2 production	[114]
C_3N_4	Cu_2O	H_2 production	[115]
C_3N_4	BiOI	Bisphenol A degradation, enhanced photocurrents	[116]
C_3N_4	In_2O_3	H_2 production, enhanced photocurrents	[117]
C_3N_4	SnS_2	RhB degradation, MO degradation, 4-NP degradation	[118]

MO methyl orange, *RhB* rhodamine B, *MB* methylene blue, *DR81* direct red 81, *4-NP* 4-nitrophenol

semiconductors (as shown in Fig. 5 with TiO_2 as example), and thus charge transfer of photogenerated electrons from the conduction band of carbon nitrides to the adjacent oxide/sulfide is often utilized to achieve efficient charge carrier separation improving photocatalytic activity, whereas photogenerated holes can transfer to the carbon nitride valence band. Such a heterojunction is classified as a type II heterojunction.

Moreover, carbon nitride exhibits a band gap of 2.7 eV in the visible-light range and can be used to sensitize large band gap semiconductors to form visible-light active semiconductor composites. Typical examples are the combination with TiO_2 [100–102], ZnO [5], or $NaTaO_3$ [103]. However, more and more reports on the combination of carbon nitride with other visible light absorbing semiconductors are reported, with some examples given in Table 3.

Apart from the combination in composites with oxides or sulfides, carbon nitride was recently used to form a composite with poly(3,4-ethylenedioxythiophene) (PEDOT):poly(styrenesulfonate) (PSS) [119]. PEDOT is known as a good hole conductor, and using PSS as the counter ion to PEDOT can increase its dispersibility in aqueous media. The resulting composite enhanced the hole transfer from the carbon nitride to the sacrificial agent triethanolamine (TEA). After the addition of platinum (Pt, 1 wt%), enhanced hydrogen production rates (up to 33 μmol h^{-1} per 0.1 g with an optimum PEDOT loading of 2 wt%, 10 vol.% TEA solution, 300-W Xe lamp, $\lambda > 400$ nm) was observed, because of the spatial separation of the reduction (Pt) and oxidation (PEDOT:PSS) reaction sites (Fig. 17).

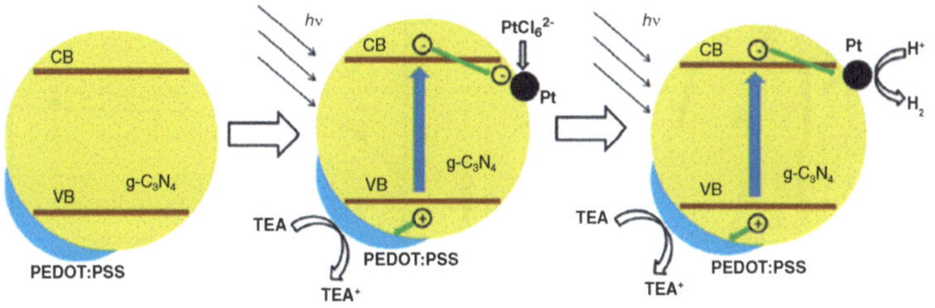

Fig. 17 Photodeposition of Pt on carbon nitride and proposed mechanism of photocatalytic H_2 generation from TEA solutions. Reproduced with permission from Xing et al. [119], ©2014 Royal Society of Chemistry

In this example, PEDOT:PSS and Pt are acting as charge mediators for holes (to the TEA) or for electrons (from the conduction band), respectively, thus acting as co-catalysts. Other electron mediators include graphene-derived carbon, such as GO or rGO (as described in Sect. 2) because they are highly conductive. Carbon nitride was recently combined with rGO and an additional co-catalyst, MoS_2, to prepare a multilayer structure of those three 2D materials [120]. Although photo-excited electrons were transferred from the carbon nitride via rGO to MoS_2 working as co-catalyst for Cr(VI) reduction, MoS_2 also strongly enhanced the overall light absorption of the composite.

Another semiconducting polymer, polyaniline (PANI), recently gained attention when forming a heterojunction with other semiconductors such as TiO_2 [121, 122]. Interestingly, the semiconductor behavior of PANI can be influenced by the surrounding medium and its pH value. Under acidic conditions, PANI is protonated and electron-poor, resulting in p-type behavior, whereas at basic pH, PANI is electron-rich and exhibits n-type behavior. At pH 3 PANI is stable, protonated, and thus positively charged, and was sequentially deposited on conducting glass with negatively charged $Ti_{0.91}O_2$ nanosheets [121]. In basic electrolyte containing methanol, the PANI/$Ti_{0.91}O_2$ multilayer electrodes showed anodic photocurrents of up to 1.5 $\mu A/cm^2$ (Fig. 18). By switching the electrolyte to hydrochloric acid containing an electron scavenger (use of methanol gave no photocurrents), cathodic photocurrents were observed. Although the nanosheets are always n-type, the amount of photocurrent could be controlled by the number of layers deposited.

3.3 *Molecular Complexes Combined with Semiconductors*

In some cases, light absorbing complexes, so-called photosensitizers (PS), are attached to semiconductor surfaces to broaden the light absorption of the resulting composite compared to the semiconductor itself. The most famous example is the dye-sensitized solar cell (DSSC) [123], in which a broad absorbing Ru-bipyridine-based dye PS is adsorbed on a nanoparticulate TiO_2 electrode, injecting electrons into the TiO_2 conduction band upon light irradiation. Inspired by DSSCs, water

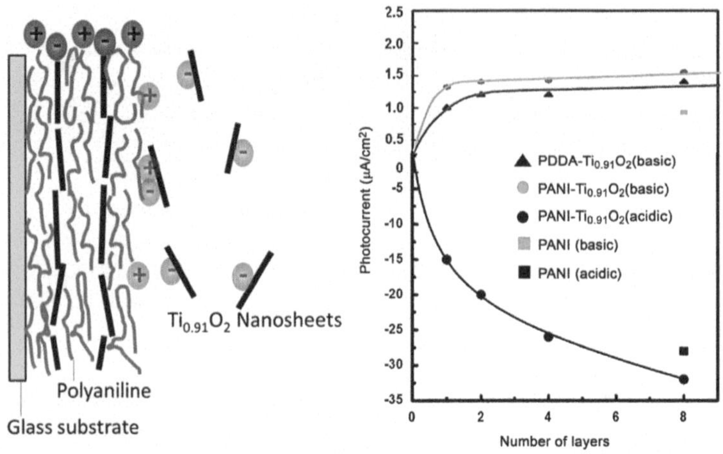

Fig. 18 *Left*: Layer-by-layer deposition of polyaniline (PANI) and $Ti_{0.91}O_2$ nanosheets. *Right*: anodic or cathodic photocurrents of PANI/$Ti_{0.91}O_2$ nanosheets depending on pH. Reproduced with permission from Seger et al. [121], ©2013 Wiley-VCH

splitting dye-sensitized photoelectrochemical cells have recently gained attention. Instead of using a redox couple to re-reduce the oxidized dyes, a water oxidation catalyst (WOC) on the TiO_2 anode, e.g., IrO_2, can be applied to generate oxygen, whereas the reduction of protons occurs on the counter cathode [124, 125]. By using a PS on p-type semiconductor films, such as NiO, together with a water reduction catalysts (WRC), hydrogen evolving photocathodes for water splitting dye-sensitized photoelectrochemical cells can be constructed. Water oxidation can then occur on the counter anode. For example, Li et al. used a donor-acceptor-type metal-free dye on NiO to construct a hydrogen-evolving p-type photocathode (Fig. 19), and used a cobalt-based WRC to generate H_2 from phosphate buffer solution [126].

Molecular WRCs or water oxidation catalysts (WOCs) are often immobilized on semiconductors to improve photocatalytic or photoelectrochemical water splitting. For example, Klepser and Bartlett anchored a molecular Fe-based molecular WOC (Fe [tetraethyl *N,N′*-bis(2-methylpyridyl-4-phosphonate)-*N,N′*-dimethylcyclohexyldiamine]Cl_2) on the surface of a WO_3 photoanode [127]. Thus, no sacrificial oxidant to regenerate the catalyst was needed anymore, and photoelectrochemical water oxidation rates were increased by 60%.

Porphyrins are another class of dyes used to sensitize semiconductors such as TiO_2. For example, Kurimoto et al. used tetra(4-carboxy)phenylporphyrinatotin (IV) (Sn(IV)TCPP) to sensitize TiO_2, performing simultaneous photocatalytic H_2 generation and cyclohexane oxygenation in water (Fig. 20) [128]. Upon exciting Sn (IV)TCPP with visible light ($\lambda > 480$ nm), electron injection into the conduction band of TiO_2 occurs. H_2 is generated at the Pt co-catalyst, whereas the oxidation reaction occurs at the dye. The maximum total yield of the oxygenation products was reached at neutral pH conditions.

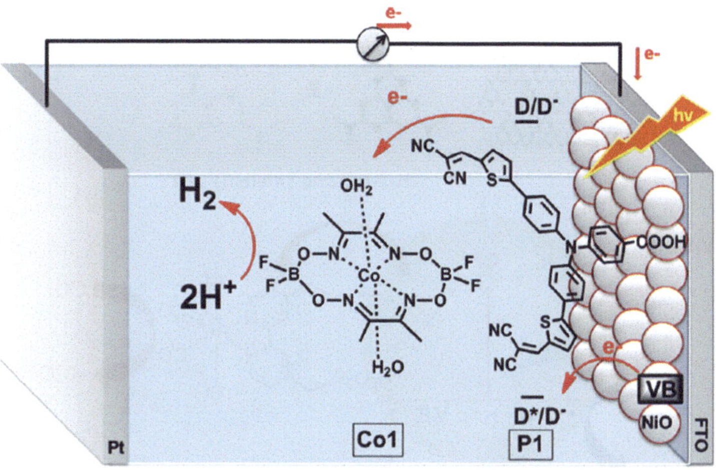

Fig. 19 Dye-sensitized photoelectrochemical cell for water splitting based on a dye-sensitized photocathode. Reproduced with permission of Li et al. [126], ©2012 Royal Society of Chemistry

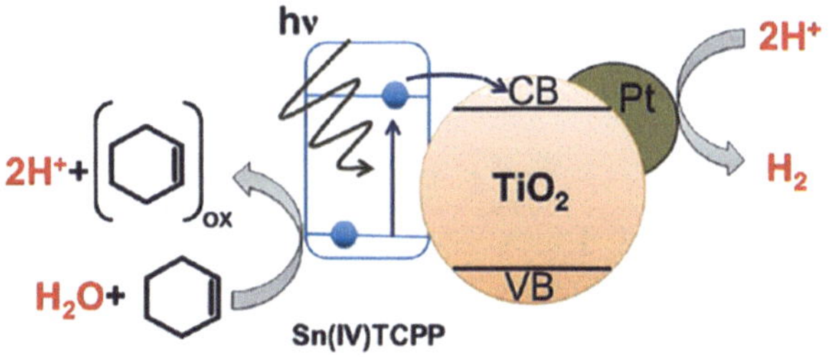

Fig. 20 Simultaneous hydrogen evolution with the photochemical oxygenation of cyclohexene on Sn(IV)TCPP-sensitized TiO_2. Reproduced with permission from Kurimoto et al. [128], ©2014 Royal Society of Chemistry

Maeda et al. immobilized the Ru-complex "*trans*(Cl)-[Ru{4,4′-$(CH_2PO_3H_2)_2$-2,2′-bipyridine}$(CO)_2Cl_2$]" acting as a reduction catalyst onto carbon nitride [129]. In this case, because carbon nitride has such high conduction band potential, electron transfer of photoexcited electron takes place from the carbon nitride to the Ru-complex (Fig. 21), from which carbon dioxide was reduced to formic acid (in a mixture of acetonitrile and triethanolamine (4:1 v/v), 400-W high-pressure Hg lamp with a $NaNO_2$ solution filter to cut-off UV light, $\lambda > 400$ nm).

Many groups try to replace the expensive Ru-based PS with complexes based on more Earth-abundant metals or even with metal-free PS for photocatalytic reactions. For example, in a completely noble metal-free approach, Karnahl et al. immobilized a series of heteroleptic copper photosensitizers on TiO_2 and showed H_2 generation in the presence of $[Fe_3(CO)_{12}]$ (5 mmol) as water reduction catalyst (WRC) (25 mg ~ 1 μmol PS, mixture of THF/TEA/H_2O [4:3:1, 10 mL]) [130]. When encapsulated with plasma-polymerized allylamine (PPAAm), desorption of the copper dye is reduced, leading to a more active and stable absorber system.

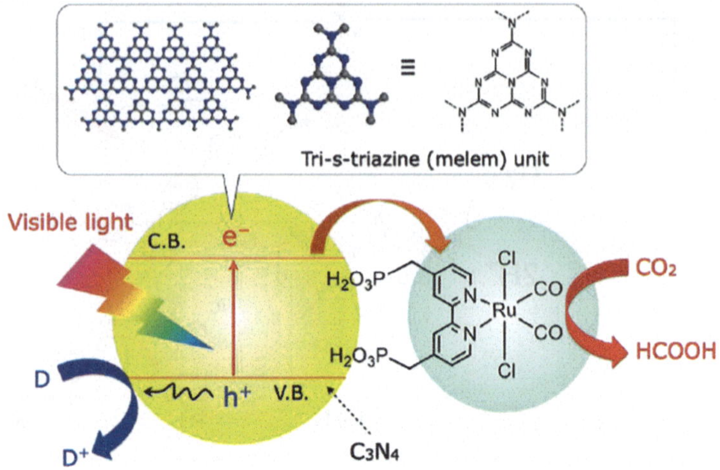

Fig. 21 Schematic illustration of photocatalytic CO_2 reduction to formic acid on the Ru-complex/C_3N_4 composite under visible light illumination, TEA as electron donor. Reproduced with permission of Maeda et al. [129], ©2014 Royal Society of Chemistry

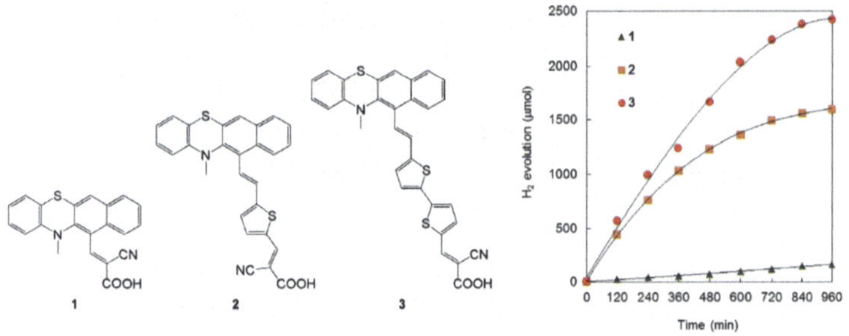

Fig. 22 Donor-spacer-acceptor dyes to sensitize Pt-TiO_2, used for hydrogen generation from TEA/water (*right*). Reproduced with permission from Watanabe et al. [131], ©2014 Royal Society of Chemistry

Watanabe et al. used metal-free donor-spacer-acceptor type dyes containing benzo[*b*]phenothiazine to sensitize Pt-TiO_2 for visible-light-driven photocatalytic hydrogen production (10 vol.% aqueous TEA (10 mL), 33.0 mg TiO_2/dye/Pt catalyst, pH 7.0, 300-W Xe lamp, $\lambda < 420$ nm), with up to 1.65% quantum efficiency at 420 nm (Fig. 22) [131]. Dyes 2 and 3 showed longer decay times in the time-resolved absorption spectra, indicating increased charge carrier lifetimes with increased spacer length, leading to increased photocatalytic activity.

In an approach to improve water oxidation in inorganic heterogeneous photocatalysis, Li et al. sensitized the visible-light-absorbing artificial photocatalysts Rh-doped Ru-$SrTiO_3$ and Ru_2S_3-CdS with the photosystem II (PSII), the natural water-oxidizing enzyme [132]. Therefore, they isolated PSII membrane fragments

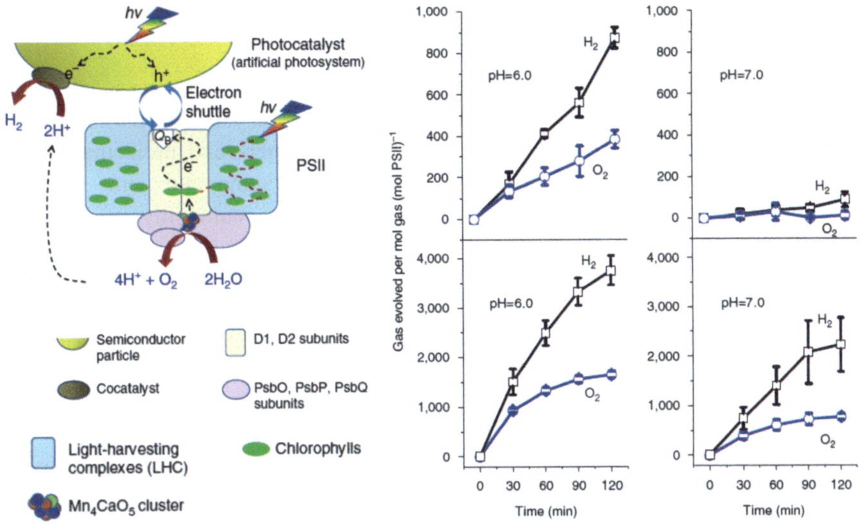

Fig. 23 *Left*: Charge carrier generation, flow and transfer in a PSII-artificial photocatalyst composite; *Right*: overall water splitting gas evolution with PSII/Rh-doped Ru-$SrTiO_3$ (*top row*) and PSII/Ru_2S_3-CdS (*bottom row*) at different pH values in *visible light*. Reproduced with permission from Wang et al. [132], ©2014 Nature Publishing Group

from spinach with a non-ionic surfactant, namely Triton-X. The PSII fragments were then mixed with the artificial photocatalysts and $Fe(CN)_6^{3-}$ electron mediator in a sodium buffer solution, and irradiated with a 300-W Xe lamp ($\lambda > 420$ nm). The resulting composite was able to split water into H_2 and O_2 without sacrificial reagents in visible light. The electrons from water oxidation by PSII are shuttled to the Rh-doped Ru-$SrTiO_3$ or Ru_2S_3-CdS photocatalyst for proton reduction via a $Fe(CN)_6^{3-}/Fe(CN)_6^{4-}$ electron mediator. At pH 6, overall water splitting was achieved with both composites in a stoichiometric ratio of H_2 and O_2, the composite Rh-doped Ru-$SrTiO_3$ showing the higher activity (Fig. 23) because of its higher hydrophilicity for PSII binding.

At pH 7, the rates decreased as a result of the pH-dependence of O_2 evolution by PSII. Under irradiation with light with wavelengths >600 nm, where only the PSII fragments were able to absorb light, only O_2 evolution was detected, showing the necessity for a hydrogen-evolving photocatalyst. Although the composite also generated H_2 and O_2 in outdoor solar light, the stability of the PSII against reactive oxygen species is still an issue to be solved.

4 Conclusion

This chapter has given a short overview about the recent strategies to create composites with semiconductors for improved photocatalysis and photoelectrochemistry efficiency, including composites with polymers and molecular complexes. Many more examples are known in the literature, and some aspects including tandem devices and tandem photoelectrodes could only be covered sparsely.

It is unfortunate that very few composites are made of p–n-heterojunctions, which show very interesting band shifting produced by Fermi level alignment. This type of composite is especially interesting for p-type materials whose conduction band is below the hydrogen reduction potential, because the combination with an n-type semiconductor can shift the band edge upwards, even enough for hydrogen generation.

Another promising strategy to form future heterojunctions might be to develop nanostructuring techniques apart from nanorods/nanotubes/nanocubes. One possible strategy could be the generation of porosity and high surface area of one semiconductor, e.g., with ordered pore systems, and then filling the nanostructural voids with the second semiconductor for a large interfacial contact between both phases or components, because an intimate contact between both components is required for efficient charge separation. This concept has already been shown for multiferroic materials [133] and for electrochemical water oxidation [134], and should also be considered for future composite preparations in photocatalysis. Moreover, as also stated, heterojunctions do not need to be limited to only two components, as more than two components or phases can further improve the charge carrier separation.

Acknowledgement R.M. gratefully acknowledges funding in the Emmy-Noether program (MA 5392/3-1) of the German Research Foundation DFG.

References

1. Marschall R (2014) Adv Funct Mater 24:2421
2. Soldat J, Marschall R, Wark M (2014) Chem Sci 5:3746
3. Kho YK, Iwase A, Teoh WY, Mädler L, Kudo A, Amal R (2010) J Phys Chem C 114:2821
4. Siedl N, Baumann SO, Elser MJ, Diwald O (2012) J Phys Chem C 116:22967
5. Zhou J, Zhang M, Zhu Y (2014) Phys Chem Chem Phys 16:17627
6. Kim ES, Nishimura N, Magesh G, Kim JY, Jang J-W, Jun H, Kubota J, Domen K, Lee JS (2013) J Am Chem Soc 135:5375
7. Mukherji A, Seger B, (Max) Lu GQ, Wang L (2011) ACS Nano 5:3483
8. Williams G, Seger B, Kamat PV (2008) ACS Nano 2:1487
9. Park H, Choi W, Hoffmann MR (2008) J Mater Chem 18:2379
10. Hou Y, Zuo F, Dagg A, Feng P (2013) Angew Chem Int Ed 52:1248
11. Ostermann R, Sallard S, Smarsly BM (2009) Phys Chem Chem Phys 11:3648

12. Devadoss A, Sudhagar P, Ravidhas C, Hishinuma R, Terashima C, Nakata K, Kondo T, Shitanda I (2014) Phys Chem Chem Phys 16:21237
13. Pilli SK, Janarthanan R, Deutsch TG, Furtak TE, Brown LD, Turner JA, Herring AM (2013) Phys Chem Chem Phys 15:14723
14. Ismail AA, Geioushy RA, Bouzid H, Al-Sayari SA, Al-Hajry A, Bahnemann DW (2013) Appl Catal B 129:62
15. Cherevan AS, Gebhardt P, Shearer CJ, Matsukawa M, Domen K, Eder D (2014) Energy Environ Sci 7:791
16. Yin H, Yu K, Song C, Huang R, Zhu Z (2014) ACS Appl Mater Interfaces 6:14851
17. Bledowski M, Wang L, Ramakrishnan A, Khavryuchenko OV, Khavryuchenko VD, Ricci PC, Strunk J, Cremer T, Kolbeck C, Beranek R (2011) Phys Chem Chem Phys 13:21511
18. Wang X, Xu Q, Li M, Shen S, Wang X, Wang Y, Feng Z, Shi J, Han H, Li C (2012) Angew Chem Int Ed 51:13089
19. Zhu Y, Liu Y, Lv Y, Ling Q, Di L, Zhu Y (2014) J Mater Chem A 2:13041
20. Li P, Xu H, Liu L, Kako T, Umezawa N, Abe H, Ye J (2014) J Mater Chem A 2:5606
21. Teoh WY, Amal R, Mädler L (2010) Nanoscale 2:1324
22. Kandiel TA, Feldhoff A, Robben L, Dillert R, Bahnemann DW (2010) Chem Mater 22:2050
23. Zhao H, Liu L, Andino JM, Li Y (2013) J Mater Chem A 1:8209
24. Kandiel TA, Robben L, Alkaim A, Bahnemann D (2013) Photochem Photobiol Sci 12:602
25. Grewe T, Meier K, Tüysüz H (2014) Catal Today 225:142
26. Riboni F, Bettini LG, Bahnemann DW, Selli E (2013) Catal Today 209:28
27. Lee SS, Bai H, Liu Z, Sun DD (2013) Appl Catal B 140–141:68
28. Abdi FF, Han L, Smets AHM, Zeman M, Dam B, van de Krol R (2013) Nat Commun 4:2195
29. Ho-Kimura S, Moniz SJA, Handoko AD, Tang J (2014) J Mater Chem A 2:3948
30. Pihosh Y, Turkevych I, Mawatari K, Asai T, Hisatomi T, Uemura J, Tosa M, Shimamura K, Kubota J, Domen K, Kitamori T (2014) Small 10:3692
31. Moniz SJ, Zhu J, Tang J (2014) Adv Energy Mater 4:1301590
32. He Z, Shi Y, Gao C, Wen L, Chen J, Song S (2014) J Phys Chem C 118:389
33. Jia Q, Iwase A, Kudo A (2014) Chem Sci 5:1513
34. Zhu Y, Chen Z, Gao T, Huang Q, Niu F, Qin L, Tang P, Huang Y, Sha Z, Wang Y (2015) Appl Catal B 163:16
35. Kim JS, Han HS, Shin S, Han GS, Jung HS, Hong KS, Noh JH (2014) Int J Hydrogen Energy 39:17473
36. Yu J, Jin J, Cheng B, Jaroniec M (2014) J Mater Chem A 2:3407
37. Li Y, Wang H, Peng S (2014) J Phys Chem C 118:19842
38. Jia T, Kolpin A, Ma C, Chan RC-T, Kwok W-M, Tsang SCE (2014) Chem Commun 50:1185
39. Chang K, Mei Z, Wang T, Kang Q, Ouyang S, Ye J (2014) ACS Nano 8:7078
40. Eley C, Li T, Liao F, Fairclough SM, Smith JM, Smith G, Tsang SCE (2014) Angew Chem Int Ed 53:7838
41. Lingampalli SR, Gautam UK, Rao CNR (2013) Energy Environ Sci 6:3589
42. Oliveira LC, Oliveira HS, Mayrink G, Mansur HS, Mansur AA, Moreira RL (2014) Appl Catal Environ 152–153:403
43. Tan C-S, Hsiao C-H, Wang S-C, Liu P-H, Lu M-Y, Huang MH, Ouyang H, Chen L-J (2014) ACS Nano 8:9422
44. Hao L-X, Chen G, Yu Y-G, Zhou Y-S, Han Z-H, Liu Y (2014) Int J Hydrogen Energy 39: 14479
45. Wang Z, Hou J, Yang C, Jiao S, Zhu H (2014) Chem Commun 50:1731
46. Zhu H, Chen Z, Wu K, Lian T (2014) Chem Sci 5:3905
47. Okano M, Sakamoto M, Teranishi T, Kanemitsu Y (2014) J Phys Chem Lett 5:2951
48. Yu S, Kim YH, Lee SY, Song HD, Yi J (2014) Angew Chem Int Ed 53:11203
49. Wang P, Li D, Chen J, Zhang X, Xian J, Yang X, Zheng X, Li X, Shao Y (2014) Appl Catal Environ 160–161:217

50. Sahai S, Ikram A, Rai S, Dass S, Shrivastav R, Satsangi VR (2014) Int J Hydrogen Energy 39:11860
51. Devarapalli RR, Kamaja CK, Shelke MV (2014) J Mater Chem A 2:13352
52. Bora DK, Braun A, Constable EC (2013) Energy Environ Sci 6:407
53. Zhu S, Yao F, Yin C, Li Y, Peng W, Ma J, Zhang D (2014) Microporous Mesoporous Mater 190:10
54. Pham M-H, Dinh C-T, Vuong G-T, Ta N-D, Do T-O (2014) Phys Chem Chem Phys 16:5937
55. Jin T, Diao P, Wu Q, Di X, Hu D, Xie Y, Zhang M (2014) Appl Catal B 148–149:304
56. Guo Y, Fu Y, Liu Y, Shen S (2014) RSC Adv 4:36967
57. McDonald KJ, Choi K-S (2011) Chem Mater 23:4863
58. Zhao D, Xiao Y, Wang X, Gao Q, Cao M (2014) Nano Energy 7:124
59. He L, Jing L, Luan Y, Wang L, Fu H (2014) ACS Catal 4:990
60. Sun Z, Xie K, Li ZA, Sinev I, Ebbinghaus P, Erbe A, Farle M, Schuhmann W, Muhler M, Ventosa E (2014) Chem Eur J 20:2022
61. Xu J-W, Gao Z-D, Han K, Liu Y, Song Y-Y (2014) ACS Appl Mater Interfaces 6:15122
62. Liu J, Liu X, Li J, Pan L, Sun Z (2014) RSC Adv 4:38594
63. Tian J, Sang Y, Yu G, Jiang H, Mu X, Liu H (2013) Adv Mater 25:5075
64. Guerrero M, Altube A, García-Lecina E, Rossinyol E, Baró MD, Pellicer E, Sort J (2014) ACS Appl Mater Interfaces 6:13994
65. Cheng W-Y, Yu T-H, Chao K-J, Lu S-Y (2014) ChemCatChem 6:293
66. Chen W-T, Jovic V, Sun-Waterhouse D, Idriss H, Waterhouse GI (2013) Int J Hydrogen Energy 38:15036
67. Li T, Li X, Zhao Q, Shi Y, Teng W (2014) Appl Catal B 156–157:362
68. Li Y, Luo S, Wei Z, Meng D, Ding M, Liu C (2014) Phys Chem Chem Phys 16:4361
69. Wang Y, Liu L, Xu L, Cao X, Li X, Huang Y, Meng C, Wang Z, Zhu W (2014) Nanoscale 6:6790
70. Liu K-I, Hsueh Y-C, Chen H-S, Perng T-P (2014) J Mater Chem A 2:5387
71. Zhu X, Zhang F, Wang M, Ding J, Sun S, Bao J, Gao C (2014) Appl Surf Sci
72. Qiu Y, Leung S-F, Wei Z, Lin Q, Zheng X, Zhang Y, Fan Z, Yang S (2014) J Phys Chem C 118:22465
73. Guo S, Han S, Chi B, Pu J, Li J (2014) ACS Appl Mater Interfaces 6:4743
74. Yu Y, Zhang P, Kuang Y, Ding Y, Yao J, Xu J, Cao Y (2014) J Phys Chem C 118:20982
75. Yazdani N, Bozyigit D, Utke I, Buchheim J, Youn SK, Patscheider J, Wood V, Park HG (2014) ACS Appl Mater Interfaces 6:1389
76. Cao X, Tian G, Chen Y, Zhou J, Zhou W, Tian C, Fu H (2014) J Mater Chem A 2:4366
77. Zhu A, Zhao Q, Li X, Shi Y (2014) ACS Appl Mater Interfaces 6:671
78. Jiao Z, Zhang Y, Chen T, Dong Q, Lu G, Bi Y (2014) Chem Eur J 20:2654
79. Shaner MR, Fountaine KT, Ardo S, Coridan RH, Atwater HA, Lewis NS (2014) Energy Environ Sci 7:779
80. Qi H, Wolfe J, Wang D, Fan HJ, Fichou D, Chen Z (2014) Nanoscale 6:13457
81. Kargar A, Jing Y, Kim SJ, Riley CT, Pan X, Wang D (2013) ACS Nano 7:11112
82. Zhong M, Ma Y, Oleynikov P, Domen K, Delaunay J-J (2014) Energy Environ Sci 7:1693
83. Qin D-D, Tao C-L (2014) RSC Adv 4:16968
84. Guo X, Zhu H, Li Q (2014) Appl Catal B 160–161:408
85. Wu Z, Wang H, Xue Y, Li B, Geng B (2014) J Mater Chem A 2:17502
86. Zamiri R, Abbastabar Ahangar H, Tobaldi DM, Rebelo A, Seabra MP, Shabani M, Ferreira JMF (2014) Phys Chem Chem Phys 16:22418
87. Guo J, Zhou H, Ouyang S, Kako T, Ye J (2014) Nanoscale 6:7303
88. Guan X, Guo L (2014) ACS Catal 4:3020
89. Hu R, Xiao X, Tu S, Zuo X, Nan J (2015) Appl Catal B 163:510
90. Liang N, Zai J, Xu M, Zhu Q, Wei X, Qian X (2014) J Mater Chem A 2:4208
91. Huang H, Wang S, Zhang Y, Chu PK (2014) RSC Adv 4:41219

92. Yang C, Tran PD, Boix PP, Bassi PS, Yantara N, Wong LH, Barber J (2014) Nanoscale 6:6506
93. Sharma D, Verma A, Satsangi VR, Shrivastav R, Dass S (2014) Int J Hydrogen Energy 39: 4189
94. Hou J, Yang C, Cheng H, Jiao S, Takeda O, Zhu H (2014) Energy Environ Sci 7:3758
95. Grinberg I, West DV, Torres M, Gou G, Stein DM, Wu L, Chen G, Gallo EM, Akbashev AR, Davies PK, Spanier JE, Rappe AM (2013) Nature 503:509
96. Saito K, Kudo A (2013) Inorg Chem 52:5621
97. Wang Q, Hisatomi T, Ma SSK, Li Y, Domen K (2014) Chem Mater 26:4144
98. Luo Y, Liu X, Tang X, Luo Y, Zeng Q, Deng X (2014) J Mater Chem A 2:14927
99. Marschall R, Wang L (2014) 225:111
100. Ma W, Han D, Zhou M, Sun H, Wang L, Dong X, Niu L (2014) Chem Sci 5:3946
101. Chen Y, Huang W, He D, Situ Y, Huang H (2014) ACS Appl Mater Interfaces 6:14405
102. Zang Y, Li L, Xu Y, Zuo Y, Li G (2014) J Mater Chem A 2:15774
103. Kumar S, Kumar B, Surendar T, Shanker V (2014) Mater Res Bull 49:310
104. Zhou S, Liu Y, Li J, Wang Y, Jiang G, Zhao Z, Wang D, Duan A, Liu J, Wei Y (2014) Appl Catal Environ 158–159:20
105. Zhang J, Wang Y, Jin J, Zhang J, Lin Z, Huang F, Yu J (2013) ACS Appl Mater Interfaces 5:10317
106. Lu M, Pei Z, Weng S, Feng W, Fang Z, Zheng Z, Huang M, Liu P (2014) Phys Chem Chem Phys 16:21280
107. Dai X, Xie M, Meng S, Fu X, Chen S (2014) Appl Catal Environ 158–159:382
108. Tian N, Huang H, He Y, Guo Y, Zhang Y (2014) RSC Adv 4:42716
109. Theerthagiri J, Senthil RA, Priya A, Madhavan J, Michael RJV, Ashokkumar M (2014) RSC Adv 4:38222
110. Sun L, Qi Y, Jia C-J, Jin Z, Fan W (2014) Nanoscale 6:2649
111. Lublow M, Fischer A, Merschjann C, Yang F, Schedel-Niedrig T, Veyan J-F, Chabal YJ (2014) J Mater Chem A 2:12697
112. Li H, Liu J, Hou W, Du N, Zhang R, Tao X (2014) Appl Catal Environ 160–161:89
113. Hou Y, Zuo F, Dagg AP, Liu J, Feng P (2014) Adv Mater 26:5043
114. Chen J, Shen S, Guo P, Wu P, Guo L (2014) J Mater Chem A 2:4605
115. Chen J, Shen S, Guo P, Wang M, Wu P, Wang X, Guo L (2014) Appl Catal B 152–153:335
116. Chang C, Zhu L, Wang S, Chu X, Yue L (2014) ACS Appl Mater Interfaces 6:5083
117. Cao S-W, Liu X-F, Yuan Y-P, Zhang Z-Y, Liao Y-S, Fang J, Loo SCJ, Sum TC, Xue C (2014) Appl Catal B 147:940
118. Zhang Z, Huang J, Zhang M, Yuan Q, Dong B (2015) Appl Catal B 163:298
119. Xing Z, Chen Z, Zong X, Wang L (2014) Chem Commun 50:6762
120. Hou Y, Wen Z, Cui S, Guo X, Chen J (2013) Adv Mater 25:6291
121. Seger B, McCray J, Mukherji A, Zong X, Xing Z, Wang L (2013) Angew Chem Int Ed 52: 6400
122. Guo N, Liang Y, Lan S, Liu L, Zhang J, Ji G, Gan S (2014) J Phys Chem C 118:18343
123. O'Regan B, Grätzel M (1991) Nature 353:737
124. Swierk JR, McCool NS, Saunders TP, Barber GD, Strayer ME, Vargas-Barbosa NM, Mallouk TE (2014) J Phys Chem C 118:17046
125. Swierk JR, Mallouk TE (2013) Chem Soc Rev 42:2357
126. Li L, Duan L, Wen F, Li C, Wang M, Hagfeldt A, Sun L (2012) Chem Commun 48:988
127. Klepser BM, Bartlett BM (2014) J Am Chem Soc 136:1694
128. Kurimoto K, Yamazaki T, Suzuri Y, Nabetani Y, Onuki S, Takagi S, Shimada T, Tachibana H, Inoue H (2014) Photochem Photobiol Sci 13:154
129. Maeda K, Kuriki R, Zhang M, Wang X, Ishitani O (2014) J Mater Chem A 2:15146

130. Karnahl M, Mejía E, Rockstroh N, Tschierlei S, Luo S-P, Grabow K, Kruth A, Brüser V, Junge H, Lochbrunner S, Beller M (2014) ChemCatChem 6:82
131. Watanabe M, Hagiwara H, Iribe A, Ogata Y, Shiomi K, Staykov A, Ida S, Tanaka K, Ishihara T (2014) J Mater Chem A 2:12952
132. Wang W, Chen J, Li C, Tian W (2014) Nat Commun 5:4647
133. Haffer S, Lüder C, Walther T, Köferstein R, Ebbinghaus SG, Tiemann M (2014) Microporous Mesoporous Mater 196:300
134. Grewe T, Deng X, Weidenthaler C, Schüth F, Tüysüz H (2013) Chem Mater 25:4926

Top Curr Chem (2016) 371: 173–214
DOI: 10.1007/128_2015_649

Published online: 6 August 2015

Catalytic Oxygen Evolution by Cobalt Oxido Thin Films

D. Kwabena Bediako, Andrew M. Ullman, and Daniel G. Nocera

Abstract The contemporary demand to generate fuels from solar energy has stimulated intense effort to develop water splitting catalysts that can be coupled to light-absorbing materials. Cobalt oxido catalyst (Co-OECs) films deposited from buffered Co^{II} solutions have emerged as arguably the most studied class of heterogeneous oxygen evolution catalysts. The interest in these materials stems from their formation by self-assembly, their self-healing properties, and their promising catalytic activity under a variety of conditions. The structure and function of these catalysts are reviewed here together with studies of molecular Co-O cluster compounds, which have proven invaluable in elucidating the chemistry of the Co-OECs.

Keywords Artificial leaf · Cluster compounds · Cobalt oxide · Electrocatalysis · Oxygen evolution reaction · Proton-coupled electron transfer · Self-assembly · Self-healing · Solar fuels · Water splitting

Contents

D.K. Bediako, A.M. Ullman, and D.G. Nocera (✉)
Department of Chemistry and Chemical Biology, Harvard University, 12 Oxford Street, Cambridge, MA, USA
e-mail: dnocera@fas.harvard.edu

1 Introduction

The way we power the planet is changing. Distributed energy generation and storage is now within scientific and technological reach and accordingly has the promise of supplanting traditional energy infrastructure. Centralized power plants are giving way to utility-scale wind and solar farms, and, in combination with smaller scale commercial and residential forms of renewable generation, a new energy ecosystem is evolving before our eyes. In Germany, the country's largest energy supplier, E.ON, announced in late 2014 that it plans to spin off the portion of the company centered around conventional fossil fuels to focus on the increasingly important renewable energy sector (http://www.eon.com/en/media/news/press-releases/2014/11/30/new-corporate-strategy-eon-to-focus-on-renewables-distribution-networks-and-customer-solutions-and-to-spin-off-the-majority-of-a-new-publicly-listed-company-specializing-in-power-generation-global-energy-trading-and-exploration-and-production.html). The impetus for this historic restructuring originates with the rise of marketable renewable technology innovations and the increased demand for individualized consumer products. There is an increasing desire for individuals to control their energy, and hence this has turned them to the Sun as their direct energy supply.

Standing at the crossroads of these drivers of change is energy storage. The intermittency of renewable energy (solar and wind) sources necessitates suitable storage. Flow and solid-state battery technologies maintain a dominant position in the energy storage marketplace [1]. However, marketable improvements in these technologies are sporadic, sprouting mainly from cost-saving measures in engineering and manufacturing. Moreover, truly outstanding improvements in these technologies are limited by the finite energy density of materials used at the battery anode and cathode [2].

Fuels possess the energy density needed for large scale global energy storage [2]. Of the various solar fuel cycles, water splitting is perhaps the most attractive chemical reaction for storing intermittent renewable energy because the source of fuel, water, is ubiquitous, and significant energy density may be achieved in the O–O and H–H bonds of oxygen and hydrogen, which greatly increases the maximum energy density [3]. The thermodynamic potential for this reaction is given in (3), and formal potentials for the constituent half-reactions are given in (1) and (2).

$$2H_2O \rightarrow O_2 + 4H^+ + 4e^-;\ E^o_{OER} = 1.23V - 0.059\ V\ \times\ pH, \quad (1)$$

$$4H^+ + 4e^- \rightarrow 2H_2;\ E^o_{HER} = 0V - 0.059\ V\ \times\ pH, \quad (2)$$

$$2H_2O \rightarrow O_2 + 2H_2;\ E^o = E^o_{HER} - E^o_{OER} = -1.23\ V. \quad (3)$$

This E° voltage defines the minimum amount of energy required to carry out the reaction, and the maximum amount of energy that can be stored and later recovered from the bonds of H_2 and O_2. Driving this reaction inevitably requires an additional voltage in excess of E°, which is termed the overpotential. Minimizing this overpotential is a critical parameter in maximizing the efficiency of a solar fuels device. Of the two half reactions (1) and (2), the four electron, four proton oxidation of water to dioxygen (i.e., the oxygen evolution reaction, OER) is the more kinetically challenging reaction, and therefore executing this reaction at any given rate typically demands a more substantial input of overpotential than the hydrogen evolution reaction, HER (2).

Existing methods used to split water commercially (mainly for the production of high purity H_2) involve the use of proton-exchange membrane (PEM) electrolyzers that operate in acidic electrolytes, alkaline electrolyzers that operate in strongly basic solution, and solid-oxide electrolyzers that operate at high temperatures of ~1,000°C [4]. In all cases, water splitting is performed under very harsh conditions, and as a result the overall balance of system cost is too high to allow these technologies to be economically viable renewable solar fuel generators [5, 6]. PEM electrolyzers incur additional costs associated with the precious metals that are required as electrodes. Moreover, extremely harsh conditions are invariably at odds with: materials stability if the catalyst is to be interfaced directly to light harvesting systems [7, 8], other processes such as CO_2 reduction (which is best performed at pH 7), the construction of inorganic/biological hybrid systems [9], and the use of distributed energy storage schemes where water is at a premium by allowing the use of "dirty" water [10, 11]. To this end, there are advantages in performing water splitting under more benign conditions, such as at intermediate (close to neutral) pH.

The focus of this review is the oxidic thin film catalysts of cobalt that we have found to be active OER catalysts at neutral and near-neutral pHs. This Co-based catalyst has been a springboard for the development of other oxidic catalysts based on Mn [12, 13] and Ni [14–16], as well as mixed-metal oxidic catalysts [17–34]. We discuss the preparation of Co-OEC and its structural interrogation using X-ray absorption spectroscopy and X-ray total scattering methods together with pair distribution function analysis. The kinetics of catalytic turnover and film growth, which form the basis for the inherent self-repair property of these catalysts, are also reviewed. We also discuss the advances in our understanding of this catalyst that have arisen from the investigation of molecular complexes.

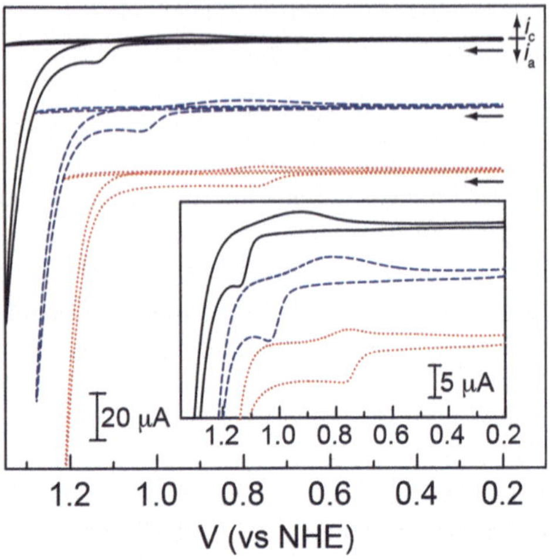

Fig. 1 Cyclic voltammograms using a glassy carbon working electrode, 50 mV/s scan rate, of aqueous 0.5 mM Co^{2+} in 0.1 M Pi electrolyte, pH 7.0 (*solid line*, *black*), 0.1 M MeP_i electrolyte, pH 8.0 (*dashed line*, *blue*), and 0.1 M Bi electrolyte, pH 9.2 (*dotted line*, *red*). Background traces in each electrolyte medium in the absence of Co^{2+} are overlaid. *Inset* shows CVs in the presence of Co^{2+} on an expanded current scale. Figure adapted from [37]

2 Self-Assembly of Co-OEC Thin-Films

A film that is catalytically active for water oxidation is formed on the surface of inert conducting electrodes upon anodic polarization at >1.0 V (vs NHE) in pH 7 phosphate (P_i)-buffered solutions containing sub-millimolar concentrations of simple Co^{2+} salts [35, 36]. Cyclic voltammograms (CVs) in buffered Co^{2+}-containing electrolytes at pH 7, 8 and 9.2 (Fig. 1) display sharp anodic waves that are well separated from a subsequent catalytic wave [35, 37], which is a result of oxygen evolution. A Faradaic efficiency of 100% indicates that OER occurs in the absence of side reactions. The initial anodic pre-feature (Fig. 1) reflects the primary process for catalyst formation; after catalyst deposition, subsequent CVs in Co-containing or Co-free electrolytes display broader waves that are representative of the redox changes within the electrodeposited film [35].

Film formation and O_2 evolution catalysis occur over a wide range of electrolyte types and pH conditions (Fig. 1). The presence of the electrolyte was observed to be crucial to film formation at low potentials as well as its long-term stability (Fig. 2). These materials stand in contrast to cobalt-based films that are formed in non-buffering electrolytes such as sulfate or nitrate, which require high metal salt concentrations and large voltages to drive deposition of films that are ultimately unstable in neutral pH electrolytes [38, 39]. Although we collectively term these oxygen-evolving catalyst films Co-OEC, they may be further distinguished as CoP_i, CoB_i, and $CoMeP_i$ depending on the electrolyte used during electrodeposition and catalysis.

Co-OEC films were shown to be permeable and conductive, because O_2 evolution activity scales with film thickness (Fig. 3) [40]. This result indicated early on that active sites are present and accessible throughout the entire thickness of the film, and not just at the outermost layer of the material. This property was exploited to prepare highly active anodes that achieved geometric current densities of 100 mA/cm^2 at an overpotential of 363 mV in pH 9.2 B_i electrolyte by extending all three dimensions of the catalyst film. Unlike commercial electrolyzer systems

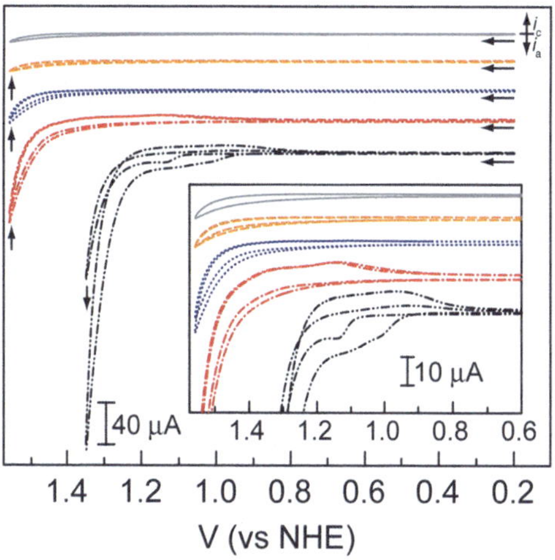

Fig. 2 Cyclic voltammogram using a glassy carbon working electrode, 50 mV/s scan rate, of 0.1 M K_2SO_4 electrolyte, pH 7.0, containing from top to bottom 0 (*solid line*), 0.5 (*dashed line*), 5 (*dotted line*), and 50 (*dot-dashed line*) mM Co^{2+}. CV traces of glassy carbon working electrode, 50 mV/s scan rate, of 0.5 mM Co^{2+} in P_i electrolyte, pH 7.0 (*dash-dotted line*) are shown for comparison. *Vertical arrows* indicate progression between the first and fifth scans taken without pause. *Inset* shows all CVs on an expanded current and potential scale. Figure adapted from [37]

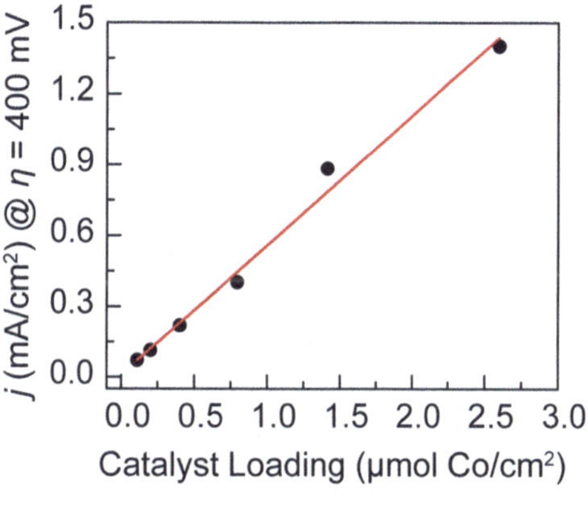

Fig. 3 CoB_i film activity (current density at 400 mV overpotential) vs catalyst loading for films deposited at 0.72 V vs Ag/AgCl in 0.1 M KB_i solution (pH 9.2). Figure adapted from [40]

that require water of high purity for sustained optimal operation, Co-OEC activities were found to be resistant to degradation in impure waters [40], making them potential targets for future cheap, distributed energy storage systems in impoverished regions. This resistance to degradation, along with the ease of preparation and good activity characteristics, prompted us to investigate the structural and mechanistic basis for Co-OEC formation and catalysis.

3 Co-OEC Structure and Valency

Insights into catalyst structure and valency are critical parameters for uncovering the mechanistic basis for catalyst activity. In particular, the construction of structure–activity relationships is a vital part of all catalyst development (homogeneous and heterogeneous), potentially revealing key insights that lead to the design of catalysts with enhanced activity. However, obtaining structural information on Co-OEC films has been impeded by their ostensibly amorphous nature within the resolution of conventional crystallographic methods [35] signifying minimal long-range structural order [41]. In this regard, they stand in contrast to the crystalline brucite, spinel, and perovskite Co compounds which have been the subject of the vast majority of heterogeneous OER mechanistic studies [42]. Although presenting challenges to structural characterization, the absence of long-range structural coherence in these films also means that they offer a unique platform for the study of water oxidation at hitherto underexplored structurally disordered heterogeneous OECs. In this section, we discuss X-ray absorption spectroscopy (XAS) and X-ray pair distribution function (PDF) studies that have shed light on complementary aspects of the local coordination environment in these catalyst systems. Together with X-ray absorption near edge structure (XANES) studies, electron paramagnetic resonance (EPR) spectroscopy methods have also been particularly useful for developing a clear picture of the valency of these catalyst films, and particularly of the resting state, which is the predominant form of the catalyst during turnover.

3.1 Structural Studies

Cobalt *K*-edge extended X-ray absorption fine structure (EXAFS) studies have been performed on CoP_i films to elucidate the structural connectivity of the catalyst [43, 44]. Figure 4 compares the Fourier transform (FT) of EXAFS spectra of two CoP_i samples collected in situ at room temperature during catalysis with that of a crystalline CoO(OH) model compound [44]. The abscissa is the apparent distance, which is shorter than the actual distance, between a given absorber–backscatterer pair of atoms produced by a phase shift. Peaks I and II at actual distances of about 1.9 and 2.8 Å reflect Co–O and Co–Co vectors, respectively. Notably, the relative intensity of the Co–Co vector is considerably greater in CoO(OH) than in either CoP_i sample, indicating a larger average number of Co–Co vectors per Co ion in CoO(OH) than in CoP_i. In addition, the EXAFS FT for CoO(OH) displays prominent peaks at distances greater than 3 Å (peaks III and IV), consistent with contributions from scattering interactions in outer shells, as expected for a crystalline compound. However, the analogous features are considerably weaker in CoP_i films, indicating a much smaller coherent domain size.

The EXAFS data shown in Fig. 4 also suggest that the deposition conditions and/or film thickness can impact the domain size (or distribution of domain sizes) in

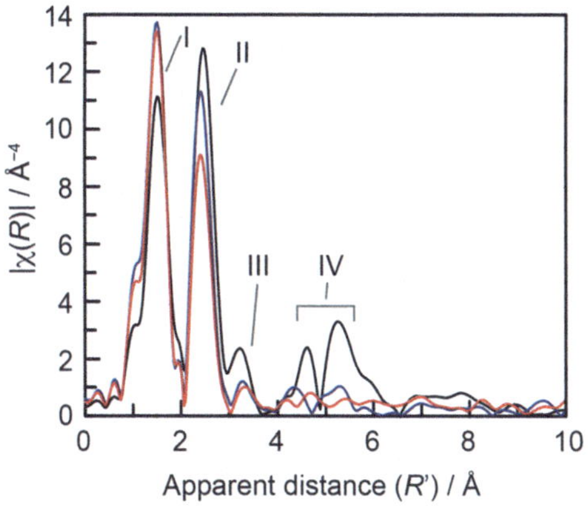

Fig. 4 Fourier transforms of EXAFS spectra for CoP_i and CoO(OH). Fourier transforms of *k*-space oscillations for "thick" CoP_i film deposited at 1.25 V (*blue*), "thin" CoP_i deposited at 1.1 V (*red*), and crystalline CoO(OH) (*black*). Figure adapted from [44]

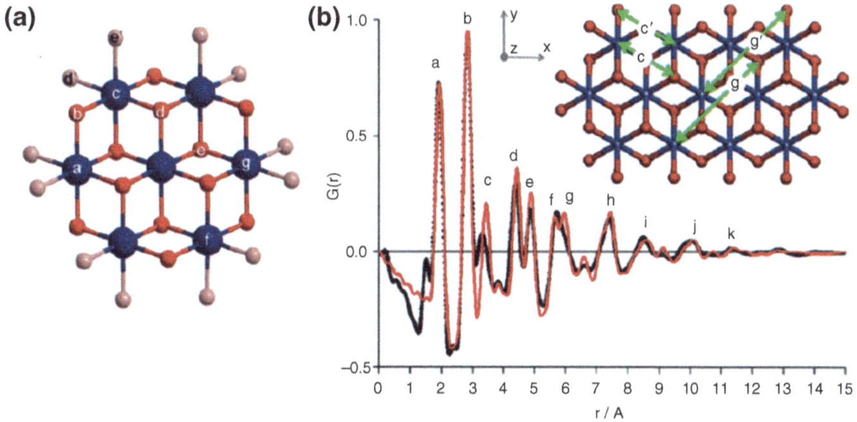

Fig. 5 (**a**) Edge-sharing molecular cobaltate cluster (MCC) obtained for "thin" CoP_i films deposited at 1.1 V, as determined from EXAFS measurements. Bridging oxo/hydroxo ligands are shown in *red*, and nonbridging oxygen ligands (including water, hydroxide, and phosphate, *light red*) complete the octahedral coordination geometry of each peripheral Co ion (*blue*). Figure adapted from [44]. (**b**) Comparison between the experimental electron pair distribution function, $G(r)$ measured for CoP_i (*black trace*) and the $G(r)$ calculated from the structural model (*inset*). Figure adapted from [48]

catalyst films. Fitting to the EXAFS spectra establishes that ordered domains in CoP_i consist of edge-sharing molecular cobaltate clusters (MCCs) with the minimal structural unit shown in Fig. 5a [44]. Differences in cluster sizes (or distributions of sizes) between the two CoP_i materials may arise as a consequence of the difference in the amount of material deposited. The absence of detectable crystallites in diffraction experiments indicated that these molecular clusters are not infinitely stacked in ordered arrays, but are to a large extent disordered relative to one another.

Attempts to incorporate cobalt–potassium vectors into the fits for this in situ cobalt *K*-edge EXAFS data were unsuccessful [44], leading to substantially increased error values or negligible effects on the fitting quality. A CoP_i catalyst was prepared with sodium phosphate used in place of potassium phosphate as the electrolyte during electrodeposition. The replacement of K^+ with Na^+ slightly increases the intensity of peaks in the FT EXAFS at apparent distances >3 Å, which is the opposite trend from what would be expected if one of these peaks had a Co–K component, because potassium is a stronger scatterer than sodium; any peaks that result from a Co–K vector in the FT of the CoP_i EXAFS data would certainly be reduced upon substitution with sodium. On the basis of these results we concluded that alkali cations are most likely coordinated to the peripheral ligands of the MCC or to interstitial lattice water molecules. Risch et al. [45] subsequently reported ex situ XAS data on frozen Co-OEC samples deposited in KP_i and $CaCl_2$ at the respective potassium and calcium *K*-edges to interrogate specifically any structural role of redox-inert cations in these catalyst films. They found that potassium binding appears largely nonspecific and disordered, in agreement with our previous in situ report. However, on the basis of Ca *K*-edge EXAFS data of films deposited in Ca^{2+}-containing electrolytes, they proposed the formation of $CaCo_3O_4$ cubanes, which are analogous to the $CaMn_3O_4$ cubane moiety found in the oxygen-evolving center of photosystem II. Nevertheless, they found that substitution of these redox-inert cations did not lead to any major differences in catalytic activity, suggesting that these ions are not directly involved in the catalytic mechanism in Co-OEC.

High energy X-ray total scattering and pair distribution function (PDF) analysis is a powerful tool for structural studies of disordered and nanocrystalline systems, because it permits consideration of a wider range of atom–atom distances that are not accessible by EXAFS [46, 47]. Tiede and co-workers employed PDF analysis to shed more light on the coherent domain size and structure of CoP_i [48]. Their results indicate that atom pair correlations persist out to about 13 Å, and PDF patterns were fit to a model consisting of a 13–14 Co ion lattice domain consisting of edge-sharing CoO_6 octahedra (Fig. 5b), in good correspondence with the in situ EXAFS results discussed above. Phosphate was found as a disordered component in the films, and minority "defect" sites consisting of complete Co_4O_4 cubanes were also proposed. In addition, the PDF data suggested that the clusters might possess distorted coordination geometries at their periphery, such as alterations in the terminal O ion positions. It was proposed that these terminal Co–O distortions could play a key role in O_2 evolution catalysis at CoP_i by facilitating the formation of terminal peroxo intermediates.

We employed an X-ray PDF analysis set to explore the influence of deposition conditions on the structure of Co-OEC films. These studies established that the nature of the buffering electrolyte (P_i and B_i) affects the intermediate structure of the catalyst [49]. The refined PDF fits and models for CoB_i and CoP_i catalysts are shown in Fig. 6a, b. Films of the CoB_i comprise three layers on average of the

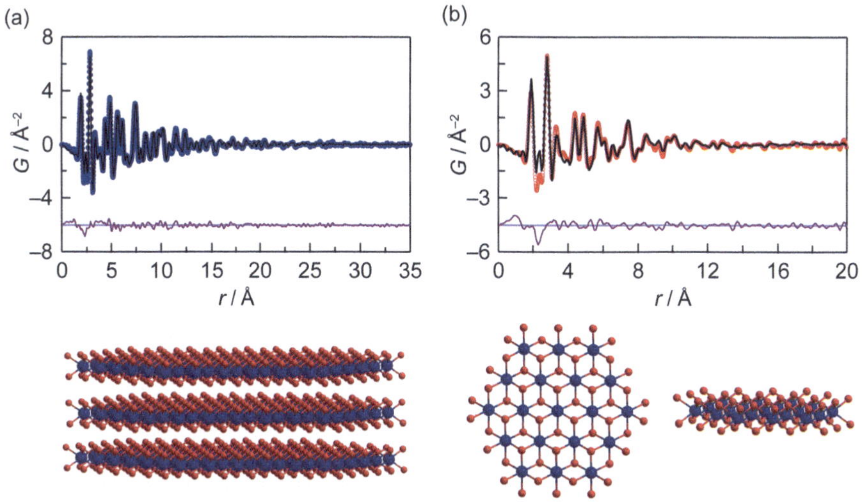

Fig. 6 *Top*: Cylindrical atomistic model fit (*black lines*) to the (**a**) CoB_i catalyst PDF data (*blue circles*) and (**b**) CoP_i catalyst PDF data (*red circles*). The difference curves are shown in *purple*, and are offset for clarity. *Bottom*: (**a**) View of the refined model for the average coherent domain in CoB_i films, and (**b**) two views of the refined model for the average coherent domain in CoP_i films. Figure adapted from [49]

MCC, and the diameter of each MCC layer is approximately 35 Å. Constraining the model to a single layer results in a poorer fit, suggesting that the coherent domains of CoB_i do indeed consist of multiple layers of MCCs with weak but significant interlayer correlations. The PDF data also reveal significant interlayer expansion relative to crystalline CoOOH, possibly to accommodate something larger than protons in the interlayer region such as weakly scattering or disordered electrolyte ions, though B_i was not identified explicitly. The refined CoP_i model and fit (Fig. 6b) indicates that the diameter of the clusters is approximately 14 Å, considerably smaller than those of CoB_i and on a par with the EXAFS data as well as the PDF studies of Tiede et al. Unlike CoB_i, modeling PDF data could not accommodate significant interlayer correlations in CoP_i, suggesting that the coherent domains of the catalyst consist of single layer MCCs whose arrangements are significantly disordered in the film.

3.2 *Catalyst Resting State Valency*

Electrocatalytic intermediates are typically formed in minor equilibrium. Though crucial for catalysis, such intermediates tend to contribute negligibly to spectroscopic data of the catalyst under steady state conditions. Insight into such intermediates is provided from the catalyst resting state, because it sheds light on the character of the catalyst prior to subsequent redox transformations. Cobalt *K*-edge XAS was used to probe the resting state oxidation state of Co in catalyst films by

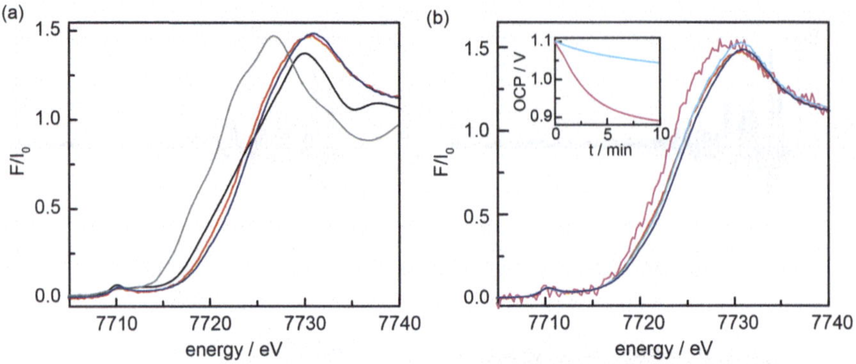

Fig. 7 XANES spectra for CoP_i samples and model compounds. (**a**) Thin 1.1 V deposited-CoP_i during operation at 1.25 V (*red*), thick 1.25 V deposited-CoP_i during operation at 1.25 V (*blue*), CoOOH (*black*), and CoO (*gray*). (**b**) Thin 1.1 V deposited-CoP_i at OCP (*magenta*) and 1.25 V (*red*), thick 1.25 V deposited-CoP_i at OCP (*cyan*) and 1.25 V (*blue*). *Inset*: open circuit potential vs time traces. Figure adapted from [44]

comparison of the XANES spectra to those of model compounds [44]. XANES spectra of CoP_i samples collected in situ during catalysis at 1.25 V are shown in Fig. 7a along with the spectra of solid samples of CoO and CoOOH. The observed edge positions are consistent with an average Co valency ≥3 for the CoP_i samples at 1.25 V, indicative of some formal Co^{IV} valency in the resting state. However, it is not straightforward to determine precise metal valency based on XANES spectra alone because of their sensitivity to geometry and the specific ligand environment [50]. However, this qualitative assignment is supported by several pieces of independent data, including CV data and the observed bond lengths in the cobalt-oxido clusters compared to bond lengths in crystalline cobaltate compounds.

As described in Sect. 2, CVs of CoP_i exhibit an anodic wave as a pre-feature to a catalytic wave. The half-wave potential of the pre-feature is observed at ~+0.95 V. The $Co(H_2O)_6^{3+/2+}$ couple is about 1.9 V [51] and the pK_a values of $Co(H_2O)_6^{3+}$ and $Co(H_2O)_6^{2+}$ are 2.9 [52] and 9.2 [53], respectively. Therefore, it is reasonable to consider that at pH 7 the proton-coupled oxidation of $Co^{2+}_{(aq)}$ could lead to a Co^{3+} film if more than one proton is transferred in the process (see Sect. 5) [53]. In addition, the $Co(OH)_2^{+/0}$ couple is estimated to be +1.1 V and the Pourbaix diagram of cobalt indicates that the thermodynamic potential for the conversion of Co_3O_4 to CoOOH occurs at ~0.75 V at pH 7 [54]. The $Co^{3+/2+}$ couple for CoP_i at 0.95 V occurs ~0.14 V negative of the catalytic wave for water oxidation, consistent with a formal oxidation state of Co^{3+}, prior to any catalytic behavior. The catalytic wave is therefore obtained only upon further oxidation of the Co^{3+} film.

The first shell Co–O distances and the nearest neighbor Co–Co distances deduced from EXAFS data [44] support the assignment of a Co valency greater than 3 by comparison to corresponding distances in alkali cobaltates of various valencies [55]. In the stoichiometric Co^{3+} oxides, CoOOH and $LiCoO_2$, the Co–Co distances are 2.85 and 2.81 Å, and the Co–O distances are 1.90 and 1.92 Å, respectively. For the series Na_xCoO_2, the corresponding distances range from 2.89 and 1.94 Å for $x = 1$ (all Co^{3+}), to 2.81 and 1.81 Å for $x = 0.3$ (i.e., formally 70% Co^{IV}). The Co–Co distance (2.82 Å) observed for the CoP_i samples under an applied potential is at the short end of the range of distances observed in cobaltates with valencies greater than 3 and the Co–O distance (1.89 Å) is in the middle of the corresponding range, suggesting a significant presence of Co^{IV} during turnover.

XANES results (Fig. 7a) indicate that there is a greater average Co valency in the CoP_i film that was deposited at 1.25 V (a potential at which water oxidation is coincident with catalyst deposition) compared to that prepared at 1.1 V (where water oxidation is negligible) [44]. Furthermore, in situ XANES spectra acquired for CoP_i films at open circuit (OC) conditions together with OC decay transients (Fig. 7b) indicate that film deposition at high potentials permits the trapping of some of the Co^{IV} centers within the film, thereby delaying their eventual discharge (reduction upon the OER) when the cell is held at OC. This trapping effect was exploited to confirm more precisely the existence of Co^{IV} centers in the resting state of CoP_i films during turnover using EPR spectroscopy.

McAlpin et al. [56] and Gerken et al. [57] conducted EPR studies of CoP_i samples freeze-quenched after deposition at a variety of potentials. Continuous wave (CW) X-band EPR spectra of frozen catalyst samples display a broad resonance at $g_{eff} \approx 5$ along with a broad derivative lineshape at $g_{eff} = 2.27$ (Fig. 8a). The $g_{eff} \approx 5$ feature was assigned to an $S = 3/2$ Co^{II} species based on EPR spectra of model Co^{II} compounds. The progressive rise of the $g_{eff} = 2.27$ feature as the deposition potential is increased along with its similarity to the g value of the signal observed in a tetracobalt cubane model compound ($[Co_4O_4(C_5H_5N)_4(CH_3CO_2)_4](ClO_4)$) possessing Co centers with formal cobalt oxidation states of III, III, III, IV (Fig. 8b), provides strong evidence for the assignment of the $g_{eff} = 2.27$ signal to low-spin Co^{IV}-containing species in the catalyst film. Moreover, the potential dependence of this signal suggests that it arises predominantly from species generated during electrocatalytic water oxidation, corroborating the finding of the in situ XANES study.

As observed in the XAS studies above (Fig. 7b), removal of the applied potential causes the open circuit potential to decay, implying a change in the average valency of the film. In correspondence with the XAS and electrochemical results, EPR spectra of films prepared identically, but with variable delay times between catalyst isolation from the film and freezing at 77 K, revealed that the Co^{IV} signal decays over the course of minutes at room temperature even in a dry, isolated film (Fig. 8c) [56].

These EPR studies permitted an estimation of the population of Co^{IV} spins in the catalyst films. By comparison to an $S = ½$ spin standard, $Cu(EDTA)(SO_4)$, it was estimated that Co^{IV} centers represented 3% and 7% of the total cobalt centers in trapped films grown at 1.14 and 1.34 V, respectively [56]. It was also shown that the Co^{IV} valency was not merely trapped in films deposited at elevated potentials, but was important to the electronic structure of the resting state of CoP_i during turnover. By preparing two catalyst films in an identical manner at 1.04 V, below the onset of

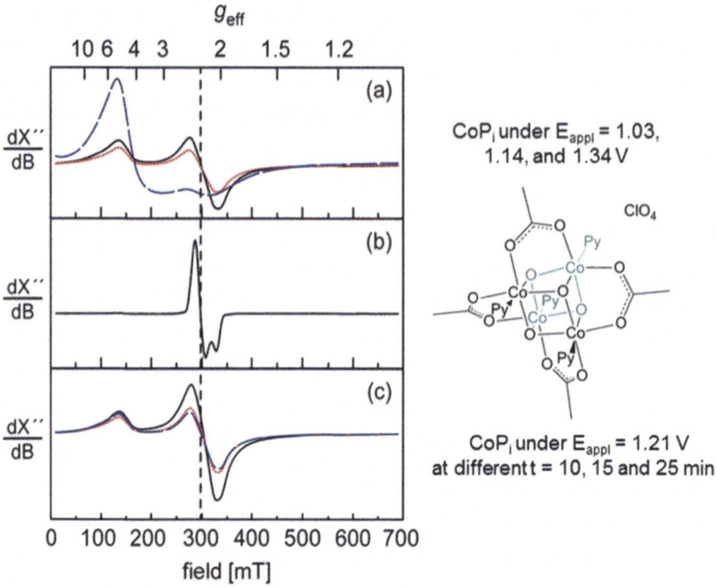

Fig. 8 CW X-band EPR spectra at $T = 5.7$ K and microwave power = 1.02 mW. (**a**) CoP$_i$ catalyst films deposited from aqueous 0.5 mM $Co(NO_3)_2$ solutions in 0.1 M P$_i$ electrolyte (pH 7.0) at 1.03 V (*blue dashed curve*), 1.14 V (*red dotted curve*), and 1.34 V (*black solid curve*). Spectra were scaled by the amount of cobalt in each sample, as measured by atomic absorption spectroscopy. (**b**) The $Co^{III}{}_3Co^{IV}$ cluster $[Co_4O_4(C_5H_5N)_4(CH_3CO_2)_4]$-$(ClO_4)$. The *vertical dotted line* indicates $g = 2.27$. (**c**) CoP$_i$ catalyst films electrodeposited at 1.21 V with 10 (*solid line*), 15 (*red-dotted line*) and 25 (*blue-dashed line*) min of delay between the termination of electrolysis and freezing of the sample at 77 K. Spectra were scaled relative to the mass of CoP$_i$ in each sample, as measured by atomic absorption spectroscopy. Figures adapted from [56]

water oxidation catalysis, it was ensured that the as-deposited film exists almost completely in its Co^{III} form (Sect. 5). Catalyst material was isolated from the one electrode immediately following deposition, whereas the other electrode was transferred to Co-free P$_i$ electrolyte solution and electrolyzed at a potential sufficient for water oxidation catalysis (>1.2 V). The catalyst film subjected to water oxidation in the absence of additional film formation in this fashion exhibited a decrease in the Co^{II} signal concomitant with an increase in the Co^{IV} signal, just as observed in Fig. 8a.

Additional multifrequency EPR studies and electronic structure calculations were performed on the Co^{IV}-containing cubane compound to investigate further the electronic and geometric structure relationships of MCCs such as those found in CoP$_i$ [58]. The magnetic parameters determined by electron nuclear double resonance (ENDOR) and spectroscopic results from electron spin echo envelope modulation (ESEEM) combined with density functional theory (DFT) calculations revealed that the unpaired spin of the cubane is highly delocalized. Each octahedrally coordinated cobalt ion was found to possess a low-spin electron configuration resulting from the anionic oxo and carboxylato ligands, and a fractional electron

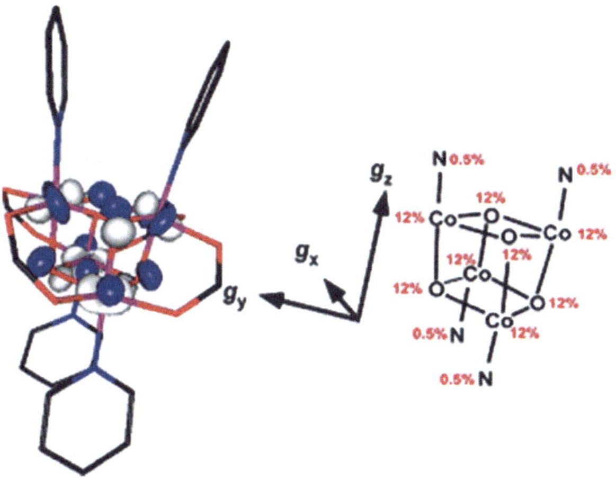

Fig. 9 Singly occupied molecular orbitals of $[Co_4O_4(C_5H_5N)_4(CH_3CO_2)_4](ClO_4)$ from DFT calculations. Calculations show the unpaired electron spin density spread almost equally over the eight core atoms of the cubane based on their Löwdin spin populations. Reproduced from [58]

hole resides on each metal center in a Co $3d_{xz,yz}$-based molecular orbital, essentially giving a $[Co^{+3.125}{}_4O_4]$ cluster (Fig. 9). However, closer inspection of the EPR spectra of the molecular model and those of CoP_i revealed features consistent with less delocalization of the unpaired electron spin in the case of CoP_i compared to the model. It was proposed that because the $Co^{III/IV}$ redox transformations in the film (Sect. 2) and at active sites during turnover (Sect. 4) are coupled to proton loss from bound OH_x ligands, the increase in ligand-field strength accompanying this proton loss could result in substantial localization of unpaired spin density [58]. This localization of the oxidized equivalent is consistent with subsequent reactivity of the Co^{IV} moieties in CoP_i compared to the molecular model. These findings implicate the importance of PCET, not only for redox/potential leveling but also for hole/radical localization, leading to efficient oxygen evolution in Co-OEC films.

The aforementioned EXAFS and X-ray PDF studies combined with the geometric structure of Co-OEC films support a structural model wherein Co-OEC is composed of cobaltate clusters of molecular dimensions. In addition, XANES and EPR studies are consistent with a proportion of Co centers achieving a formal oxidation state of +4 at potentials sufficient for water oxidation catalysis, indicating a $Co^{III/IV}$ mixed valence resting state during oxygen evolution. Together with the mechanistic studies discussed in the following section, a cohesive picture of water oxidation by Co-OEC films is unveiled.

4 Mechanistic Insights into O_2 Evolution Catalysis at Co-OEC Thin Films

The preceding sections address the electrosynthesis of Co-OECs and the structural and electronic characterization of its resting state. Against this backdrop, we discuss electrokinetic studies that provide insight into the mechanism of oxygen evolution mediated by this thin-film catalyst over a wide pH range. Because of the extreme kinetic demands of the OER, the efficiency and conditions required for this reaction are key determinants of the overall viability of energy storage via water splitting. Therefore, the continued development of effective OER catalysts and elucidation of their mechanisms stand as central scientific and technological challenges in energy conversion. For porous catalyst films of this type, the relationship between film thickness (catalyst loading) and film activity is of central importance to the search for solar water splitting at high efficiency. This is because the overpotential is an extrinsic parameter that depends on the current density desired. Hence, in addition to making more active catalyst materials, low overpotentials may be realized by preparing thicker films where a higher active site density (per projected geometric area) may be accessed. This allows faster rates to be achieved with thicker films (Fig. 3), but demands that charge hopping between active sites is well matched to catalytic rates. Thus the mechanisms of catalytic turnover and charge hopping are equally important design elements. These topics are addressed in the following sections.

4.1 Active Site Electrokinetics

Steady-state electrokinetic data were acquired over a range of electrolyte conditions between pH 4 and 12 in order to uncover the electrochemical rate law for O_2 evolution [59]. Tafel plots, which describe the variation of the steady state current density with the potential or overpotential supplied at the electrode [60], provide key mechanistic insights for electrochemical reactions that are not mass-transport limited. Under these conditions, the current density measured at steady state is directly proportional to the rate of the underlying reaction [60]. It was shown that this condition applies to the electrokinetic studies of CoP_i; Tafel plots collected at 1,000 and 2,000 rpm of CoP_i films on a Pt rotating disk electrode exhibited similar slopes (61 mV/decade) to those collected on a stationary FTO electrode (62 mV/decade) in a stirred solution, indicating that the observed currents are not limited by mass transport to the catalyst films over the sampled current/potential range. In addition, the Tafel data were shown to be reproducible between sequential potential sweeps and insensitive to the direction of potential sweep, indicating that the film is not appreciably altered over the course of Tafel data collection. Furthermore, measurable steady-state current is observed upon application of an overpotential of >0.2 V to CoP_i films. At these overpotentials, any current arising from the

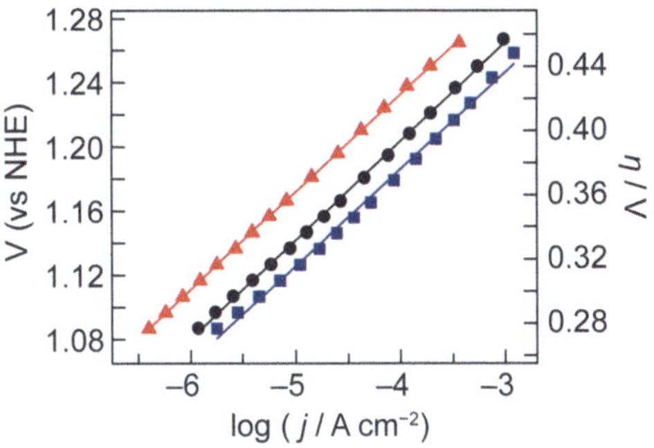

Fig. 10 Tafel plots, $V = (V_{appl} - iR)$, $\eta = (V - E^{\circ})$, for CoP_i catalyst films grown with passage of 6 (*red-filled triangle*), 24 (*filled circle*), and 60 (*blue-filled square*) mC/cm^2. Adapted from [59]

oxygen reduction back reaction can be safely ignored and, thus, kinetic information about the OER can be deduced directly from the steady state current density measurements.

Tafel slopes were found to be similar over the same range of current densities for CoP_i films of varying thickness (Fig. 10), and the activity of the film increases markedly as the film thickness is increased from 10 to 90 nm [59]. All three film thicknesses exhibited similar lower limit turnover frequencies (TOF) of ~2×10^{-3} s^{-1} per Co at an overpotential of 410 mV. This TOF estimate assumed that every Co center in the film is catalytically active. Because the vast majority of the cobalt centers in the film are expected to play a purely structural role, this is likely a gross underestimate of the real TOF of active sites. However, in the absence of a direct measure of the electrocatalytically active surface area or the active site density, this value is offered purely for the purposes of comparing different film thicknesses. The similarity in TOF amongst the films of varying thickness highlights the porous and conductive nature of the catalyst films, because the active sites are accessible throughout the film and not merely at the film/solution interface. Moreover, the independence of Tafel slope with film thickness indicated that the reaction kinetics are not influenced by barriers to electron transport or mass transport within the film – factors that can severely skew the activation-controlled electrokinetic data. These barriers can be particularly problematic for oxide films formed by the anodization of a metal substrate, where a compact barrier oxide layer between the buried metal and the catalytic hydrous oxide can pose significant challenges for charge transport [61–63].

The Tafel slope of nearly 59 mV/decade observed for the thin CoP_i films indicated that a single mechanism is dominant over the potential range investigated [60]. The observed Tafel slope defines a transfer coefficient (usually denoted α) of 1, consistent with a one-electron pre-equilibrium from the resting state followed by a turnover-limiting chemical step. It is important to note here that α (which is a purely experimental parameter equal to the reciprocal Tafel slope in dimensionless form) is distinct from the fundamental symmetry factor (usually denoted β) characterizing the position of the activated complex along the reaction coordinate in a microscopic one-electron transfer reaction. In multielectron catalysis, the transfer coefficient can take any value between 0 and the total number of electrons

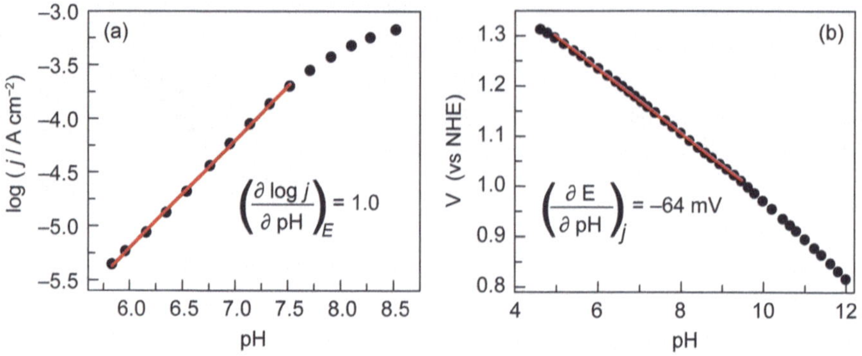

Fig. 11 (**a**) pH dependence of steady-state catalytic current density at constant potential ($E = 1.18$ V) for a catalyst film functioning in 0.1 M P_i electrolyte. Attenuated current densities are observed for pH > 8, because under these conditions P_i ($pK_a = 7.2$) cannot maintain a constant local pH. (**b**) pH dependence of steady-state electrode potential at constant current density ($j_{anodic} = 30\,\mu A/cm^2$) for a catalyst film operated in 0.1 M P_i electrolyte. Maintaining a low current density ensures that the buffering capacity of P_i is not overwhelmed even when the bulk pH is adjusted outside of the ideal P_i buffering range. Adapted from [59]

transferred per turnover, n (for the OER $n = 4$ and therefore $0 \leq \alpha \leq 4$). Therefore, for any multielectron reaction, the lowest Tafel slope attainable is $2.3 \times RT/nF$ [60]. On the other hand, β by definition must lie between 0 and 1, because it represents the ratio between the changes in the electrochemical free energy of activation and the electrochemical free energy of the reaction. For an irreversible one-electron reaction, α (which is always the parameter that can be measured directly) is taken as numerically equal to β [60].

Examination of the pH dependence of the reaction rate by both potentiostatic (Fig. 11a) and galvanostatic (Fig. 11b) techniques is consistent with an inverse first-order dependence on proton activity, indicative of the loss of a single proton in an equilibrium step prior to the turnover-limiting process [59]. Although phosphate is neither reactant nor product in the overall water oxidation reaction, it can serve as a proton acceptor in reactions involving PCET transformations. Notwithstanding, a zeroth-order dependence of phosphate was observed over a >1.5 decade range of phosphate concentration, indicating that proton transfer to phosphate is not involved in the turnover-limiting step and therefore proton transfer to bulk solution is also not turnover-limiting because phosphate is the most likely proton acceptor in P_i electrolyte. Moreover, turnover-limiting proton transfer to surface-bound phosphate is unlikely provided that phosphate has not saturated the catalyst film over the potential range studied. It was, however, noted that turnover-limiting internal proton transfers, as might be required for oxo-hydroxo exchange between catalyst active sites, could not be ruled out.

To a first approximation, these electrokinetic studies define an electrochemical rate law described by the following expression [59]:

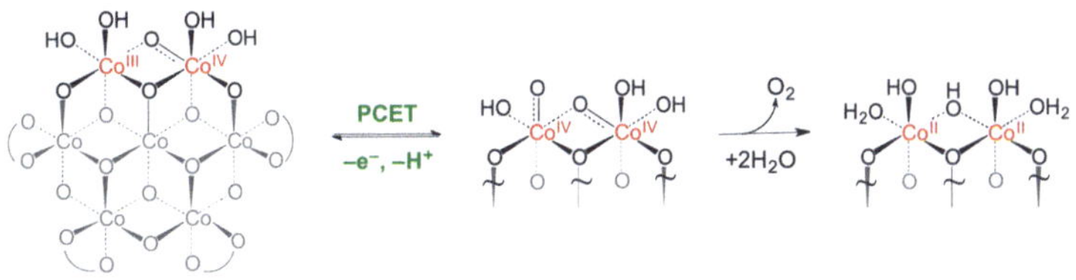

Fig. 12 Proposed pathway for OER by CoP_i. A PCET minor equilibrium proceeded by a turnover-limiting O–O bond forming step is consistent with current dependencies on proton and electron equivalencies. *Curved lines* denote phosphate, or OH_x terminal or bridging ligands. Reproduced from [59]

$$j = k_0^{\mathrm{OER}}(a_{\mathrm{H}^+})^{-1}\exp\left[\frac{FE}{RT}\right], \tag{4}$$

where k_0^{OER} is a potential-independent constant. This constant is proportional to the exchange current density and increases with catalyst loading as observed in Fig. 12. The exact mechanistic basis for the proportionality between this constant and film thickness is addressed in Sect. 4.2. Rearrangement of the logarithmic form of (4) yields a Tafel slope, $b = \partial \mathrm{E}/\partial \log(j)$, of 59 mV/decade that is also consistent with the experimental data shown in Fig. 11 and is equivalent to a transfer coefficient of $\alpha = RT/bF = 1$. Equation (4) is consistent with a mechanistic sequence involving a reversible one-electron, one-proton minor equilibrium step followed by a turnover-limiting chemical step [59]:

$$\mathrm{A} \rightleftharpoons \mathrm{B} + e^- + \mathrm{H}^+, \tag{5}$$

$$\mathrm{B} \rightarrow \mathrm{C}. \tag{6}$$

The assignment of chemical species to A, B, and C in (5) and (6) was aided by knowledge of the oxidation state changes in the film as a function of the applied potential. As described in Sect. 3, CV features of CoP_i are consistent with a predominant Co oxidation state of >+3 during catalysis. Accordingly, the pre-equilibrium step ($\mathrm{A} \rightleftharpoons \mathrm{B}$) was assigned to a formal $\mathrm{Co^{III}} \rightleftharpoons \mathrm{Co^{IV}}$ redox transition that is coupled to a proton transfer. It was suggested that the electrokinetic behavior described by (4) was in accord with $\mathrm{Co^{IV}}$ (i.e., B) as the minor component of this pre-equilibrium. Because the surface coverage of B, denoted θ_{B}, must be less than 10% for the pre-equilibrium to exhibit the observed Nernstian dependence on potential at the highest potential (1.27 V) that Tafel data were collected, the maximum fractional surface coverage of $\mathrm{Co^{IV}}$ ($\theta_{\mathrm{B,max}}$) must be 10%, and θ_{B} is expected to diminish by an order of magnitude per ~60 mV decrease in potential [60].

As discussed in Sect. 3, EPR spectra of CoP_i films subjected to prolonged electrolysis at 1.14 V, a value sufficient for water oxidation, exhibit a $\mathrm{Co^{IV}}$ signal estimated to arise from 3% of the cobalt centers in the film [56]. Noting the sizeable dead time (~25 min) between termination of electrolysis and freezing of the EPR sample, it is to be expected that the population of Co^{IV} in the film during electrolysis at 1.14 V would be significantly in excess of the residual 3% observed by EPR. Correspondingly, the in situ XANES data [44] discussed in Sect. 3 are consistent with an appreciable population of Co^{IV} when the films are held at potentials sufficient for water oxidation catalysis. The large population of Co^{IV} observed by EPR and in situ XANES studies, therefore, indicated that the catalyst resting state, A, is comprised of $Co^{III/IV}$ mixed valence clusters. The pre-equilibrium redox process ($A \rightleftharpoons B$) was therefore assigned to further oxidation of the $Co^{III/IV}$ mixed valence clusters preceding a turnover-limiting chemical process [59].

Based on the pH titration data shown in Fig. 11b, it was inferred that the reactant in the equilibrium step defined in (5) possessed a pK_a greater than 12 [59]. Indeed, additional studies under more alkaline conditions undertaken by Gerken et al. [57] extend the pK_a of this species beyond 14. By comparison to the pK_as of known Co aqua species [52] and to potentiometric titrations of crystalline cobalt oxide compounds [64, 65], we deduced that the reactant in (5) was most likely a [Co^{III}–OH] moiety present in a $Co^{III/IV}$ mixed valence cluster (species A), which is oxidized in a PCET equilibrium to a [Co^{IV}–O] center (species B).

Films prepared in isotopically labeled $^{18}OH_2$-enriched electrolyte and operated in non-isotopically labeled water were shown to extrude $^{32}O_2$ and $^{34}O_2$ and, to a lesser extent, $^{36}O_2$ [59]. The observation of $^{34}O_2$ and $^{36}O_2$ was ascribed to the participation of μ-oxo/μ-hydroxide moieties in oxygen production given that (1) the terminal Co–$^{18}OH_x$ moieties of the films used in the ^{18}O experiments are subject to exchange with bulk solvent over the time required to purge the gas-tight electrochemical cell used in these measurements and (2) the μ-oxo/μ-hydroxide moieties in the catalyst clusters exchange at a much slower rate. Importantly, although these results are consistent with the participation of μ-oxo/μ-hydroxo sites in O–O bond formation as a plausible mechanistic pathway in the OER, it does not necessarily imply that this pathway is the *predominant* route for O_2 formation; the slow extrusion of ^{18}O could simply reflect a minor pathway for O_2 evolution that contributes minimally to the observed electrokinetics. Because of the ambiguities associated with interpreting these ^{18}O-labeling results, it was noted that studies identifying the rate and nature of oxo-exchange in these films are needed before specific conclusions can be drawn. Once these exchange rates are identified, experiments that permit more rapid switching between isotopically enriched and non-isotopically enriched electrolytes (along with in-line analysis of products) could be leveraged to shed more light on the mechanism of O–O bond formation.

The proposed reaction pathway for the overall transformation is shown in Fig. 12 using the MCC model derived from EXAFS [44] and X-ray PDF analysis [48] (Sect. 3.1). Given the pre-equilibrium in (5) and the observed 60 mV/decade Tafel slope, the [Co^{IV}–O] product must participate in a turnover-limiting chemical step. The electrochemical rate law is consistent with a chemical step that does not involve proton transfer to bulk electrolyte, and as such, it was concluded that the O–O bond is formed in this irreversible chemical process.

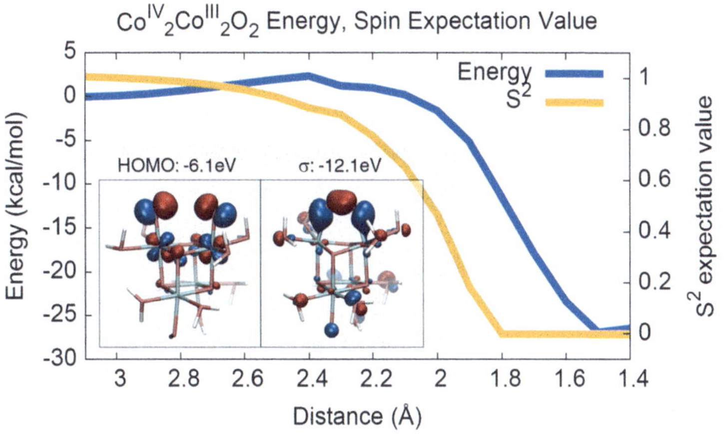

Fig. 13 DFT energies (*left Y*-axis, in *blue*) and $\langle S^2 \rangle$ (*right Y*-axis, in *orange*) as a function of O–O distance for a tetracobalt cubane model catalyst in the $[Co^{III}{}_2Co^{IV}{}_2]$ state. *Left inset*: highest occupied molecular orbital (HOMO) at $r_{O-O} = 3.0$ Å. *Right inset*: σ bonding orbital at $r_{O-O} = 1.4$ Å. Reprinted from [66]. Copyright 2011 American Chemical Society

Wang and Van Voorhis [66] explored the catalytic mechanism of O_2 evolution at a cobalt cubane model compound using DFT . They computed the energetics and barriers for the steps leading up to and including the O–O bond formation reaction using an explicit solvent model within a hybrid quantum mechanics/molecular mechanics (QM/MM) framework. Their studies on this model compound indicated that the formation of two cofacial terminal Co^{IV}–O generated a stable O–O bond with a low activation barrier (Fig. 13) and a low thermodynamic driving force. Based on the low barrier for O–O bond formation, it was proposed that the turnover-limiting step in CoP_i could be the addition of a water molecule or an intramolecular proton transfer, both of which would be consistent with the electrokinetic data. However, the authors noted one important deficiency of this cubane model relative to CoP_i: the model lacks any μ_2-(hydr)oxo centers, which are present in the MCC structure (Figs. 5 and 6), and which are implicated in O–O bond formation by the ^{18}O-labeling studies described above. We note that alternative turnover-limiting steps could also be the scission of the Co–O bond following O–O bond formation or hydrogen atom abstraction from H_2O by a Co oxyl (Co^{III}–$O^{\bullet}$). This computational study revealed that the localization of the hole density was severely dependent on the precise OH center deprotonated in the proton-coupled oxidation, providing further evidence for the role of PCET in hole/radical localization, which was suggested by the EPR studies [58] discussed in Sect. 3.2.

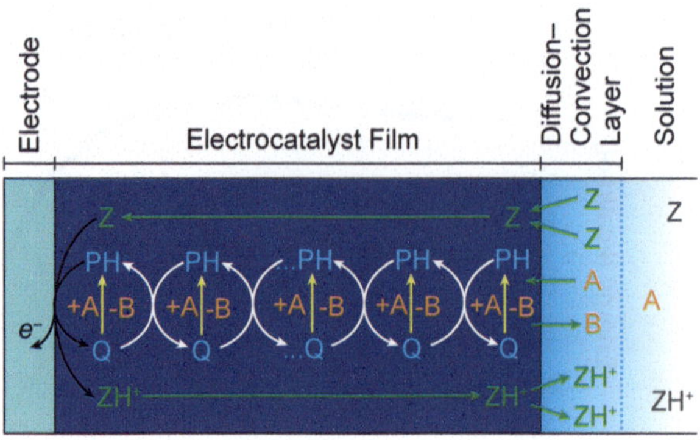

Fig. 14 Electrocatalytic oxidation of the substrate A into the products B in the presence of an acid-base pair ZH^+/Z by means of an immobilized $PH/Q+e^-$ catalyst couple. Adapted from [67]

4.2 *Proton-Electron Transport-Catalysis Mechanism*

In addition to the catalytic reaction, a major controlling factor in overall observed activity is electron transport through the film to the electrode to regenerate the catalyst [67]. Because protons are generated during the catalytic reaction, proton transfer – and therefore proton-coupled electron transfer [68–76] – reactions play a key role in both catalysis and transport. Using the active site turnover mechanism discussed in Sect. 4.1 [59], we set out to provide a more complete electrochemical rate law than the simplified version in (4).

The general scheme for the analysis is displayed in Fig. 14. Here "PH" represents the resting $Co^{III}(OH)/Co^{IV}(O)$ state (Fig. 12, left) and "Q" represents the pre-turnover-limiting $[Co^{IV}(O)]_2$ state, colloquially referred to as the "active form" of the catalyst (Fig. 12, middle). The substrate (A) and product (B) are H_2O and O_2, respectively. The acid-base pair, ZH^+/Z, is $H_2PO_4^-/HPO_4^{2-}$, $B(OH)_3/B(OH)_4^-$, or H_3O^+/H_2O (under appropriate electrolyte conditions).

The permeation of electrolyte into the film and the fact that the hopping involves no net charge transport (e^- plus H^+) ensures that transport is not driven by a gradient in the electric field across the film; instead proton coupled electron hopping proceeds down a gradient in electrochemical potential, akin to the description used for polymer electrode coatings – the so-called "redox conductors" [77–79]. In this model, proton coupled electron hopping through the film may be likened to a linear diffusion transport obeying Fick's law that is modified by a catalytic rate constant term, k_{cat}:

$$D_{H,e}\frac{d^2C_{PH}}{dx^2}+k_{cat}C_Q=0, \tag{7}$$

$$D_{H,e}\frac{d^2C_Q}{dx^2}-k_{cat}C_Q=0. \tag{8}$$

Here, x is the distance between any specified site in the film and the electrode substrate, and "C" represents the concentrations of the subscript species.

At the electrode, the following expression of the Nernst law is appropriate at the electrode surface ($x = 0$):

$$\frac{(C_{\mathrm{Q}})_{x=0} \times (C_{\mathrm{H^+}})_{x=0}}{(C_{\mathrm{PH}})_{x=0} \times C^0} = \exp\left[\frac{F}{RT}\left(E - E^o_{\mathrm{Q+H^+/PH}}\right)\right], \tag{9}$$

where C^0 is a normalizing concentration of 1 M.

A detailed analysis of the implications of these diffusion–reaction equations leads to an equation for the Tafel plot at high buffer concentration (where there is no pH gradient at the film–electrolyte interface):

$$\log j = \log j_0 + \frac{F}{RT\ln 10}\eta, \tag{10}$$

where j_0 is the nominal exchange current density (current density at overpotential, $\eta = 0$). Increasing film thickness increases j_0 in accordance with the following relation:

$$\log j_0 = \log\left[F\frac{C^0}{C^0_{\mathrm{H^+}}}C_{\mathrm{cat}}k_{\mathrm{cat}}d_f^{\mathrm{opt}}\tanh\left(\frac{d_f}{d_f^{\mathrm{opt}}}\right)\right] + \frac{F\left(E_A^{\mathrm{eq}} - E^0_{\mathrm{Q+H^+/PH}}\right)}{RT\ln 10}, \tag{11}$$

with an optimal film thickness defined as

$$d_f^{\mathrm{opt}} = \sqrt{\frac{D_{H,e}}{k_{\mathrm{cat}}}}. \tag{12}$$

C_{cat} represents the total concentration of catalyst active sites and E_A^{eq} is the thermodynamic potential for the reaction catalyzed (OER).

Equations (10)–(12) describe Tafel plots that possess a slope of 59 mV and whose exchange current densities change progressively with film thickness as shown in Fig. 15a. Below an "optimal" film thickness value, d_f^{opt}, the current density response increases in proportion to film thickness, but beyond d_f^{opt} the response displays a plateau. The variation of the exchange current density with the film thickness is shown in Fig. 15b where the maximal value reached by j_0, denoted $j_0^{\max}$, may be expressed by

$$j_0^{\max} = F\frac{C^0}{C^0_{\mathrm{H^+}}}C_{\mathrm{cat}}\sqrt{k_{\mathrm{cat}}D_{H,e}}\exp\left[\frac{F\left(E_A^{\mathrm{eq}} - E^0_{\mathrm{Q+H^+/PH}}\right)}{RT}\right]. \tag{13}$$

This reveals that a high film activity can be achieved by improving the intrinsic activity of the catalyst (increasing k_{cat}) and/or by increasing the film thickness (d_f),

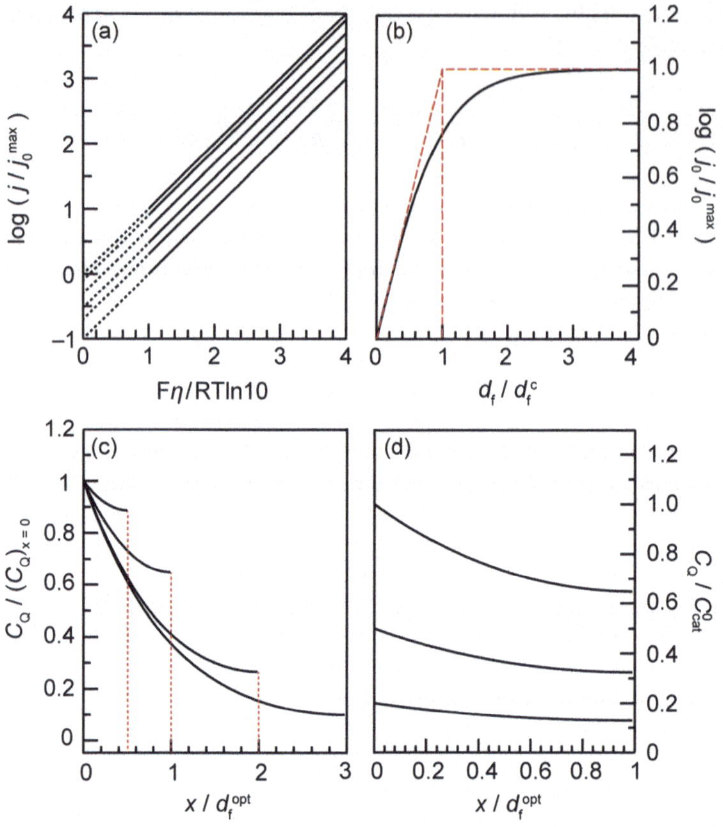

Fig. 15 Electrokinetic profile at large buffer concentrations. (**a**) Tafel plot trends predicted for a series of increasing thicknesses, from *bottom* to *top*. (**b**) Variation of the current density with the film thickness (relative to the optimal film thickness (d_f^{opt}), at $\eta = 0$. j_0 and j_0^{max} are defined in (11)–(13). For d_f^{opt} see (12). (**c**) Concentration profile of oxidized catalyst form, Q (relative to the concentration of Q at the electrode–film interface), in the case of mixed control by the turnover-limiting reaction and the diffusion-like proton–electron hopping under pure kinetic conditions for a series of films with thicknesses equal to $0.5\,d_f^{opt}$, d_f^{opt}, $2\,d_f^{opt}$, and $3\,d_f^{opt}$. (**d**) Concentration profile of Q (relative to total catalyst concentration in the film) as a function of increasing electrode potential (from *bottom* to *top*) for a film of thickness equal to d_f^{opt}. Figures adapted from [67]

but only up to a point. Once $d_f \times \sqrt{k_{cat}} > \sqrt{D_{H,e}}$, the film activity no longer improves with thickness. A crucial implication is, to the extent that catalyst film development focuses on improving intrinsic activity, one must also address the issue of proton-coupled electron transport (represented by the effective diffusion coefficient, $D_{H,e}$), in order to extract maximal benefit from a more active catalyst.

The underlying reason for the plateau in activity is described by Fig. 15c where the concentration of the oxidized form of the catalyst (Q) is plotted as a function of the distance from the electrode substrate (for films of different thicknesses). Beyond

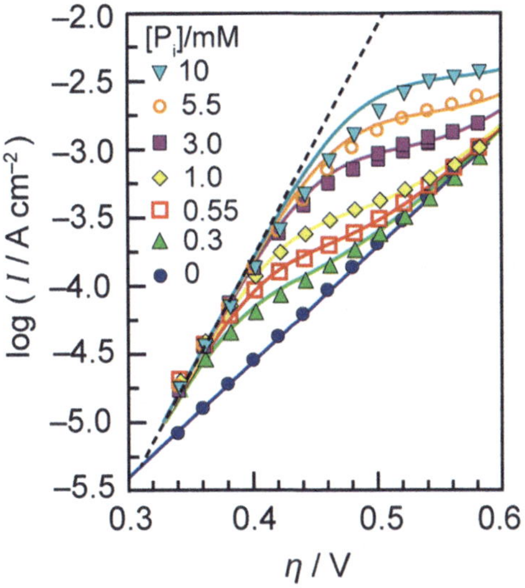

Fig. 16 Tafel plots of a CoP_i film (ca. 200 nm thick) deposited onto a platinum rotating disk electrode and operated at 2,500 rpm in varying concentrations of NaP_i pH 6 with 1 M $NaClO_4$ as supporting electrolyte. The slopes of the straight lines are F/RT ln 10 (*dashed black*) and $F/2RT$ ln 10 (*solid blue*). Adapted from [67]

d_f^{opt} the total integrated concentration of Q over the electrode surface ceases to increase appreciably with film thickness, leading to no net gain in film activity. The role of the actual electrode potential is then only to define the concentration of oxidized catalyst at the immediate film–electrode interface. Increasing the electrode potential shifts the position of the PH/(Q+H^+) equilibrium at the interface. This consequently leads to greater values of C_Q throughout the film (Fig. 15d) and therefore a greater current density (Fig. 15a).

This mechanistic framework was tested against the electrokinetic data acquired over a wide range of buffering conditions (including non-buffered electrolytes) (Fig. 16). We found that the model was fully consistent with proton-coupled electron transport turnover in Co-OEC films. More broadly, these concepts are readily applicable to systems comprising molecular catalysts that are coated onto electrode surfaces [80, 81] or to the increasing number of crystalline oxides that become amorphous and electrochemically porous during turnover [82, 83].

Having discussed the structural and mechanistic basis for the catalytic behavior of these materials, we turn our attention to the fundamental underpinnings of the self-healing behavior of these systems.

5 Catalyst Life Cycle: Growth and Repair

All catalytic systems to our knowledge require the catalyst to be repaired. In deactivation, the catalyst is transformed to an inactive species, which then must be regenerated, usually by a chemical process that lies outside the catalytic cycle.

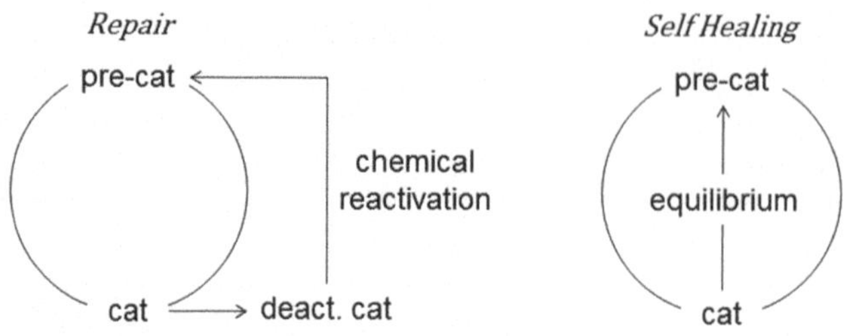

Fig. 17 The difference between a catalyst that is repaired vs self-healing. The CoP_i catalyst is unique among all catalysts in that it is self-healing

The CoP_i catalyst is self-healing, i.e., the catalyst may be regenerated by an equilibrium within the catalytic cycle. These two processes are compared in Fig. 17.

Self-healing in CoP_i is established from the interplay of the potential at which OER occurs vs the potential at which the catalyst nucleates and grows. Accordingly, the kinetics of the oxidative self-assembly process are critical to developing a coherent picture of the full life cycle of the catalyst from nucleation to dissolution and, ultimately, self-repair. Mechanistic insights into catalyst formation are now discussed.

The separation in potentials between water oxidation and catalyst formation [35, 37] (Sect. 2) were exploited to isolate the kinetics for the latter [53]. Chronoamperograms (Fig. 18) with step potentials spanning the initial rising portion of the anodic pre-feature wave discussed in Sect. 2 revealed that the deposition process is not merely a diffusion-controlled oxidation of Co^{2+} ions. The initial decay in current immediately after the potential step is followed by a rise in current produced by the formation of catalyst nuclei on the surface [84] which, in effect, increases the surface area of the electrode. Quasi three-dimensional growth of these catalyst particles gives rise to a rapid increase in current until the reactant-depleted diffusion zones surrounding each growing catalyst particle begin to overlap, causing the onset of semi-infinite linear diffusion and an associated $t^{-1/2}$ decay of the current [85]. Importantly, the step potential has a marked impact on the time, t_{max}, and current density, j_{max}, at which the chronoamperogram exhibits its characteristic peak. These characteristic features established that Co-OEC formation occurs via nucleation, followed by diffusion-limited growth [53].

Examination of the chronoamperometric transients indicated that the nucleated growth process was characterized by progressive nucleation, as opposed to instantaneous nucleation. In the instantaneous case, all potential nucleation sites, N_0, on the surface are assumed to give rise to nuclei at time zero following the potential step. In contrast, for progressive nucleation the number of occupied nucleation sites, N, increases with time, leading to a nucleation rate defined by [86]

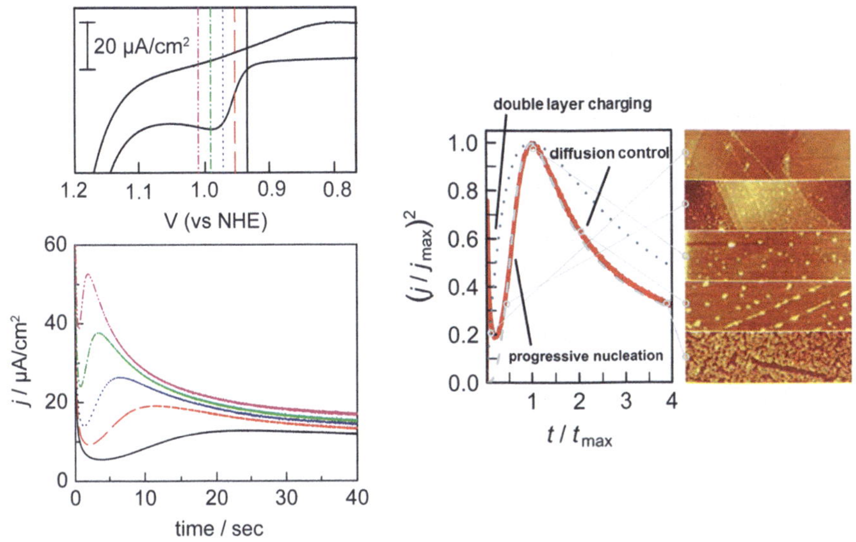

Fig. 18 *Top left*: Cyclic voltammogram (scan rate = 5 mV/s) and (*bottom left*) potential step chronoamperograms of a freshly polished glassy carbon disk electrode in 0.4 mM Co^{2+}, and 0.02 M MeP_i, 1.97 M KNO_3 electrolyte at pH 7.5. Chronoamperometry data recorded with a step voltage of 1.01 (*magenta, dot-dashed line*), 0.99 (*green, dot-dashed line*), 0.97 (*blue, dotted line*), 0.95 (*red, dashed line*), and 0.93 V (*black, solid line*) following a 100-s pulse at 0.75 V (not shown). Step voltages used to collect chronoamperograms are indicted with vertical lines overlaid on the CV trace. *Right*: Comparison of simulated nucleation transients (*gray dashed line*) with the experimental data (*red line*) modeled by (14) for a progressive growth mechanism with $n = 4$. The *dotted line* is the predicted growth curve for an instantaneous nucleation growth mechanism. Atomic force microscopy (AFM) images of Co-OEC on the electrode surface are shown for various points along the experimental curve. Adapted from [53]

$$\frac{\partial N}{\partial t} = k_N \left(\frac{c}{c_0} \right)^n (N_0 - N), \tag{14}$$

where k_N is the nucleation rate constant, and the term $N_0 - N$ represents the number of unoccupied nucleation sites on the electrode surface. In (14), c and c_0 represent the local and bulk concentrations of all solution species upon which the nucleation rate depends, with the exponent, n, characterizing the overall reaction order. Comparison of simulated nucleation transients with the experimental data indicated that Co-OEC nucleation was best modeled by an $n = 4$ curve (Fig. 18), highlighting the complexity of the initial catalyst formation process relative to simple metal deposition/plating.

Atomic force microscopy (AFM) imaging of Co-OEC early on during deposition on highly-ordered pyrolytic graphite (HOPG) was also explored to study catalyst formation (Figs. 18 and 19) [53]. The number of nuclei was observed to increase dramatically at times between 0.2 and $1 \times t_{max}$, after which relatively few new nuclei are formed. Instead, the already formed nuclei grow and coalesce to coat a

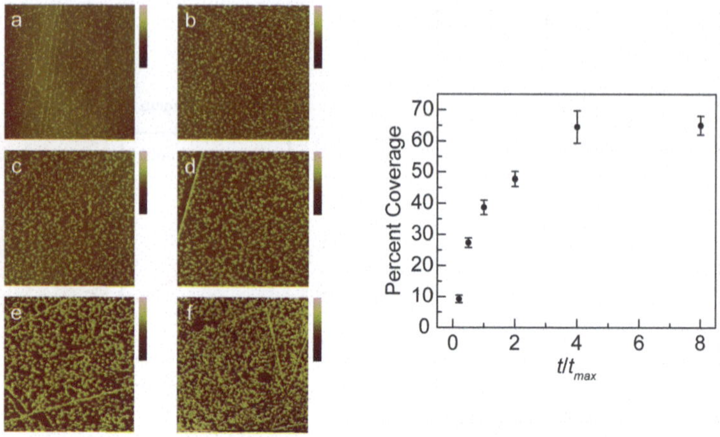

Fig. 19 *Left*: Representative 5 × 5 μm^2 AFM images of a highly oriented pyrolytic graphite electrode after being subjected to potential step polarization from 0.75 to 0.97 V for (**a**) 0.2, (**b**) 0.5, (**c**) 1, (**d**) 2, (**e**) 4, and (**f**) 8 × t_{max} (~10 s). *Bars* to the right of each image indicate the depth with full scale values of (**a**) 20, (**b**) 30, (**c**) 50, (**d**) 75, (**e**) 75, and (**f**) 50 nm. Electrolyte conditions: 0.4 mM Co^{2+} and 0.02 M MeP_i, 1.97 M KNO_3 electrolyte at pH 7.5 (2 M ionic strength). *Right*: Plot displays coverage percentage of catalyst vs the normalized duration of potential step polarization, t/t_{max}. Reproduced from [53]

greater proportion of the surface between 1 and 4 × t_{max}. Besides confirming the progressive nature of catalyst nucleation, these AFM studies highlight a convenient method for patterning semiconductor surfaces with catalyst islands. The appropriate choice of (photo)potential would enable modulation of the t_{max} parameter for any desired surface, and controlling the deposition time relative to this t_{max} would permit control over the coverage of the catalyst islands on the conducting or semiconducting surface. More recently, a high throughput, scalable, and low cost method for patterning periodic sub-micron structures of CoP_i over large areas of silicon has been developed, called reactive interface patterning promoted by lithographic electrochemistry (RIPPLE) [87]. Through an interplay of electrochemical and hydrodynamic effects, the RIPPLE method allows for catalyst to be deposited with tunable and high spatial resolution using cyclic voltammetry; the number of patterns scales with N–1 CV scans. In addition to planar substrates, three-dimensional surfaces and their vertical sidewalls may be patterned. Patterns of sufficient fidelity are realized that optical light control may also be achieved [88].

Activation-controlled current densities for steady state catalyst growth were determined as a function of the applied potential by extrapolating current densities measured at variable rotation rates on a rotating disk electrode (RDE) to infinite rotation speed (Fig. 20a). Tafel plots of steady state catalyst deposition displayed a Tafel slope of 60 mV/decade (Fig. 20b) [53]. The observed 60 mV/decade slope implies a catalyst assembly mechanism involving a one-electron reversible equilibrium preceding a chemical rate-limiting step for catalyst formation [60].

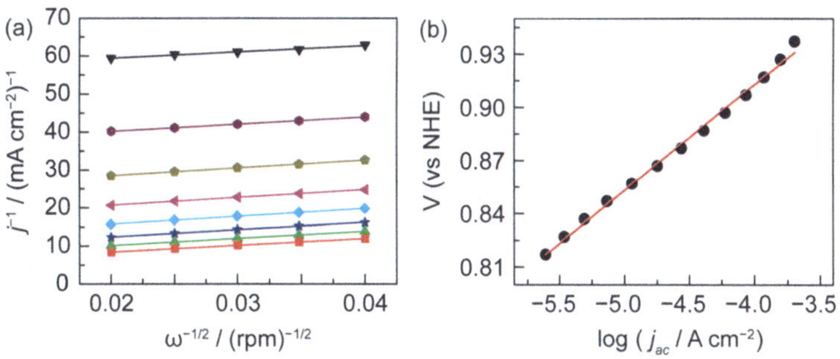

Fig. 20 (**a**) Koutecký–Levich plots Co-OEC catalyst film formation from 0.4 mM Co^{2+}, 0.02 M MeP_i, 1.97 M KNO_3 electrolyte, pH 7.5 at applied potentials of (from *top* to *bottom*) 0.87, 0.88, 0.89, 0.90, 0.91, 0.92, 0.93, and 0.94 V. (**b**) Tafel plot of Co-OEC catalyst film formation from 0.4 mM Co^{2+}, 0.02 M MeP_i, 1.97 M KNO_3 electrolyte, pH 7.5 onto a Pt rotating disk electrode. Activation controlled current density values (j_{ac}) were derived from Koutecký–Levich analysis of plots in (**a**). The Tafel slope is 60 mV/decade. Reproduced from [53]

Additional electrokinetic data was acquired as a function of Co^{2+} concentration, pH, and buffer (methyl phosphonate, MeP_i) concentration. In each case, Tafel plots relating potential to the activation-controlled current density were obtained over the range of electrolyte conditions. Interpolation of these plots permitted determination of the respective reaction orders in Co^{2+} concentration, proton activity, and buffer concentration [53]. These studies uncovered a first-order dependence on Co^{2+} concentration, indicating the reversible one-electron equilibrium defined by the 60 mV/decade Tafel slope involves one Co^{2+} ion, thereby excluding the possibility of multi-nuclear solution species participating in the one-electron equilibrium in question. It was also observed that the rate of deposition possesses an inverse third-order dependence on proton activity. This result suggested that the coupling of three proton transfers to the single cobalt one-electron transfer is what permits access to Co^{3+} at these low potentials. The high degree of proton coupling in the oxidation of the Co^{2+} precursor implicates a critical role for the MeP_i buffering species. Notwithstanding its role as a PCET proton acceptor, MeP_i may coordinate to Co centers in a conflicting inhibitory role. Interestingly, an inverse second-order dependence in MeP_i is observed for buffer concentrations in excess of ~30 mM. At intermediate buffer strengths, an inverse first-order dependence in MeP_i is found, and below 2 mM MeP_i, a plateau is observed implying that, at these concentrations, MeP_i is no longer explicitly represented in the rate law for electrodeposition. On the basis of Co^{2+}/MeP_i binding calculations, one half of the second-order behavior was explained as caused by coordination of the buffer to dissolved Co^{2+}, whereas the underlying first-order behavior, was attributed to buffering species that were bound to clusters of the growing catalyst film. Thus, it was proposed that equilibrium dissociation of buffering species from the surface was a critical prerequisite for incorporation of newly oxidized Co centers. These findings are summarized in the electrochemical rate law for catalyst formation [53]:

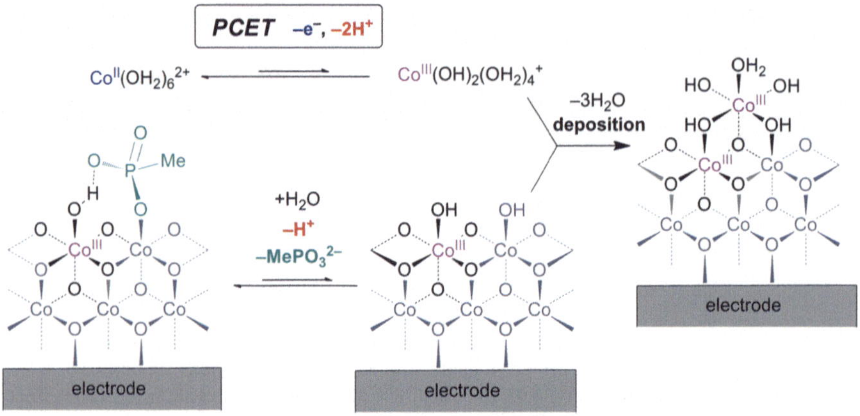

Fig. 21 Proposed mechanism of Co-OEC catalyst film formation. Adapted from [53]

$$v = k_0^{\mathrm{dep}}\left[\mathrm{Co}^{2+}\right](a_{\mathrm{H}^+})^{-3}[\mathrm{MeP_i}]^{-1}\exp\left[\frac{FE}{RT}\right], \tag{15}$$

where k_0^{dep} is a potential-independent constant which is proportional to the exchange current density for the electrodeposition process.

Within the framework of the Co-OEC structural model discussed in Sect. 3, it was proposed that growth of the MCCs, and thus of the catalyst film itself, proceeds via attachment of new Co fragments to the exposed edges of these clusters. The proposed mechanistic model for Co-OEC film growth is shown in Fig. 21 [53]. A portion of the expanding cobaltate cluster is shown to present the surface active sites from which growth is proposed to occur. In this mechanistic model, surface and solution phase reactions exist in equilibrium prior to the rate-limiting phase transfer of Co to the growing catalyst surface. The exposed edge sites were proposed to be directly coordinated to MeP_i buffering species in the resting state. Further studies on model complexes (Sect. 6) suggest that the mode of inhibition by MeP_i may more likely involve H-bonding to the surface of the clusters. Thus, the $Co(OH_2)_6^{2+}$ solution precursor is proposed to undergo a one electron-proton minor equilibrium PCET reaction to form a $Co(OH)_2(OH_2)_4^+$ intermediate. Simultaneously, a surface equilibrium involving $MePO_3^{2-}$ dissociation and deprotonation of a surface aquo species takes place. This heterogeneous equilibrium forms a surface intermediate poised for rate-limiting binding of the solution-based Co $(OH)_2(OH_2)_4^+$ intermediate to effect catalyst growth by one Co center [53].

The above mechanistic insights into catalyst formation are critical for explaining the self-repair process of Co-OEC films. We had shown previously through ^{57}Co radiolabeling of CoP_i films that, when held at open circuit, cobalt slowly leached out of films into bulk solution. Yet upon application of a potential bias sufficient to trigger O_2 evolution, Co was taken up from solution and redeposited on the catalyst

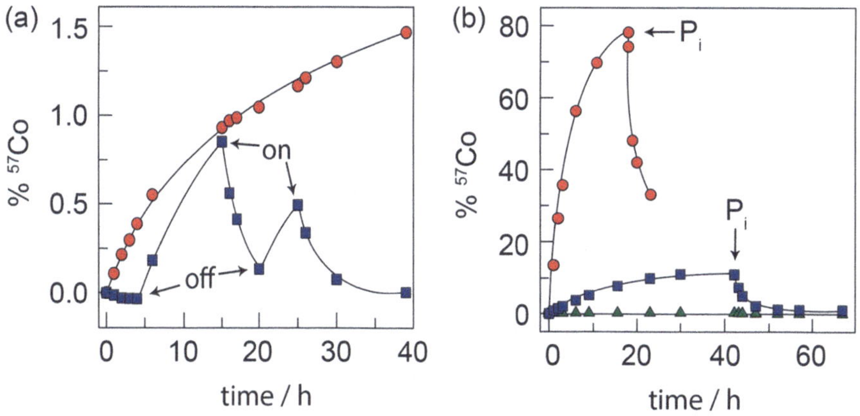

Fig. 22 (**a**) Percentage of ^{57}Co leached from a CoP_i catalyst film on an electrode: with a potential bias of 1.3 V (NHE) (*blue squares*) turned on and off at the times designated and held at open circuit potential (*red circles*). (**b**) Percentage of ^{57}Co leached from Co-X (where X is a non-buffering electrolyte such as SO_4^{2-} or NO_3^-) films on an electrode under a potential bias of 1.3 V (*red circles*) and 1.5 V (*blue squares*) (NHE) and held at open circuit potential (*green triangles*). Phosphate was added at the time points indicated by the *arrows*. Adapted from [89]

film (Fig. 22a) [89]. Moreover, it was shown that phosphate was a critical ingredient to enabling this self-healing capacity (Fig. 22b) [89].

The mechanistic basis for Co-OEC self-healing is shown in Fig. 23a. The electrodeposition process displays an inverse third-order dependence on proton activity (see (5)) [53] compared to the inverse first-order dependence on proton activity found for the OER mediated by Co-OEC [59] (see (4) and (11)). Thus, as the pH is lowered, the potential necessary for catalyst formation increases much faster than the corresponding rise in potential required for water oxidation catalysis as shown in Fig. 23a. Extrapolation of the kinetic profile for catalyst assembly reveals a crossover point at ca. pH 5.2. This self-healing behavior arises because, at pH > 5.2, the potentials necessary to sustain catalyst film formation and growth are well below the potentials required for water oxidation catalysis. Thus, upon application of a potential sufficient to generate O_2, ample driving force exists to redeposit any Co^{2+} in solution that may have leached from the film in between catalyst operation cycles. This repair process is only possible in the presence of a good proton acceptor such as P_i, B_i, or MeP_i because these buffers not only maintain a stable local pH at the electrode in order to prevent local acidification that can trigger film dissolution, but they are also crucial for permitting the high degree of proton coupling inherent in the oxidative deposition process itself (see (15)) [53].

Direct measurement of catalyst corrosion agrees with the results of Fig. 23a. Whereas no dissolution is observed in pH > 6 P_i electrolyte under an applied bias, catalyst corrosion is observed at pH 5 and below, even in the presence of P_i and under anodic polarization [53]. This predicted dependence of the pH regime of catalyst stability on Co^{2+} concentration (Fig. 23b) is in line with the observation in the literature that adding 1 mM Co^{2+} to the electrolyte solution during catalyst

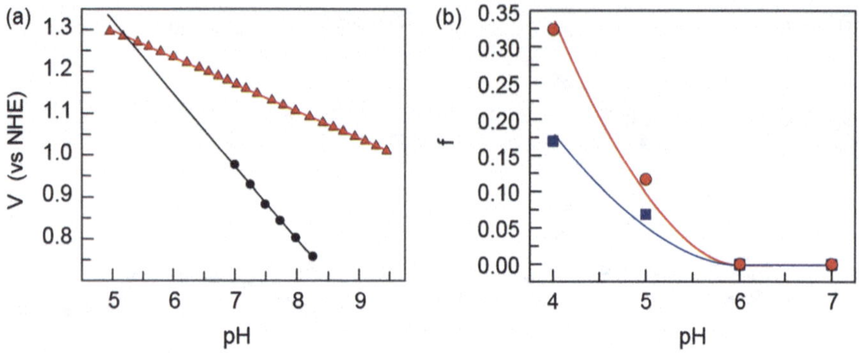

Fig. 23 (**a**) pH dependence of the potential for catalyst film formation (*black filled circle*) and oxygen evolution (*red-filled square*) at $j_{ac} = 30\ \mu A/cm^2$. (**b**)Fraction (f) of catalyst film dissolved after 1 h vs pH for electrodes poised at 1 (*blue-filled square*) and 0 (*red-filled square*) mA/cm^2 in 0.04 M Britton–Robinson buffer. *Lines* are presented to guide the eye. Adapted from [53]

operation for oxygen evolution was critical to maintain functional stability at pH 3.7 [57, 90]. The much lower potentials required for Co-OEC deposition in buffered electrolytes at intermediate pH, relative to those required for O_2 evolution also provide an explanation for the observation that some molecular compounds that are designed to be water oxidation catalysts form oxidic films upon decomposition of the precursor compound [91–94]. This issue is discussed in further detail in the next section. As we discuss below, it is critical that ligand scaffolds targeted for homogeneous water oxidation catalysts both preclude the dissociation of free Co ions into solution and maintain their integrity under the highly oxidizing conditions during the OER.

6 Molecular Cobalt Complexes as Models for Co-OEC

Unlike the amorphous solids of Co-OECs, where structure at the most precise atomistic level is undetermined, inorganic molecular complexes may be structurally characterized precisely using X-ray diffraction. Accordingly, we turned our attention to the use of multinuclear cobalt molecules, especially those with oxidic ligands (O^{2-}, OH^-, and OH_2), as models for the Co-OEC. We sought to use spectroscopic handles of multinuclear molecular cobalt complexes to provide direct comparisons to Co-OEC. For example, in Sect. 3.2 it was shown how EPR studies on the molecular cobalt cubane $Co_4O_4(OAc)_4(py)_4^+$ provided evidence that the EPR signal that arises from Co-OEC at anodic potentials is produced by a delocalized Co^{IV} valency.

Following in this theme, this section is organized to provide an overview of our research efforts in this area of cobalt model complexes. The aim is not to provide a review of the molecular water oxidation catalysts, cobalt-containing or otherwise, but to highlight the results obtained from studying molecules that have refined our thinking with regard to the mechanisms germane to the activity of Co-OEC.

6.1 *Co_4O_4 Cubanes Model Complexes: Co(III)/Co(IV) Self-Exchange and Film Conductivity*

The Co_4O_4 cubanes have proven to be very useful models of the Co-OEC. A number of molecules in this class have been reported, and they are distinguished by the ligands, such as pyridines, bipyridines, acetates, or benzoates, which stabilize the metal-oxo core [95–100]. The $Co_4O_4(OAc)_4(py)_4$ cubane, **1**, and the derivatives thereof, first synthesized by Das and coworkers [95], were not only shown to be useful for EPR studies, but have also been explored as discrete water oxidation catalysts [101–103]. Though some evidence of electrochemical water oxidation was presented in these studies, the primary mode of characterization of the water oxidation activity was through the use of a photochemical assay, utilizing $Ru(bpy)_3^{2+}$/persulfate as the photosensitizer/sacrificial oxidant. Because two oxidizing species (i.e., $Ru(bpy)_3^{3+}$ and $SO_4^{-\bullet}$) are created following the absorption of a photon, it is difficult to draw mechanistic conclusions from this photochemical assay as the driving force is not well defined. Therefore, we decided to explore the mechanism by electrochemical methods, which permit precise control over the electrode potential, and thus the driving force of the reaction.

It was discovered during these studies that the reported oxygen evolution activity could not be reproduced. Further investigation revealed that the earlier reports had used samples of **1** that had not been purified by silica gel chromatography, which was found to be essential for removing Co^{II} impurities from the as-synthesized material [104]. The Co^{II} impurity was found to be a precursor to the formation of the active catalyst, Co-OEC. Figure 24 shows a comparison of the electrochemical and photochemical responses of purified and crude samples of **1**. In both experiments, the oxygen evolution activity decreases significantly once the impurity is removed.

The lack of significant water oxidation activity for **1** suggests the need for a solvent accessible coordination site on the periphery of the metal cluster. For the Co-OECs, the edge sites of the cobaltate clusters provide these points of access and are the likely active sites for catalysis. However, because the metal ligand bonds in the oxidized cubane $\mathbf{1}^+$ are inert because of the low-spin electronic structure of the Co^{III} and Co^{IV} atoms, the complex is resistant to protonolysis and hydrolysis. This leads to a stable molecule, but the peripherally coordinated ligands present a barrier to the formation of Co^{IV}–O intermediate, which is presumably a key catalytic intermediate. Further research is needed to develop novel ligand architectures that can stabilize a Co_4O_4 core and at the same time provide open coordination sites for water or hydroxide ligands capable of participating in redox-leveling PCET events.

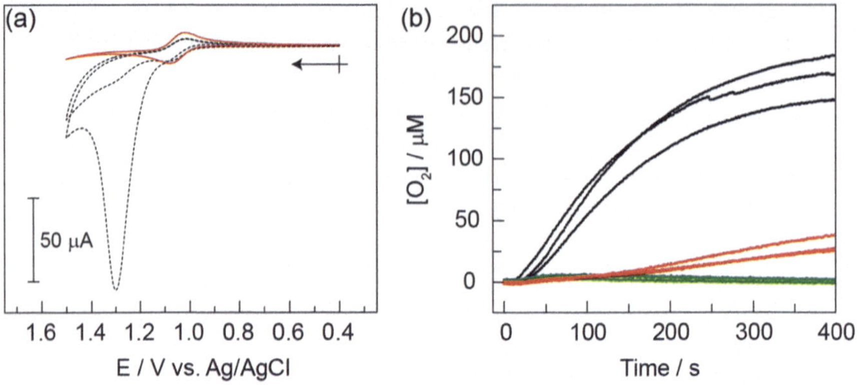

Fig. 24 (**a**) Background corrected CVs of crude (*black-dashed line*) and purified (*red-solid line*) samples of **1** (0.852 mg/mL) in 0.2 M KP_i buffer, pH 7. (**b**) Solution $[O_2]$ measurements during illumination of (*black-solid line*) crude and (*red-solid line*) purified samples of **1** (0.33 mM), and (*green-solid line*) without added **1**. Photochemical reactions were performed in the presence of 0.5 mM $Ru(bpy)_3{}^{2+}$, 35 mM $Na_2S_2O_8$, and 100 mM KP_i pH 7 buffer. Adapted from [104]

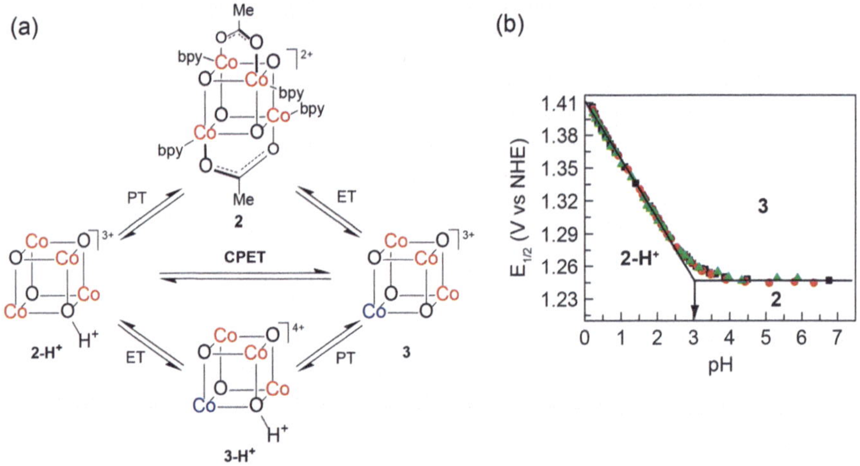

Fig. 25 (**a**) PCET square scheme for the oxidation of the molecular model cubane, **2-H**$^+$. Though the hole in the oxidized cubane is delocalized, we depict the Co^{IV} centers (*blue*) and Co^{III} centers (*red*) solely for illustrative purposes. For clarity, the ligands are not depicted for **2-H**$^+$, **3**, and **3-H**$^+$. (**b**) Pourbaix diagram of cubane **2** constructed from CV scans at 0.1 V s^{-1}. The *green*, *red*, and *black dots* correspond to three separate sets of data. Labels depicted the zones of thermodynamic stability for **2**, **2-H**$^+$, and **3**. Adapted from [105]

The Co_4O_4 core has provided a platform for the study of the PCET properties associated with charge transfer through the films. The molecule $Co_4O_4(OAc)_2(bpy)_4{}^{2+}$, **2**, first synthesized by Christou and coworkers [97], also contains a Co_4O_4 core, but differs in overall charge from the cubane of Das, **1**, because of the replacement of two acetate ligands by neutral bipyridines (Fig. 25a).

Cubane **2** exhibits a reversible $Co^{IV}Co^{III}_3/Co^{III}_4$ couple with an $E_{1/2} = 1.25$ V vs NHE at pH > 4. Below pH 4, the $E_{1/2}$ value becomes progressively more positive, exhibiting a slope of –55 mV/pH unit at pH < 2 (Fig. 25b). The electrochemical response in the low pH region of the Pourbaix diagram is indicative of a one-electron, one-proton transfer of a protonated species, **2-H**$^+$, with a p$K_a = 3.1$, to a deprotonated, oxidized cubane, **3**. Because a second plateau region was not observed at high potentials, the pK_a of the protonated form of **3** must be pH < 0.

As charge transport through Co-OEC films is thought to proceed via electron and proton hopping between Co^{III}–OH to Co^{IV}–O species (Sect. 4.2), we sought to isolate the kinetics of this PCET event by interrogating the mechanism by which **2-H**$^+$ is converted to **3** [105]. We were interested in comparing the bidirectional reaction, in which the electron and proton are transferred to separate chemical species, to the unidirectional reaction, in which they are transferred to the same species. By varying the scan rate in cyclic voltammogram experiments, the heterogeneous ET rate constant was extracted by comparing the simulated CV overlaid with the experimental data. Using this method, it was found that the rate constant for the PCET reaction at pH 1 was $k_{PCET} = 0.17$ cm s^{-1}, an order of magnitude smaller than the simple ET rate constant measured at pH 4, $k_{ET} = 2$ cm s^{-1}. The lack of an observable kinetic isotope effect (KIE) at pD = 1 indicated that the PCET reaction at the electrode surface is stepwise, proceeding by an equilibrium proton transfer followed by a rate-limiting electron transfer (PTET). This stepwise pathway was also found to be manifested in the bidirectional PCET reaction between $[Ru(bpy)_3]^{2+}$ and **3** at pH or pD = 1. In this homogeneous reaction, **3** was reduced to **2-H**$^+$ by $[Ru(bpy)_3]^{2+}$ with a driving force of ~100 mV. Stopped-flow kinetic measurements resulted in near identical second-order rate constants (~1×10^6 M^{-1} s^{-1}) in H_2O and D_2O. As in the heterogeneous reaction, the lack of an observable KIE indicated a stepwise PTET mechanism.

The unidirectional PCET self-exchange reaction between **3** and **2-H**$^+$ was interrogated further using NMR line broadening analysis. The rate constant for this reaction at pH 1, $k_{SE1} = 5.6 \times 10^4$ M^{-1} s^{-1}, was found to be only a factor of ~5 slower than the simple self-exchange ET reaction at pH 4, $k_{SE4} = 3.0 \times 10^5$ M^{-1} s^{-1}. As a comparison, the expected rate constant at pH 1 for the stepwise self-exchange PTET reaction constant can be calculated using k_{SE4} and the acid dissociation constant of **2-H**$^+$ according to

$$k_{SE1} = k_{SE4}\frac{K_a}{[H^+]} = 3.0 \times 10^5\ M^{-1}\ s^{-1}\frac{10^{-3.1}M}{10^{-1}M} = 2400\ M^{-1}\ s^{-1}. \tag{16}$$

Because the calculated rate constant is less than one twentieth of the experimentally determined value, a concerted proton-electron transfer, CPET, mechanism is implicated for this unidirectional PCET event. This supposition was confirmed by the measurement of a significant KIE = 4.3 for the self-exchange reaction at pH 1, which suggests that proton tunneling is involved in the rate-limiting step. Taken together, these results indicate that the unidirectional pathway provides an activation barrier significantly lower in energy than the intermediates of the step-wise reactions.

The $Co^{III/IV}$ self-exchange reaction is directly relevant to Co-OEC catalysis. During OER, the catalyst is poised at a potential to maintain the Co^{IV} oxidation state. Hole propagation through the film may occur by self-exchange with Co^{III}. In this way the oxidizing hole equivalents may move from the electrode surface through the film. The fairly high self-exchange $Co^{III/IV}$ rate constant measured for the cubane indicates that the Co-OEC film may be sufficiently conductive by a hole hopping mechanism.

6.2 Co_7 Clusters: Co^{II}-Co^{III} Self-Exchange and Co-OEC Deposition

Co-OEC deposition and growth is driven by the oxidation of aqueous Co^{2+} to Co^{3+}. When considering the mechanism of how this oxidation is coupled to deposition, we were confronted by the unresolved issue regarding the anomalously high observed self-exchange rate constant for the $Co(OH_2)_6^{3+/2+}$ couple [106, 107]. Because of the low-spin to high-spin transformation that takes place when $Co(OH_2)_6^{3+}$ is reduced to $Co(OH_2)_6^{2+}$, a large (~0.2 Å) change in inner-sphere bond lengths is observed. Consequently, Marcus theory predicts a very slow self-exchange rate constant for this reaction. However, experiment does not bear this out, as a k_{SE} of 5 M^{-1} s^{-1} has been measured [108], which is six orders of magnitude higher than predicted by theory [109].

We sought to address this issue by interrogating this redox couple in the context of a relevant molecular framework [110]. Such a framework is afforded by a heptanuclear cluster, **4**, shown in Fig. 26. The heptanuclear core of this molecule closely resembles the minimal cobaltate cluster unit of the Co-OEC, as determined from EXAFS studies on thin films (Fig. 5a).

Complex **4** has, as have a number of heptanuclear cobalt clusters, previously been synthesized as a potential single-molecular magnet, [111, 112]. Cluster **4** is distinguished from other known clusters by the μ_3-OH bridging units surrounding the central cobalt atom. This oxidic ligand environment is precisely that of Co-OEC and additionally provides a similar ligand field stabilization for the central cobalt atom as is found for cobalt hexaaqua complexes. Chemical oxidation of **4** with 1 equiv. of Ag^+ leads to an oxidized complex, **5**, which was structurally characterized using X-ray diffraction. Bond distance analysis in conjunction with magnetic measurements confirmed that the central cobalt's oxidation state was Co^{III} with a low-spin electronic configuration. The crystal structure also revealed hydrogen bond interactions between the triflate anions and the protons of the μ_3-OH ligands (Fig. 26), and NMR measurements established that this interaction was strong enough with **5**, but not **4**, to be maintained in solution.

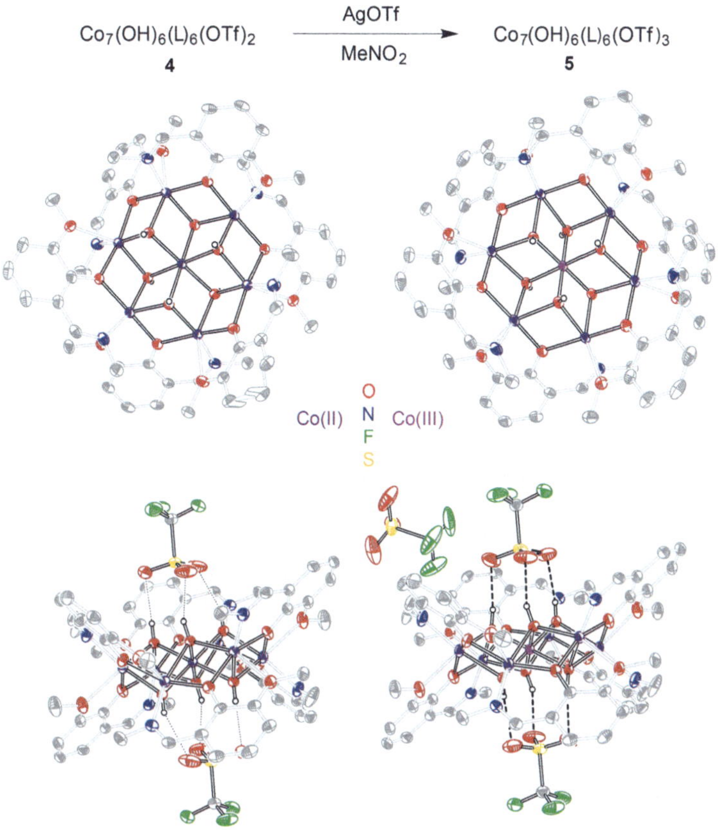

Fig. 26 *Top*: Synthetic scheme for the oxidation of **4** to **5**. *Middle*: Top down view of co-crystallized (*left*) **4** and (*right*) **5** with anions removed for clarity. *Bottom*: Side-on view of co-crystallized (*left*) **4** and (*right*) **5** with triflate anions shown. Adapted from [110]

The kinetics of the self-exchange electron transfer (ET) between **4** and **5** were measured using an isotope scrambling experiment. The second-order rate constant was found to be 1.53×10^{-3} M^{-1} s^{-1} at 40°C and to depend inversely on triflate concentration. This observation defined a mechanism for the self-exchange which involved an equilibrium dissociation of a hydrogen-bonded anion from **5** prior to rate-determining electron transfer (Fig. 27).

The slow observed rate constant is important in the context of the anomalous self-exchange rate constant for $Co(OH_2)_6^{3+/2+}$ electron transfer. For the $Co(OH_2)_6^{3+/2+}$ self-exchange reaction, an inner-sphere ET mechanism involving a water bridge was hypothesized to account for the fast rate of electron exchange. Because the two exchanging cobalt atoms, at the center of the clusters in **4** and **5**, are shielded from each other by the surrounding cobalt atoms and ligands, an inner sphere ET between these centers is precluded. Because a slow self-exchange ET rate constant was measured in this model study – consistent with the large bond

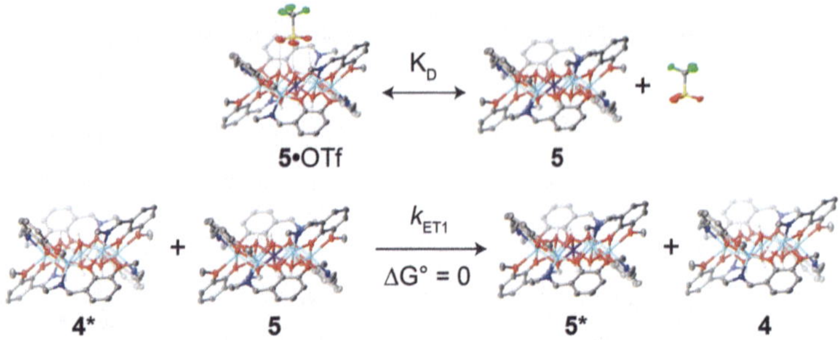

Fig. 27 Electron transfer scheme for the self-exchange reaction between model clusters **4** and **5**. K_D is the dissociation constant of the hydrogen bonded triflate anion from the oxidized cluster, **5**. The *asterisks* signify the isotopic label used to monitor the exchange reaction. Adapted from [110]

distance changes associated with a low-spin Co^{III} to high-spin Co^{II} transformation – it was concluded that an inner-sphere water bridge mechanism can account for the anomalous k_{SE} for the $Co(OH_2)_6^{3+/2+}$ exchange reaction.

Taken together, these results, in combination with others discussed above, have led to a new proposed mechanism of Co-OEC deposition. The original proposed mechanism, shown in Fig. 21, is consistent with the experimental rate law (15), which is distinguished by an inverse dependence on the $MePO_3^{2-}$ anion and a potential-dependent equilibrium ET. However, in an initial analysis, it was proposed that the $MePO_3^{2-}$ anion was bound as an inner-sphere ligand to a surface exposed Co^{III}. It is known that ligand substitution at Co^{III} atoms is usually slow because of the inert nature of d^6, low spin complexes, and therefore it is unlikely that this type of bond cleavage could exist in a fast equilibrium step. A more plausible explanation for the equilibrium dissociation of the buffer anion, which is consistent with the model studies above, involves a hydrogen-bonded $MePO_3^{2-}$ species, which rapidly associates and dissociates with surface edge site of the Co-OEC. Once the anion dissociates, a site is opened for solution-based Co^{2+} to bind before subsequent oxidation to Co^{III}.

The precise mechanism by which Co^{II} is oxidized to Co^{III} during steady state film growth can now also be re-evaluated. Based on the studies described in Sect. 4.2, it is now known that ET occurs by hole hopping through the film. This is in contrast to the initial proposal, which involved an ET from Co^{II} in solution directly to the underlying electrode. Moreover, a film-mediated ET mechanism is implicated because film growth proceeds efficiently even when the thickness of the film exceeds length scales that could allow electrons to tunnel from the solution directly to the electrode (i.e., 5–10 nm). The issue then becomes what is the nature of the ET process that supports catalyst deposition.

The EPR studies on the Co-OEC films demonstrated that at 1.03 V (vs NHE) a signal for Co^{IV} can be observed (Fig. 8a). This potential is only 100–200 mV more positive than the potential region investigated by Tafel analysis for the deposition mechanism (Fig. 20b). Therefore, these data suggest that the minor equilibrium ET leads to the oxidation of the Co^{III} film to a minor Co^{IV} species – the same species as

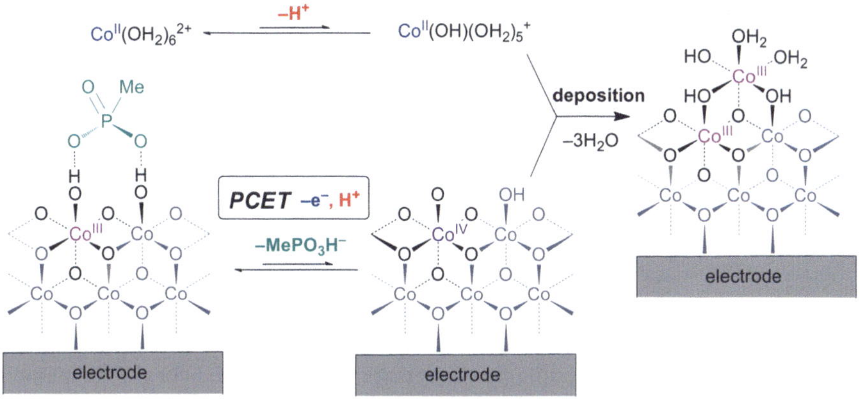

Fig. 28 New proposed mechanism of Co-OEC film formation

the resting state of the Co-OEC catalyst during catalytic turnover – which then "diffuses" to the surface of film in a series of self-exchange ET events. The diffusion process is analogous to the one described in Sect. 4.2 for the catalytic intermediate Co_2^{IV} form of the catalyst (Fig. 14), but instead involves a lower oxidation state of the film. Because the hopping mechanism can be thought of as a PCET self-exchange reaction between small domains in the Co-OEC films, $Co^{III/IV}$ self-exchange studies with molecular model complexes, such as the cubanes, **2-H**$^+$ and **3**, offer helpful insights. Studies of these Co_4 cubanes showed that the $Co^{III/IV}$ CPET self-exchange reaction is indeed rapid (possessing a rate constant on the order of 10^5 M^{-1} s^{-1}) and revealed that intrafilm $Co^{III/IV}$ hole hopping can indeed sustain the facile film growth. As a point of comparison, an exchange reaction between Co^{II} and Co^{III} in an oxidic ligand field (as in the case of the Co_7 compounds **4** and **5**) is eight orders of magnitude slower (~10^{-3} M^{-1} s^{-1}).

A revised model for the mechanism of Co-OEC growth is presented in Fig. 28. The dissociation of a hydrogen-bonded $MePO_3^{2-}$ is coupled to proton loss from the film. This increases the electron-donating ability of the inner-sphere hydroxide ligands, which in turn makes the surface-exposed Co^{III} species more susceptible to oxidation. The oxidation of the exposed Co^{III} to Co^{IV} is coupled to a deprotonation in a PCET event, and, in solution, a $Co(OH_2)_6^{2+}$ species is in minor equilibrium with its deprotonated form, $Co(OH)(OH_2)_5^+$. Following the interaction of the solution species with the surface of the film, likely through an inner-sphere bridging species, a comproportionation of the Co^{IV} and Co^{II} species into two Co^{III} species leads to the growth of the film by one cobalt atom. The rate-determining chemical step is likely the inner-sphere ET comproportionation. This is predicted to be a slow event because a Co^{II} species is being oxidized to Co^{III}, leading to large changes in bond lengths. Importantly, because this ET does not involve the electrode directly, its influence on the Tafel slope is indistinguishable from a "chemical" step, which is usually considered to involve the formation or cleavage of chemical bonds.

7 Outlook

The electrodeposition and OER catalysis of molecular cobaltate clusters may be established by self-assembly from solution. They are comprised of disordered Co–O clusters with sizes within the molecular to nanoscale range. The Co-OECs are the first self-healing catalysts of any type and they function at neutral pH. Catalysis occurs at active sites throughout the porous films and as a result proton-electron transport is a crucial determinant of the overall activity of these materials. A mixed valence $Co^{III/IV}$ resting state is revealed from spectroscopic studies and Co^{IV} intermediates are implicated in the catalytic cycle. Molecular model compounds have shed light on the kinetics of proton-coupled self-exchange reactions within the film as well as the buffer-binding process at electrolyte-exposed sites. The interaction of buffer oxoanions with the edges of the clusters during growth appears to be the basis for the electrolyte-dependent domain sizes observed in these materials. As such, these systems may be ideal for the exploration of redox leveling in electrocatalysis. Yet this requires the elucidation of precisely how deposition conditions direct catalyst structure. The exact details of the rate-limiting step (possibly O–O bond formation) are yet to be revealed, and additional isotopic labeling studies are required (in concert with a determination of aquo and hydroxo exchange rates in these materials) in order to understand this step more definitively. Analogous catalyst films based on Ni [14–16] and Mn [12, 13] have also been developed. These systems are also highly promising OECs under slightly alkaline (Ni) and acidic (Mn) conditions. The lessons learned from the CoP_i system offers encouragement that new reaction pathways and redox tuning may be realized by the self-assembly of mixed-metal films with a rationally controlled mesostructure.

References

1. Yang Z, Zhang J, Kintner-Meyer MCW et al (2011) Chem Rev 111:3577–3613
2. Cook TR, Dogutan DK, Reece SY, Surendranath Y, Teets TS, Nocera DG (2010) Chem Rev 110:6474
3. Lewis NS, Nocera DG (2006) PNAS 103:15729
4. Grimes CA, Varghese OK, Ranjan S (2008) Light, water, hydrogen: the solar generation of hydrogen by water photoelectrolysis. Springer, New York
5. Lewis NS (2007) Science 315:798
6. Turner JA (1999) Science 285:687
7. Pijpers JJH, Winkler MT, Surendranath Y, Buonassisi T, Nocera DG (2011) Proc Natl Acad Sci USA 108:10056
8. Reece SY, Hamel JA, Sung K, Jarvi TD, Esswein AJ, Pijpers JJH, Nocera DG (2011) Science 334:645
9. Torella JP, Gagliardi CJ, Chen JS, Bediako DK, Colón B, Way JC, Silver PA, Nocera DG (2015) Proc Natl Acad Sci USA 112:2337
10. Nocera DG (2009) Inorg Chem 48:10001
11. Nocera DG (2009) ChemSusChem 2:387
12. Huynh M, Bediako DK, Nocera DG (2014) J Am Chem Soc 136:6002

13. Huynh M, Bediako DK, Liu Y, Nocera DG (2014) J Phys Chem C 118:17142
14. Dincă M, Surendranath Y, Nocera DG (2010) Proc Natl Acad Sci USA 107:10337
15. Bediako DK, Lassalle-Kaiser B, Surendranath Y, Yano J, Yachandra VK, Nocera DG (2012) J Am Chem Soc 134:6801
16. Bediako DK, Surendranath Y, Nocera DG (2013) J Am Chem Soc 135:3662
17. Gerken JB, Chen JYC, Massé RC, Powell AB, Stahl SS (2012) Angew Chem Int Ed 51:6676
18. Gerken JB, Shaner SE, Massé RC, Porubsky NJ, Stahl SS (2014) Energy Environ Sci 7:2376
19. Suntivich J, Gasteiger HA, Yabuuchi N, Nakanishi H, Goodenough HB, Shao-Horn Y (2011) Nat Chem 3:546
20. Li X, Walsh FC, Pletcher D (2011) Phys Chem Chem Phys 13:1162
21. Suntivich J, Gasteiger HA, Yabuuchi N, Nakanishi H, Goodenough HB, Shao-Horn Y (2011) Science 334:1383
22. Trotochaud L, Ranney JK, Williams KN, Boettcher SW (2012) J Am Chem Soc 134:17253
23. Subbaraman R, Tripkovic D, Chang K-C, Strmcnik D, Paulikas AP, Hirunsit P, Chan M, Greeley J, Stamenkovic V, Markovic NM (2012) Nature Mat 11:550
24. Smith RDL, Prévot MS, Fagan RD, Zhang Z, Sedach PA, Siu MKJ, Trudel S, Berlinguette CP (2013) Science 340:60
25. Louie MW, Bell AT (2013) J Am Chem Soc 135:12329
26. McCrory CCL, Jung S, Peters JC, Jaramillo TF (2013) J Am Chem Soc 135:16977
27. Lu Z, Wang H, Kong D, Yan K, Hsu P-C, Zheng G, Yao H, Liang Z, Sun X, Cui Y (2014) Nature Commun 5:4345
28. Trotochaud L, Young SL, Ranney JK, Boettcher SW (2014) J Am Chem Soc 136:6744
29. Hunter BM, Blakemore JD, Deimund M, Gray HB, Winkler JR, Müller AM (2014) J Am Chem Soc 136:13118
30. Li C, Han X, Cheng F, Hu Y, Chen C, Chen J (2015) Nature Commun 6:7365
31. Friebel D, Louie MW, Bajdich M, Sanwald KE, Cai Y, Wise AM, Cheng M-J, Sokaras D, Weng T-C, Alonso-Mori R, Davis RC, Barger JR, Nørskov JK, Nilsson A, Bell AT (2015) J Am Chem Soc 137:1305
32. Zakaria MB, Hu M, Pramanik M, Li C, Tang J, Aldalbani A, Alshehri SM, Malgras V, Yamauchi Y (2015) Chem Asian J. doi:10.1002/asia.201500245
33. Du J, Chen C, Cheng F, Chen J (2015) Inorg Chem 54:5467
34. Gong M, Dai H (2014) Nano Res 8:239
35. Kanan MW, Nocera DG (2008) Science 321:1072
36. Kanan MW, Surendranath Y, Nocera DG (2009) Chem Soc Rev 38:109
37. Surendranath Y, Dinca M, Nocera DG (2009) J Am Chem Soc 131:2615
38. Suzuki O, Takahashi M, Fukunaga T, Kuboyama J (1968) US Pat 3 399 966
39. Shafirovich VY, Strelets VV (1978) Nouv J Chim 2:199
40. Esswein AJ, Surendranath Y, Reece SY, Nocera DG (2011) Energy Environ Sci 4:499
41. Liu Y, Nocera DG (2014) J Phys Chem C 118:17060
42. Surendranath Y, Nocera DG (2011) Prog Inorg Chem 57:505
43. Risch M, Khare V, Zaharieva I, Gerencser L, Chernev P, Dau H (2009) J Am Chem Soc 131:6936
44. Kanan MW, Yano J, Surendranath Y, Dincă M, Yachandra VK, Nocera DG (2010) J Am Chem Soc 132:13692
45. Risch M, Klingan K, Ringleb F, Chernev P, Zaharieva I, Fischer A, Dau H (2012) ChemSusChem 5:542
46. Egami T, Billinge SJL (2003) Underneath the Bragg-peaks: structural analysis of complex materials. Plenum, Oxford
47. Billinge SJL, Kanatzidis MG (2004) Chem Commun 2004:749
48. Du P, Kokhan O, Chapman KW, Chupas PJ, Tiede DM (2012) J Am Chem Soc 134:11096
49. Farrow CL, Bediako DK, Surendranath Y, Nocera DG, Billinge SJL (2013) J Am Chem Soc 135:6403

50. Bianconi A (1988) In: Koningsberger DC, Prins R (eds) X-Ray absorption: principles, applications, techniques of EXAFS, SEXAFS and XANES. Wiley, New York, pp 573–662
51. Brunschwig BS, Chou MH, Creutz C, Ghosh P, Sutin N (1983) J Am Chem Soc 105:4832
52. Sisley MJ, Jordan RB (2006) Inorg Chem 45:10758
53. Surendranath Y, Lutterman DA, Liu Y, Nocera DG (2012) J Am Chem Soc 134:6326
54. Chivot J, Mendoza L, Mansour C, Pauporte T, Cassir M (2008) Corrosion Sci 50:62
55. Strobel P, Muguerra H, Hebert S, Pachoud E, Colin C, Julien M (2009) J Sol State Chem 182:1872
56. McAlpin JG, Surendranath Y, Dincă M, Stich TA, Stoian SA, Casey WH, Nocera DG, Britt RD (2010) J Am Chem Soc 132:6882
57. Gerken JB, McAlpin JG, Chen JYC, Rigsby ML, Casey WH, Britt RD, Stahl SS (2011) J Am Chem Soc 133:14431
58. McAlpin JG, Stich TA, Ohlin CA, Surendranath Y, Nocera DG, Casey WH, Britt RD (2011) J Am Chem Soc 133:15444
59. Surendranath Y, Kanan MW, Nocera DG (2010) J Am Chem Soc 132:16501
60. Gileadi E (1993) Electrode kinetics for chemists, chemical engineers, and materials scientists. Wiley-VCH, New York, pp 127–184
61. Lyons MEG, Brandon MP (2008) Int J Electrochem Sci 3:1425
62. MacDonald JJ, Conway BE (1962) Proc R Soc Lond 269:419
63. Damjanovic A, Jovanovic B (1976) J Electrochem Soc 123:374
64. Trasatti S (1994) In: Ross PN, Lipkowski J (eds) Electrochemistry of novel materials. VCH, New York, Chap 5
65. Daghetti A, Lodi G, Trasatti S (1983) Mater Chem Phys 8:1
66. Wang L-P, Van Voorhis T (2011) J Phys Chem Lett 2:2200
67. Bediako DK, Costentin C, Jones EC, Nocera DG, Savéant J-M (2013) J Am Chem Soc 135:10492
68. Cukier RI, Nocera DG (1998) Annu Rev Phys Chem 49:337
69. Hammes-Schiffer S (2010) Chem Rev 110:6937
70. Savéant J-M (2012) Energy Environ Sci 5:7718
71. Mayer JM (2004) Annu Rev Phys Chem 55:363
72. Huynh MHV, Meyer TJ (2007) Chem Rev 107:5004
73. Costentin C (2008) Chem Rev 108:2145
74. Hammarström L, Styring S (2008) Phil Trans R Soc B 363:1283
75. Reece SY, Nocera DG (2009) Annu Rev Biochem 78:763
76. Hammes-Schiffer S (2009) Acc Chem Res 42:1881
77. Savéant J-M (1991) J Electroanal Chem 302:91
78. Chidsey CED, Feldman BJ, Lundgren C, Murray RW (1986) Anal Chem 58:601
79. Ofer D, Crooks RM, Wrighton MS (1990) J Am Chem Soc 112:7869
80. Jurss JW, Concepcion JC, Norris MR, Templeton JL, Meyer TJ (2010) Inorg Chem 49:3980
81. Chen Z, Concepcion JJ, Hull JF, Hoertz PG, Meyer TJ (2010) Dalton Trans 39:6950
82. May KJ, Carlton CE, Stoerzinger KA, Risch M, Suntivich J, Lee Y-L, Grimaud A, Shao-Horn Y (2012) J Phys Chem Lett 49:3980
83. González-Flores D, Sánchez I, Zaharieva I, Klingan K, Heidkamp J, Chernev P, Menezes PW, Driess M, Dau H, Montero ML (2015) Angew Chem Int Ed 54:2472
84. Scharifker B, Hills G (1983) Electrochim Acta 28:879
85. Bard AJ, Faulkner LR (2001) Electrochemical methods: fundamentals and applications. Wiley, New York, Chap 5
86. Zheng M, West AC (2004) J Electrochem Soc 151:C502
87. Kempa TJ, Bediako DK, Jones EC, Lieber CM, Nocera DG (2015) J Am Chem Soc 137:3739
88. Kempa TJ, Bediako DK, Kim SK, Park HG, Nocera DG (2015) Proc Natl Acad Sci USA 112:5309
89. Lutterman DA, Surendranath Y, Nocera DG (2009) J Am Chem Soc 131:3838
90. Gerken JB, Landis EC, Hamers RJ, Stahl SS (2010) ChemSusChem 3:1176

91. Stracke JJ, Finke RG (2011) J Am Chem Soc 133:14872
92. Limburg B, Bouwman E, Bonnet S (2012) Coord Chem Rev 256:1451
93. Hong D, Jung J, Park J, Yamada Y, Suenobu T, Lee Y-M, Nam W, Fukuzumi S (2012) Energy Environ Sci 5:7606
94. Crabtree RH (2012) Chem Rev 112:1536
95. Chakrabarty R, Bora SJ, Das BK (2007) Inorg Chem 46:9450
96. Chakrabarty R, Sarmah P, Saha B et al (2009) Inorg Chem 48:6371
97. Dimitrou K, Folting K, Streib WE, Christou G (1993) J Am Chem Soc 115:6432
98. Dimitrou K, Brown AD, Folting K, Christou G (1999) Inorg Chem 38:1834
99. Dimitrou K, Brown AD, Christou G et al (2001) Chem Comm 2001:1284
100. Ama T, Okamoto K, Yonemura T et al (1997) Chem Lett 26:1189
101. McCool NS, Robinson DM, Sheats JE, Dismukes GC (2011) J Am Chem Soc 133:11446
102. Berardi S, La Ganga G, Natali M et al (2012) J Am Chem Soc 134:11104
103. Zhang B, Li F, Yu F, Wang X, Zhou X, Li H, Jiang Y, Sun L (2014) ACS Catal 4:804
104. Ullman AM, Liu Y, Huynh M, Bediako DK, Wang H, Anderson BL, Powers DC, Breen JJ, Abruña HD, Nocera DG (2014) J Am Chem Soc 136:17681
105. Symes MD, Surendranath Y, Lutterman DA, Nocera DG (2011) J Am Chem Soc 133:5174
106. Macartney DH, Sutin N (1985) Inorg Chem 24:3403
107. Endicott JF, Durham B, Kumar K (1982) Inorg Chem 21:2437
108. Habib HS, Hunt JP (1966) J Am Chem Soc 88:1668
109. Marcus RA, Sutin N (1985) Biochim Biophys Acta 811:265
110. Ullman AM, Nocera DG (2013) J Am Chem Soc 135:15053
111. Zhou Y-L, Zeng M-H, Wei L-Q et al (2010) Chem Mater 22:4295
112. Murrie M (2010) Chem Soc Rev 39:1986

Top Curr Chem (2016) 371: 215–252
DOI: 10.1007/128_2015_642

Published online: 20 June 2015

Surface Plasmon-Assisted Solar Energy Conversion

Georgios Dodekatos, Stefan Schünemann, and Harun Tüysüz

Abstract The utilization of localized surface plasmon resonance (LSPR) from plasmonic noble metals in combination with semiconductors promises great improvements for visible light-driven photocatalysis, in particular for energy conversion. This review summarizes the basic principles of plasmonic photocatalysis, giving a comprehensive overview about the proposed mechanisms for enhancing the performance of photocatalytically active semiconductors with plasmonic devices and their applications for surface plasmon-assisted solar energy conversion. The main focus is on gold and, to a lesser extent, silver nanoparticles in combination with titania as semiconductor and their usage as active plasmonic photocatalysts. Recent advances in water splitting, hydrogen generation with sacrificial organic compounds, and CO_2 reduction to hydrocarbons for solar fuel production are highlighted. Finally, further improvements for plasmonic photocatalysts, regarding performance, stability, and economic feasibility, are discussed for surface plasmon-assisted solar energy conversion.

Keywords CO_2 reduction · Plasmonic catalysis · Solar energy for fuels · Water splitting

Contents

G. Dodekatos, S. Schünemann, and H. Tüysüz (✉)
Max-Planck-Institut für Kohlenforschung, Kaiser-Wilhelm-Platz 1, 45470 Mülheim an der Ruhr, Germany
e-mail: tueysuez@kofo.mpg.de

1 Introduction

Solar energy – as a clean and abundant energy source – has great potential for tackling the upcoming shortage of energy resources. The increasing energy demand and the great dependence of society on fossil fuels will eventually result in depleted energy supplies and advancing environmental pollution. The main issue of solar energy utilization is the efficient conversion to electricity or chemical energy. Furthermore, because of the natural fluctuation of this renewable energy, energy storage becomes another issue which has to be solved. In this regard, the conversion of solar energy to chemical fuels has attracted great interest. Hydrogen, a clean and energy dense compound, is a promising solar fuel. After the discovery of water splitting using TiO_2 photoanodes by Fujishima and Honda [1], much effort has been devoted to solar energy conversion to chemical energy through water splitting for hydrogen generation [2]. The idea of artificial photosynthesis for conducting thermodynamic uphill reactions to store energy in molecular bonds (and in doing so, mimicking nature) opens a path for overcoming the aforementioned environmental and energy problems.

In principle, photon absorption in semiconductors results in electron excitation with concomitant generation of holes, which can both be used to drive chemical reactions on the surface of the semiconductor. Generally, two main drawbacks have to be considered when semiconductors are used as photocatalysts. First, most semiconductors suffer from high bandgap energies and hence poor visible light harvesting (solar light consists of 44% visible light). Second, low quantum efficiencies hamper high photocatalytic performances; that is, generated charge carriers in the semiconductor recombine and the energy is lost in heat instead of driving catalysis on the surface. Many attempts such as transition metal doping [3], anion doping [4], dye sensitization [5], and heterojunctions have been made to overcome these drawbacks [6, 7].

Recently, the utilization of plasmonic metal nanoparticles for photocatalysis has gained great interest for boosting the performance of semiconductors [8]. The term plasmonic photocatalysis was coined by Awazu et al. in 2008, where the beneficial effect of combining plasmonic metals with semiconductor photocatalysts was demonstrated [9]. Plasmonic metals with their exceptional properties have resulted in various applications reported in the literature ranging from the well-known effect for surface-enhanced Raman spectroscopy (SERS) [10, 11], photovoltaics [12], optical sensors [13], catalytic sensors [14], biotechnology [15, 16], and heterogeneous catalysis [17] among others. In that sense, plasmonic photocatalysts for solar fuel production is the next step for exploiting the great properties of these materials. The utilization of localized surface plasmon resonance (LSPR) offers a great opportunity for tackling the limiting factors of photocatalysts for efficient performance under visible light irradiation. Besides the use of plasmonic photocatalysts for solar fuel production in the form of H_2 via photo(electro)chemical water splitting [8, 18–30], they have also been used for H_2 production from organic compounds [31–42] and CO_2 reduction [43–49]. Furthermore, their applications in water purification [50–67] and organic synthesis [68–71] were also reported. Gold and silver nanoparticles (NPs) in combination with semiconductor photocatalysts are mainly investigated in the literature because of their outstanding extinction cross sections, i.e., the sum of absorption cross section and scattering cross section [72].

In this chapter we first show the basic concept of the LSPR effect in noble metal NPs and discuss the mechanisms for enhancing the photocatalytic activity of semiconductors. In the second half of this chapter we present some applications of plasmonic photocatalysts for solar energy conversion to chemical energy through water splitting to H_2 and CO_2 conversion to hydrocarbons.

2 Principles of Localized Surface Plasmon Resonance

In this section, the basic principle of the localized surface plasmon resonance (LSPR) effect is briefly discussed. A short introduction is necessary in order to understand the working process and to use this knowledge for understanding plasmonic photocatalysis, which is elucidated later. More elaborated and detailed introduction to the LSPR phenomenon is reported elsewhere [72–74].

Reducing the size of metals to the nanoscale changes the optical properties of these materials tremendously. Electromagnetic fields with suitable frequencies are capable of inducing coherent oscillations of free electrons in the conduction band of metal nanoparticles. When the electron cloud is shifted relative to the particle, Coulomb forces from the positive nuclei act as a restoring force, which results in oscillations of the electron cloud as schematically illustrated in Fig. 1.

LSPR occurs when the oscillation frequency of the electron density and the incident light are in resonance. The oscillation of the surface plasmon causes high absorption and scattering-cross sections, together with great enhancement of the local electromagnetic field near the particle's surface [75]. For some metals, including gold and silver, the absorption maximum lies within the visible range; this makes them suitable for visible light-driven photocatalysis. The absorption spectra of metal nanoparticles (NPs) can be tuned over a wide range within the visible light spectrum by varying different parameters, including the dielectric environment and morphology, dimensions, and type of metal of the plasmonic

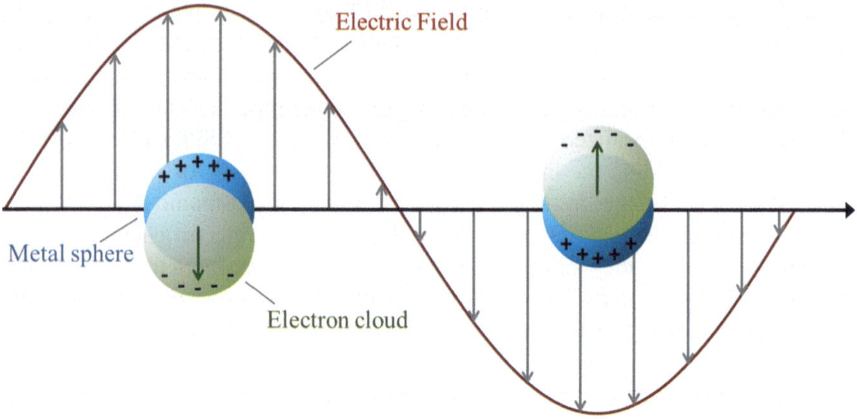

Fig. 1 Schematic plasmon oscillation for a metal sphere induced by the electric field of incident light

NP. All these factors influence not only the absorbance intensity but also the peak position and width. The dependence of the surface plasmon band position on the dielectric environment was studied experimentally [76–78] and theoretically by computer simulation [77, 79, 80]. For spherical particles, the extinction cross section can be calculated using Mie theory, which is derived from Maxwell's electromagnetic theory. The extinction cross section C_{ext} is given by

$$C_{\text{ext}} = \frac{2\pi}{k^2}\sum_{l=1}^{\infty}(2l+1)Re(a_l + b_l), \tag{1}$$

where $k = 2\pi/\lambda$ and the coefficients a_1 and b_1 are given by

$$a_l = \frac{m^2 j_l(x)[xj_l(x)]' - j_l(x)[mxj_l(mx)]'}{m^2 j_l(mx)\left[xh_l^{(1)}(x)\right]' - h_l^{(1)}(x)[mxj_l(mx)]'}, \tag{2}$$

$$b_l = \frac{j_l(x)[xj_l(x)]' - j_l(x)[mxj_l(mx)]'}{j_l(mx)\left[xh_l^{(1)}(x)\right]' - h_l^{(1)}(x)[mxj_l(mx)]'}, \tag{3}$$

where $x = ka$ (a being the sphere radius), $m^2 = \varepsilon_1/\varepsilon_0$, ($\varepsilon_1$ and ε_0 being the dielectric functions of the NP and the surrounding medium, respectively), $j_l(x)$ is the spherical Bessel function, and $h_l^{(1)}(x)$ is the spherical Hankel function [81]. Solving (1) for spheres with different diameters shows that the extinction maximum of NPs red-shifts (i.e., shifts to longer wavelengths) and broadens for larger particles and surrounding media with higher refractive indices [81].

The surface plasmon peak of the metallic particles depends on several parameters, including their composition, dimensions, size, and shape. Chen et al. [82]

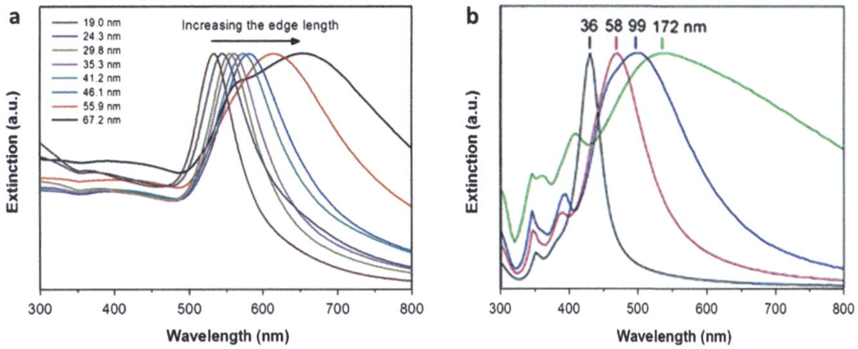

Fig. 2 (**a**) Size-dependent resonance wavelength of Au rhombic dodecahedra with different edge lengths. Reprinted with permission from [87]. Copyright 2013, Royal Society of Chemistry. (**b**) UV–vis extinction spectra of suspended Ag nanocubes with different edge lengths. Reprinted with permission from [88]. Copyright 2010, American Chemical Society

investigated the shift of the plasmon peak of gold nanoparticles with various shapes as a function of the refractive index of the environment. The authors found a linear dependence of the plasmon peak on the refractive index for all shapes studied. A linear response of the surface plasmon resonance was also found by discrete dipole approximation (DDA) [83]. Interestingly, the response to changes in the refractive index of the surrounding medium was found to be very different for different shapes. In Chen's study, Au nanospheres exhibited the smallest refractive index sensitivity whereas Au nanobranches exhibited the largest refractive index sensitivity. The strong dependence of the surface plasmon peak position on the surrounding medium is frequently used in biosensors. Further details on this topic are outside the scope of this chapter, but can be found in several reviews [13, 84, 85].

The size of plasmonic particles can easily be adjusted during synthesis, thus allowing for tuning of their light absorption abilities [86]. Figure 2 shows extinction spectra of Au rhombic dodecahedra [87] (Fig. 2a) and Ag nanocubes [88] (Fig. 2b). Both spectra show two major changes in the extinction coefficient as a function of edge length. First, increasing the edge length leads to a bathochromic shift (red shift) of the surface plasmon. Second, increasing the edge length leads to broader peaks. These two features are typically observed for plasmonic particles of all shapes [13, 89–91].

Another way to tune the LSPR to obtain the desired properties is to vary the shape of the plasmonic particles. In contrast to spherical NPs, nanorods possess two plasmon peaks because of their lower symmetry. One peak belongs to oscillations along the major axis whereas the other peak belongs to oscillations perpendicular to the major axis [91]. By varying the size of the nanorods along the major axis, the position of the plasmon absorption peak belonging to oscillations along the major axis shifts its spectral position, whereas the other peak (belonging to oscillations perpendicular to the major axis) is shifted only slightly [91].

3 Mechanisms of Plasmonic Photocatalysis

In many studies, plasmonic catalysts supported on semiconductors were reported to be highly active and selective for a wide range of reactions including selective oxidation of alcohols [71, 92], water splitting [8, 18, 29, 31, 93], and pollutant degradation [55, 58, 67, 94] under visible light irradiation. The mechanistic principles responsible for the high selectivities and activities under visible light illumination are the subject of controversial discussions in the literature and remain inconsistent within the field. Fundamentally, six different mechanisms have been discussed which include (1) direct electron transfer (DET), (2) local electromagnetic field enhancement (LEMF), (3) resonant energy transfer, (4) plasmonic heating, (5) light scattering, and (6) dipole–dipole interactions [8, 23, 95–101].

The type of mechanism for a certain reaction strongly depends on the design of the photocatalysts. For example, scattering only plays a limited role in small metal NPs, although separating the metal NPs from the semiconductor with an insulator prohibits the DET mechanism. Most attention is focused on the DET and the LEMF mechanism, but experimental evidence was also found for the other mechanisms. Herein, a brief introduction into the mechanistic details regarding the beneficial effect of LSPR towards the photocatalytic performance of semiconductors is given. The reader is referred to previous reviews discussing the proposed mechanisms in more detail [17, 100].

It should be mentioned that the so-called direct photocatalysis with plasmonic metal nanoparticles, where metal nanoparticles act both as light absorber and catalytic sites, is not the focus in this review. In contrast to indirect plasmonic photocatalysis, where plasmonic metal NPs enhance the performance of nearby semiconductors, in direct photocatalysis the photocatalytic reactions are catalyzed directly on the surface of metal nanoparticles. Hence, different reaction pathways are expected for direct and indirect plasmonic photocatalysis. For true direct plasmonic photocatalysis by metal nanoparticles, only (if at all) optically inactive supports should be used for preparation of the catalyst. Because there is an overlap in the literature regarding direct and indirect plasmonic photocatalysis for metal particles supported on various oxide supports, some reported results in this review might also be assigned to direct plasmonic photocatalysis. Nonetheless, it has to be emphasized that the main subject here is the synergistic effect of plasmonic metal nanoparticles and optically active semiconductors that can be used for solar energy conversion to fuel. The reader is referred to previous reviews covering the principle and applications of direct plasmonic photocatalysis [102–104].

3.1 Direct Electron Transfer

The direct electron transfer (DET) mechanism describes the injection of so-called hot electrons into the conduction band (CB) of semiconductors [8]. The conduction

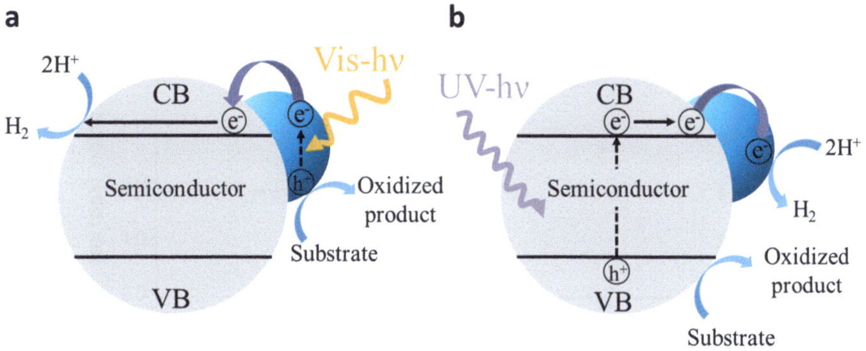

Fig. 3 (**a**) Charge carrier separation within the semiconductor under UV light irradiation and subsequent electron transfer to a metal NP. (**b**) Excitation of the surface plasmon of a metal NP by visible light and subsequent electron transfer via DET to the CB of a semiconductor

band electrons of the metal NP are excited through LSPR in the visible region of the electromagnetic spectrum. The remaining positive holes are capable of oxidizing different substrates, which is a key step for (selective) oxidation reactions. Under deaerated conditions in aqueous solution, the injected electrons reduce water to H_2 (Fig. 3a). This is in contrast to conventional photocatalysis (Fig. 3b), where photogenerated electrons in the semiconductor are transferred to the metal NPs, which act as electron sink. The beneficial charge separation effect resulting from the Schottky junction between semiconductor and metal NP [97] differs mechanistically from the effect of LSPR of plasmonic metal NPs [31]. Depending on the wavelength of the incident light, the electrons are either generated in the metal NP and subsequently transferred to the CB of the semiconductor (LSPR absorption wavelength, usually visible light) or transferred from the semiconductor's CB to the metal NP (irradiation suitable for bandgap excitation, usually UV light). Many research groups assigned their experimental observations to the direct electron transfer mechanism [26, 31, 50, 105–126].

The first experimental evidence for direct electron transfer from Au NPs to TiO_2 was found by Tian and Tatsuma in 2005 via photoelectrochemical analysis [127]. Spectroscopic measurements were conducted on Au/TiO_2 film photoanodes under white light irradiation in an N_2-saturated electrolyte containing ethanol as sacrificial electron donor. The observed increase in absorbance at 680 nm during irradiation was attributed to electrons in the CB of TiO_2 injected from Au NPs. Control experiments with bare TiO_2 photoanodes showed no absorbance changes at 680 nm. Later on, Tian's group reported similar results with visible light active Pt/TiO_2 nanocomposites [128]. Furthermore, reduced absorbance of the plasmon resonance of the Au NPs during white light irradiation in anaerobic conditions confirmed a decrease in electron density in the Au NPs because of electron transfer to TiO_2. This observation was further confirmed by Zheng et al. in 2011 using Au/TiO_2 microspheres, where the LSPR absorbance almost vanishes after 10 min visible light irradiation in N_2-saturated aqueous solution [129]. Femtosecond

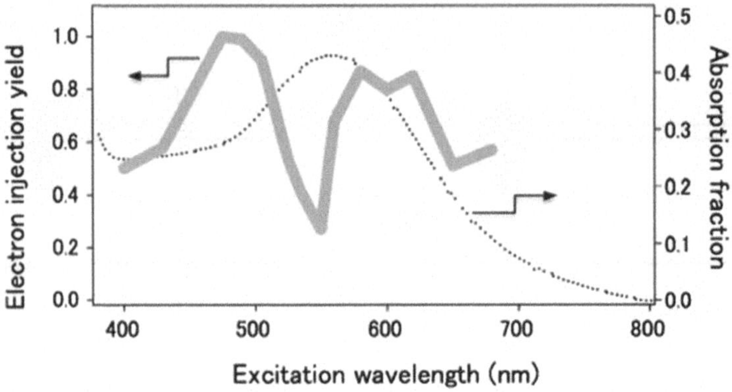

Fig. 4 Electron injection yield from gold NPs to TiO_2 NPs as a function of excitation wavelength (*gray solid line*) and absorption spectrum of the same film (*dotted line*). Reprinted with permission from [132]. Copyright 2013, Elsevier

transient absorption spectroscopy with simultaneous probing in the infrared (IR) region performed by Furube et al. in 2007 gave further indication of an electron transfer mechanism [130]. It was shown that the electron transfer from Au NPs to TiO_2 upon optical excitation of the plasmon band at 550 nm was accomplished within 240 fs (the time resolution of the apparatus). Later, more precise experiments showed that the electron transfer occurred in 50 fs or faster [131]. Interestingly, action spectrum analysis of Au/TiO_2 shows two maxima of the electron injection yield at different wavelengths deviating from the LSPR absorption maximum. It was suggested that two different electron injection pathways into TiO_2 occur, namely direct electron transfer and local electromagnetic field enhancement (Fig. 4) [132]. These results are in contrast to the results reported from Tian et al. [133] and Nishijima et al. [108] but in good agreement with the measurements of Priebe et al. [134].

Long et al. showed a poor match between the LSPR absorption of Au NPs on TiO_2 and photoelectrochemical measurements [38]. It was concluded that only high energy charge carriers excited at around 420 nm were able to migrate to TiO_2. Further evidence for DET was found by electron spin resonance (ESR) analysis. Visible light irradiation of Au/P25 (P25: anatase/rutile mixture = 4:1) in an oxygen atmosphere results in strong ESR signals assigned to peroxo-type oxygen anions ($O–O^-$) formed at the Au/TiO_2 interface [92]. However, no ESR signals were detected when Au/P25 was not irradiated and only weak signals were detected for analogous Au/rutile and Au/anatase samples. These results indicate that the contact between Au NPs and both TiO_2 phases is necessary for efficient O_2 reduction. Priebe et al. furthermore correlated the ESR signal intensities of conduction band electrons at O vacancies of Au/P25 with the irradiation wavelength and showed a matching response to the LSPR of Au [134]. Long et al. tracked the ESR signal intensity of Ti^{3+}-surface states for Au/TiO_2 (anatase nanosheets) under visible light irradiation and also assigned their results to an electron injection mechanism [38].

The previous examples with Au and TiO_2 demonstrate the concept of DET. Certainly, the applicability of the DET mechanism is not only limited to the case of Au/TiO_2, but depends on the electronic structure of the metal/semiconductor combination. Figure 5 illustrates the band edge positions of some common semiconductor materials used for photocatalysis and the position of the work function for typical metals used for plasmonic photocatalysis. The figure also shows the redox potential for some important redox couples. The oxidation of water to form molecular oxygen and protons and the reduction of protons to form molecular hydrogen (i.e., overall water splitting) are key reactions for solving the energy problem, as hydrogen and oxygen can be used in fuel cells to generate electrical energy without forming environmentally harmful products or greenhouse gases such as CO_2. As seen in Fig. 5, overall water splitting with metal NPs supported on semiconductors under light illumination with wavelengths greater than the bandgap of the semiconductor is not possible. This is because the work function of metals suitable for visible light plasmonic photocatalysis lies below the electrochemical potential for the reduction of protons to produce molecular hydrogen and above the potential of the water oxidation half reaction to produce molecular oxygen [17]. However, semiconductors decorated with metal NPs show increased activity for overall water splitting under illumination with wavelengths smaller than their bandgap energy. This is because of an improved charge carrier life time, as the CB electrons are driven to the metal NP, which reduces the charge carrier recombination (Fig. 3b) [135]. Furthermore, metal NPs themselves possess strong catalytic activity and act as active sites for H_2 evolution [2].

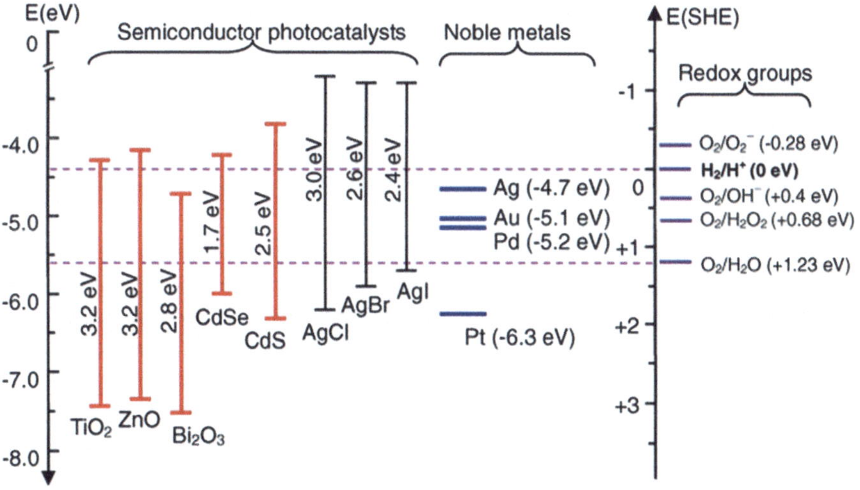

Fig. 5 Band positions of common semiconductors in photocatalysis, positions of the work function of metals used for plasmonic catalysis, and various redox potentials for different substrates vs standard hydrogen electrode (SHE). Adapted with permission from [17]. Copyright 2013, American Institute of Physics

The excited electrons in the CB of suitable semiconductors such as TiO_2 that are transferred from the surface plasmon states of metal NPs lie sufficiently high to allow for the reduction half reaction to produce hydrogen. When a suitable hole scavenger such as methanol [136], ethanol [137], or EDTA [31] is added to the reaction suspension, hydrogen formation can be observed [8, 18, 31].

3.2 Local Electromagnetic Field Enhancement

Another proposed mechanism for plasmonic photocatalysis is the enhanced charge carrier separation in a semiconductor material by strong local electromagnetic fields created by LSPR near the metal NP surface, called local-electromagnetic-field enhancement (LEMF). This is categorized as a near-field effect in contrast to the far-field effect assigned to scattering effects (see below). Besides the DET mechanism, the LEMF mechanism as source for the photocatalytic activity enhancement of semiconductors is also discussed extensively in the literature [9, 20, 21, 24, 30, 43, 93, 121, 138–144].

When a metal NP absorbs light with wavelengths near the LSPR, a local electromagnetic field is produced near the surface of the metal NP. Resulting from the strong electromagnetic field, the rate of charge carrier separation in the semiconductor is strongly increased, which leads to an increased rate for photocatalytic reactions [8, 12, 145]. The intensity of the electromagnetic field induced by surface plasmon resonance is strongly localized to the metal NP surface and decays exponentially perpendicularly from the metal NP/semiconductor interface [145].

The intensity of the electromagnetic field can be calculated using Maxwell's classic electromagnetism theory employing the finite-difference time domain method (FDTD). FDTD simulations show that the electromagnetic field around Ag nanocubes with an edge length of 75 nm is increased by a factor of 1,000 compared to the incident field. An even higher increase of the field strength was calculated for two Ag cubes separated by 1 nm where one cube is rotated by 45° (Fig. 6) [8].

The calculations showed that LSPR can enhance the local electric field by a factor of up to 10^6 at hotspots where two plasmonic particles are in close proximity to each other [8]. In an early study on the impact of the LEMF on photocatalytic performance from 2011, Ingram et al. investigated photoelectrochemical water splitting with Au spheres and Ag nanocubes deposited on nitrogen-doped TiO_2 (N-TiO_2) [20]. Under visible light irradiation, a tenfold higher photocurrent was observed with Ag/N-TiO_2 compared to N-TiO_2, whereas Au/N-TiO_2 had only a small effect on the photocurrent. The authors concluded that the intense electromagnetic field generated through LSPR of the metal NPs resulted in higher photoelectrochemical performance. The lack of overlap between the Au LSPR and the absorbance region of N-TiO_2 was used as an explanation for the low activity. On the other hand, the Ag nanocube's LSPR matches with the absorbance

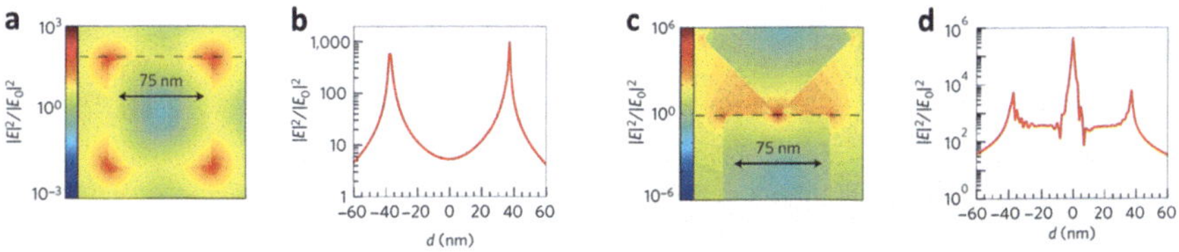

Fig. 6 (**a**) Spatial distribution of the electric field intensity at the LSPR peak wavelength, from an FDTD simulation of a 75-nm Ag nanocube and (**b**) electric field enhancement along the *dashed line* in (**a**). (**c**) Spatial distribution of the electric field intensity at the LSPR peak wavelength, from a FDTD simulation of two 75-nm Ag nanocubes separated by 1 nm (one Ag cube is rotated by 45°) and (**d**) electric field enhancement along the dashed line in (**c**). Reprinted with permission from [8]. Copyright 2011, Nature Publishing Group

of the semiconductor so that the intense electric field generated was able to induce charge separation in the semiconductor. Furthermore, the organic stabilizer (polyvinylpyrrolidone) coated around Au and Ag NPs and the limited direct contact of NPs on N-TiO_2 prohibited a direct electron transfer. Previously, the same group investigated the electron transfer from Ag NPs to N-TiO_2 via absorption studies at 680 nm under illumination at 365 nm. As mentioned before, electron transfer to TiO_2 results in an absorption peak at 680 nm. The lack of this peak confirmed that no electron transfer occurred [53]. Later on, Ingram et al. used a predictive model for the determination of the relative rate enhancement of the photocatalytic decomposition of methylene blue (MB) based on the overlap of the LSPR and semiconductor absorbance spectra [146]. Liu et al. investigated the photocatalytic activity for photoelectrochemical water splitting of TiO_2 and Au/TiO_2 under UV and visible light irradiation [18]. The authors found the photocurrent increased by a factor of 66 under visible light illumination ($\lambda = 633$ nm) for Au/TiO_2 compared to bare TiO_2. FDTD simulations of the electric field of the catalyst revealed hotspots which reach electric field intensities 1,000× higher than the incident electromagnetic field. In good agreement with the FDTD simulation discussed above, Liu et al. found that hotspots are observed when two gold NPs are in close proximity to each other (Fig. 7a–c) [18].

The presence of hotspots at nearly touching plasmonic particles is a recurring phenomenon which was also observed by other research groups [147–149]. Zhang et al. could show that, by varying the shell thickness of the SiO_2 interlayer and the TiO_2 layer for Ag@SiO_2@TiO_2 core-shell photocatalysts, an optimum structure could be obtained for the highest photocatalytic activity [150]. After comparing thicknesses of 2, 4, 8, and 20 nm for the SiO_2 interlayer and 2, 10, 20, and 40 nm for the TiO_2 shell thickness, the highest activity was observed for 2 nm SiO_2 interlayer and 20 nm TiO_2 shell exhibiting 31× larger photocatalytic activity for MB degradation under visible light irradiation compared to pristine P25. Clearly, the short distance between the Ag core and the TiO_2 shell enabled efficient charge generation in TiO_2 via LEMF. The enhancement of the electromagnetic field does not exclusively depend on the interparticle distance of two plasmonic particles but on a number of different factors such as the particle size and shape, the surrounding

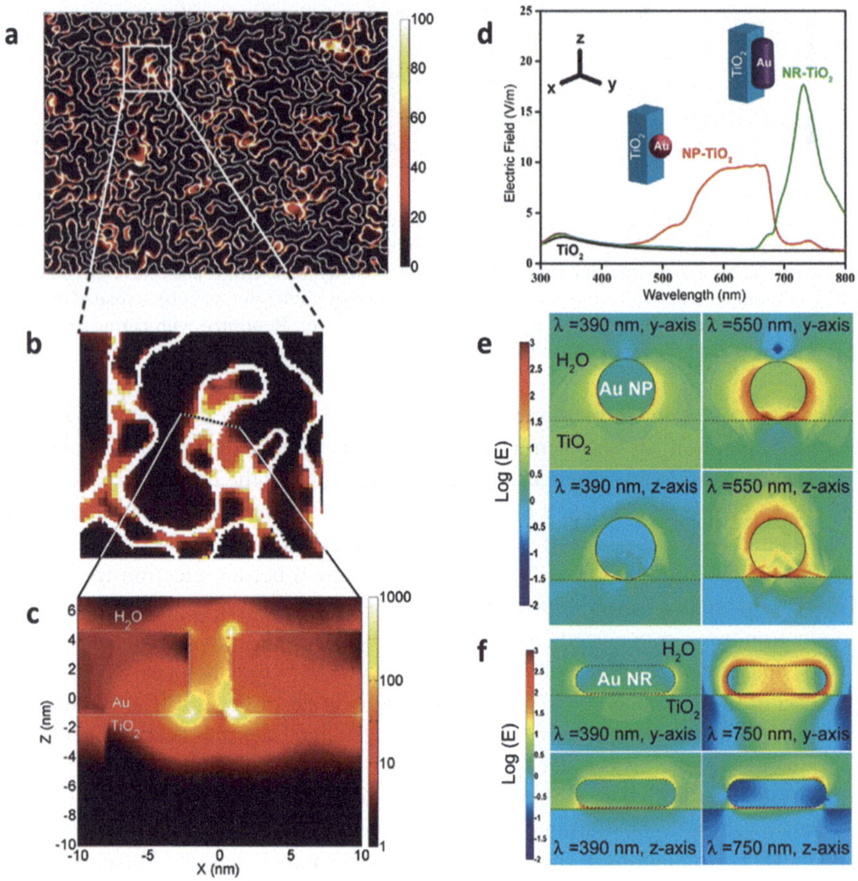

Fig. 7 (**a–c**) FDTD calculation of the electromagnetic field enhancement of Au-TiO_2 interface. Reprinted with permission from [18]. Copyright 2011, American Chemical Society. (**d**) Electric field intensity for bare TiO_2, NP-TiO_2 and nanorod TiO_2 as a function of incident light wavelength (FDTD simulation). (**e**, **f**) Electric field on the *y–z* plane for NP-TiO_2 and NR-TiO_2. Light is incident along the *y* or *z* axis. Reprinted with permission from [27]. Copyright 2013, American Chemical Society

dielectric environment [75, 147], and the incident wavelength [28]. Certainly, the strong dependence of the LEMF on these factors can be used to design photocatalysts with enhanced activity. Recently, Pu et al. developed a catalytic system for photoelectrochemical water oxidation [28]. Their photocatalyst was designed from two different gold nanoparticles, spheres and rods, which absorb light and thus enhance the local electromagnetic field in their vicinity at different wavelengths (Fig. 7d–f).

Because a broadened absorption within the visible range of the electromagnetic spectrum can be achieved by employing two different shapes of Au NPs, this

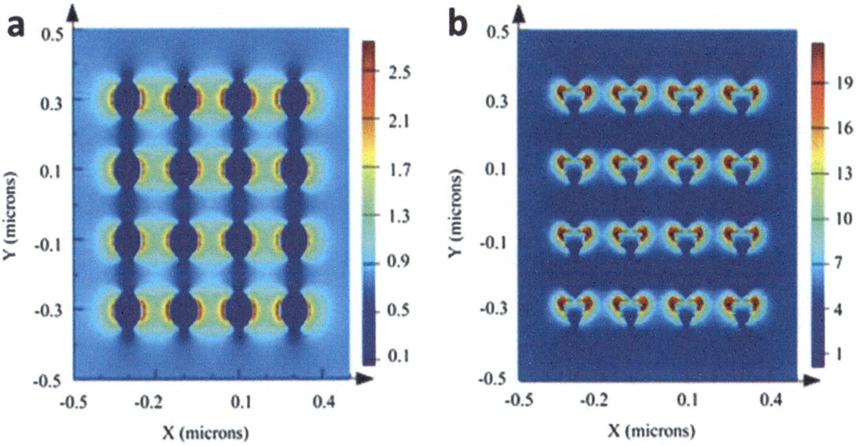

Fig. 8 FDTD simulation of the spatial distribution of the electric field intensity of (**a**) circular Ag and (**b**) triangular Ag at 550-nm irradiation. Reprinted with permission from [151]. Copyright 2014, Elsevier

constellation allows for a higher activity towards photoelectrochemical water oxidation throughout the entire UV–vis spectrum [28].

Another example of delicate structural tuning of a plasmonic catalyst is given by Zhang et al., who compared the photocurrent density of Ag/ZnO photoelectrodes with different shapes of Ag NPs [151]. It was found that the photocurrent density drastically increased by a factor of 3.1 when Ag nanoprisms instead of Ag nanospheres were deposited on a ZnO surface. FDTD simulations (Fig. 8) revealed that the electromagnetic field around the tips of Ag triangles is enhanced by a factor of 7.8 compared to the electric field around the edge of spherical particles at 550-nm irradiation, which causes the strongly increased photocurrent density [151].

3.3 *Resonant Energy Transfer*

The prerequisite for DET is the direct contact between the metal and semiconductor. In 2012, Cushing et al. reported visible light activity even when the metal NP and the semiconductor are separated by an insulating layer of SiO_2 which prevents DET. Therefore, Cushing et al. [96] proposed the resonant energy transfer (RET) mechanism to explain the enhanced photocatalytic activity of Au@Cu_2O core shell materials and Au@SiO_2@Cu_2O sandwich structures. It was found that the Au@SiO_2@Cu_2O sandwich structure with an insulating SiO_2 layer between Au NPs and the Cu_2O semiconductor was even more active than the Au@CuO_2 core shell material, even though DET is prohibited by the insulating SiO_2 layer. The experimental findings were explained by resonant energy transfer, in which electron-hole pairs are generated in the semiconductor by dipole–dipole

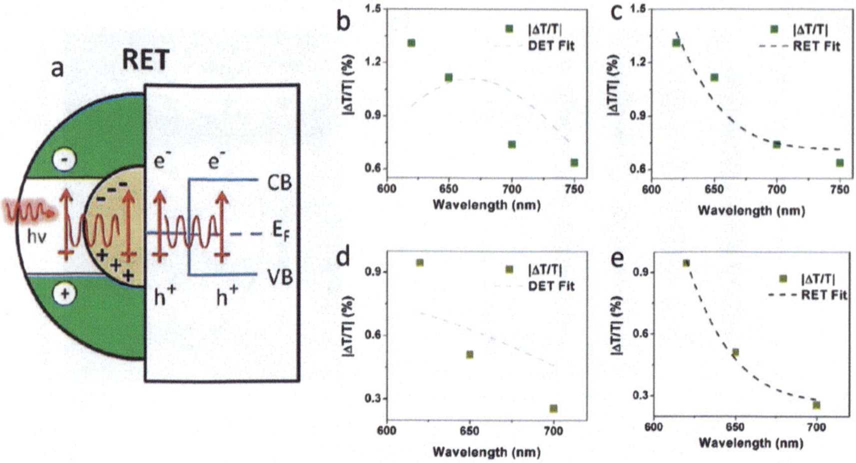

Fig. 9 (**a**) Resonant energy transfer (RET) from the metal LSPR dipole to the electron-hole pair in the semiconductor shell. Wavelength-dependent signal amplitudes for Au@Cu_2O NPs and fit with (**b**) DET and (**c**) RET models. DET (**d**) and RET (**e**) models and experimental results from Au@SiO_2@Cu_2O NPs. Reprinted with permission from [96]. Copyright 2012, American Chemical Society

interactions between metal NPs and semiconductor (Fig. 9a). This mechanism allows the generation of electron-hole pairs in the CB and VB of the semiconductor when using light with energies below that of the bandgap as a consequence of the high overlap integral of the metal NP's plasmon resonance and the Cu_2O band gap. Further evidence for the RET mechanism was found by transient-absorption measurements at different wavelengths near the absorbance maximum of the Au surface plasmon while spectroscopically probing the number of free charge carriers created by plasmonic energy transfer in Cu_2O, which is directly proportional to the relative change in transmission ($|\Delta T/T|$) at 800 nm. The charge carrier density follows the overlap integral between Au LSPR and Cu_2O density of states (RET Fit) (Fig. 9c, e). In contrast, when assuming DET as the predominant charge carrier generation mechanism, the expected change in transmission (DET Fit) does not match the experimental results (Fig. 9b, d), which suggest that RET is the predominating mechanism. The same group also found evidence for the RET mechanism using Ag@Cu_2O core shell systems [152].

3.4 Plasmonic Heating

It has long been known that the efficient absorption of visible light of various metal NPs in small volumes is a source of localized heat on the nanoscale [153–155]. Responsible for the heat generation on the nanoscale is the local electromagnetic field enhancement [156–158]. Even though local heating to temperatures as high as 600 °C was observed [159], plasmonic heating is rarely discussed as the responsible mechanism for the increased photocatalytic activity found for plasmonic photocatalysts [154]. Baffou et al. [158] studied the formation of microbubbles in water when plasmonic Au NPs were irradiated at wavelengths corresponding to their surface plasmon. In this study, a local temperature of up to 220 °C by plasmonic heating was observed. It was shown that plasmonic heating allows important industrial reactions such as ethylene epoxidation, CO oxidation, and NH_3 oxidation to run at lower temperatures compared to conventional catalysis, which makes the photocatalytic process more energy efficient [160]. For example, low-energy laser irradiation of plasmonic particles can be used to provide the necessary heat for reforming an ethanol-water mixture [161]. The localization of the heat allows miniaturization of heterogeneous catalysis by lab-on-a-chip applications. Plasmonic heating provides the necessary heat in the reactor, although the chip and the fluid lines remain at room temperature [161]. The facile control of the reaction temperature enabled by plasmonic heating can also be used to change product distributions. Wang et al. controlled the temperature of Au NPs in the range from 30 to approximately 600 °C by varying the laser intensity, which allowed for tuning of the product distribution, namely CH_4 and CO, from the reaction of CO_2 and H_2 (the reverse of steam reforming) [159].

It should be mentioned that plasmonic heating is not limited to catalytic applications. Before scientists showed interest in plasmonic heating for catalytic applications, it was utilized in biomedical treatments such as plasmonic photothermal therapy, which uses the local heat near the metal NPs to photo-damage, for example, cancer cells [153].

Direct or indirect temperature measurement with the required high spatial resolution is difficult. However, numerical calculations of the heat evolved by plasmonic heating are possible [156]. Baffou et al. [156] calculated the heat evolved by plasmonic heating for a large range of different morphologies such as spheres, rods, triangles, and connected disks with identical volumes by employing Green's dyadic method. It was found that nanorods are more efficient heaters than nanospheres, and that thinner structures are more efficient plasmonic heaters than bulky particles (Fig. 10). Seemingly, particles with a high surface area/volume ratio are more efficient for plasmonic heating. The results show that heat evolves mainly in the outer part of particles facing the incoming light. Bulk regions and regions opposing the incoming light are less efficient plasmonic heaters, which can be explained by shielding effects from the plasmonic particles. Consequently, particles with a low surface area (e.g., spheres) are less effective plasmonic heaters than particles with a high surface area, considering the same volume (e.g., disks) [156]. The heat generation power can be enhanced by improving the electric fields inside the plasmonic particles, which is in contrast to LEMF where the electric fields have to be optimized outside the plasmonic particle. Thus, hotspots of the electromagnetic field do not inevitably occur when thermal hotspots are present and vice versa [154].

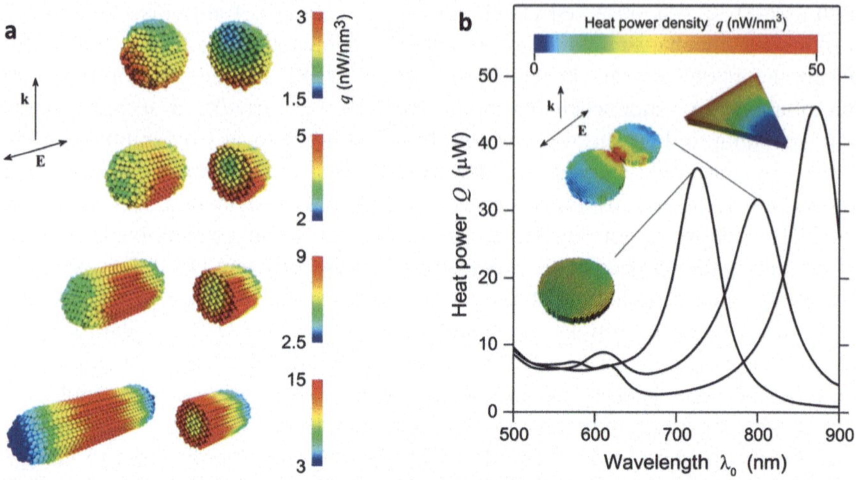

Fig. 10 (**a**) Three-dimensional mapping of the heat power density computed for different nanoparticles at their respective plasmon resonance with cross sections and the vector of the incoming light. (**b**) Calculated spectra of the heat generated in structures deposited on a planar glass surface immersed in water. The *three insets* represent the 3D heat power density computed at the main plasmon resonance of the particle. Reprinted with permission from [156]. Copyright 2009, AIP Publishing LLC

3.5 Scattering

Scattering is another phenomenon that needs to be discussed as a mechanism of plasmonic catalysis. Scattering occurs predominantly for large (>40 nm) metal NPs [89]. Excited surface plasmons can decay either non-radiatively by transferring the energy to hot electrons or radiatively by resonant scattering. Non-radiative decay is characteristic for smaller particles (<40 nm) whereas radiative decay of the surface plasmon by scattering resonant photons is characteristic for larger particles [162]. Illustratively, the activity enhancement caused by scattering at plasmonic particles can easily be understood as an elongation of the distance a photon travels through the catalyst because of scattering at multiple NPs (Fig. 11a). The extended path length through the catalyst increases the chance for a photon to be absorbed by the semiconductor and thus to create electron-hole pairs. The scattering properties of metal NPs strongly depend on the size and also on the shape of the nanoparticle (Fig. 11b, c). Enhanced photocatalytic activity is most often described to be the result of increased scattering on large Ag particles [8, 9, 150, 164–166].

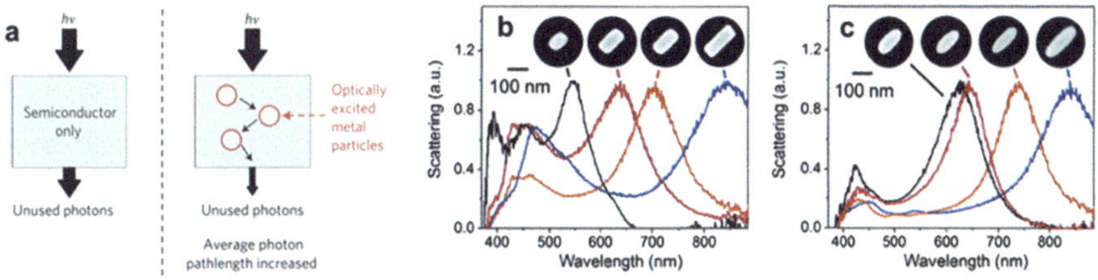

Fig. 11 (**a**) Increased photon path length caused by scattering. Reprinted with permission from [8]. Copyright 2011, Nature Publishing Group. (**b**) Normalized scattering spectra of nanorods with different aspect ratios and SEM images of the corresponding particles. (**c**) Normalized scattering spectra of nanorice and SEM images of the corresponding particles. Reprinted with permission from [163]. Copyright 2007, American Chemical Society

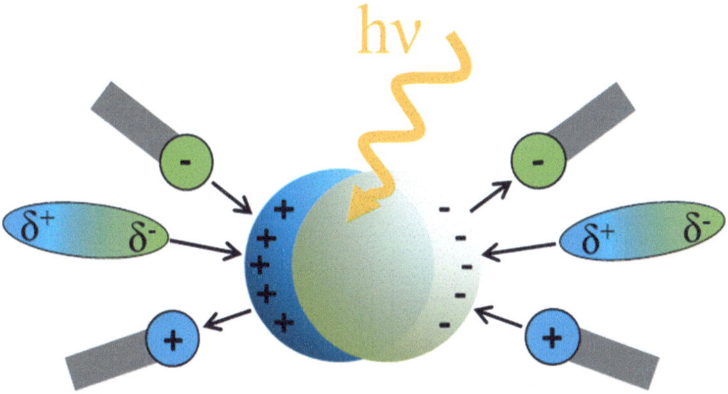

Fig. 12 Light induced dipole of a metal NP and dipole–dipole interaction with polar, nonpolar, and ionic compounds

3.6 Dipole–Dipole Interactions

The light induced oscillation of the surface plasmon results in a fast-varying dipole (Fig. 1). Thus, polar molecules such as alcohols, thiols, aldehydes, etc., are attracted to the metal surface by dipole–dipole interactions, which increase their local concentration in proximity of the plasmonic particles, i.e., the catalytically active center. Furthermore, the dipole of the plasmonic particle also polarizes nonpolar molecules and selectively attracts and repels ions (Fig. 12) [17].

4 Surface Plasmon-Assisted Solar Energy Conversion

4.1 Water Splitting

Since Liu et al. [18] first demonstrated the use of plasmonic photocatalysts as photoanodes for photoelectrochemical water splitting, much effort has been made to achieve visible light active photoelectrodes for this purpose. In general, various enhancement factors for photoelectrochemical water splitting are reported by employing plasmonic nanostructures on semiconductor electrodes. Beginning with the work of Liu et al. in 2011, they showed a 5-fold enhancement at $\lambda = 532$ nm and a 66-fold enhancement at $\lambda = 633$ nm for anodic TiO_2 films with deposited Au NPs compared to neat anodic TiO_2 electrodes (Fig. 13a) [18]. This was attributed to the creation of high intensity electric fields through the LEMF mechanism (Sect. 3.2). Wang et al. demonstrated about threefold enhanced photocurrent by the fabrication of Au/TiO_2/Au nanosheets grown on Ti foil compared to neat TiO_2 nanosheets under visible light irradiation [24]. Abdi et al. showed a 2.5-fold higher photocurrent under 1 sun illumination for $BiVO_4$ photoanodes by depositing

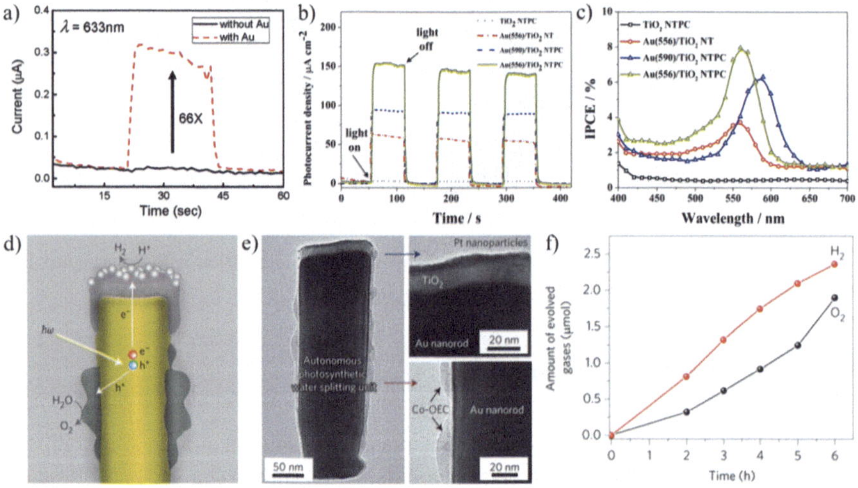

Fig. 13 (**a**) Photocurrent measurement for anodic TiO_2 with and without Au NPs under $\lambda = 633$ nm irradiation for 22 s. Reprinted with permission from [18]. Copyright 2011, American Chemical Society. (**b**) Photocurrent measurements at 1.23 V vs RHE under $\lambda > 420$ nm irradiation with 60 s light on/off cycles for TiO_2 nanotubes (NTs), TiO_2 nanotube photonic crystals (NTPCs) with and without Au NPs with different LSPR absorption wavelengths (556) and (590). (**c**) Corresponding IPCE measurements revealing the performance enhancements in the visible light region in accordance to the LSPR absorption. Reprinted with permission from [27]. Copyright 2013, American Chemical Society. (**d**, **e**) Schematic cross-section and TEM images of an individual unit of an autonomous water splitting device consisting of an Au nanorod coated with Pt NPs decorated TiO_2 and an Co-based oxygen evolution catalyst (OEC). (**f**) H_2 and O_2 evolution over time under white light irradiation (AM 1.5 solar sun filter) for the system described in (**d**) and (**e**). Reprinted with permission [117]. Copyright 2013, Nature Publishing Group

$Ag@SiO_2$ core-shell NPs on the surface of the semiconductor [167]. Li et al. incorporated a hematite nanorod array into an Au nanohole array pattern on FTO (fluorine-doped tin oxide) glass as photoanode. They found an 18× enhancement in internal photon to current efficiency (IPCE) at 650 nm (1 sun illumination) compared with hematite on bare FTO because of the LSPR of the Au nanohole array [168]. The same group reported in 2014 an IPCE enhancement factor of 7 at 650 nm (1 sun illumination) for $CdS/Au/TiO_2$ sandwich structures compared with analogous samples without Au NPs [124]. Kong et al. were able to show an 83-fold enhancement in photocurrent generation under visible light irradiation for mesoporous iron oxide nanopyramid arrays coated with a layer of Au NPs compared to electrodes without Au NPs [169].

Another approach to enhance the photocatalytic performance of plasmonic photocatalysts is the combination with photonic crystals (PCs). PCs exhibit, because of a periodic change of the dielectric constant, a band of forbidden wavelengths, which prevents the propagation of photons with certain energies (similar to the electronic bandgap of a semiconductor). At the edges of the photonic bandgap, the propagation of photons is slowed down, which results in higher photon absorption and hence higher activity. Wang and co-workers could show remarkably increased photocurrents for photoelectrochemical water splitting with Au/TiO_2 nanotubes where a PC layer was seamlessly connected on top of the nanotube array (NTPC) [27]. The performance was significantly increased for the sample with matching Au LSPR band and photonic bandgap (Au(556)/TiO_2 NTPC compared to Au(590)/TiO_2 NTPC, Fig. 13b, c) and showed a 2.5× higher photocurrent compared to the analogous sample without photonic crystal structure (Au(556)/ TiO_2 NT).

It should be noted that the LSPR band and the photonic bandgap can be tuned by changing the size and morphology of the metal NPs (for LSPR band) and by the size of the inverse opals in the structure (for the photonic bandgap). Hence, it is possible to tune the structure of materials to have light harvesting efficiency over a broad range of the solar spectrum. Liu, Kang et al. fabricated a bi-layer photoanode consisting of Au NPs deposited on a TiO_2 nanorod array with a TiO_2 PC layer on top [170]. They could show a roughly 2× higher photocurrent under visible light irradiation and 1.6× higher photocurrents under simulated sunlight irradiation for the photoanode with PC layer compared to the sample without PC layer. Zhan et al. found enhanced water splitting activities by using different Au/TiO_2 electrode structures [171]. Three structures were investigated: Au NPs deposited onto the surface of a 300 nm thick TiO_2-layer (Au-on-TiO_2), Au NP-layer in between two 150 nm thick TiO_2-layers (Au-in-TiO_2), and two Au NP-layers embedded in between three 100 nm thick TiO_2-layers (3D Au-embedded TiO_2). The 3D Au-embedded TiO_2 showed the highest LSPR absorbance followed by Au-in-TiO_2 and Au-on-TiO_2. The group found a fivefold enhancement for the photocurrent under visible light irradiation for 3D Au-embedded TiO_2 and a threefold enhancement for Au-in-TiO_2 compared to Au-on-TiO_2, which was in good agreement with the observed Au LSPR absorbance. Erwin et al. reported higher efficiencies for plasmon-enhanced photocatalytic water splitting capability of Au–Ag bimetallic nanostructured cubes compared to simple Au nanospheres [172]. It was demonstrated that significantly fewer particles (by a factor of 245) were needed for the Au–Ag bimetallic nanostructures compared to Au nanospheres for achieving similar photocurrents. Nanomaterials were deposited together with P25 on FTO glass and used as photoanodes. The nanostructures were coated with a SiO_2 shell. The photocurrent enhancements were attributed to the LEMF mechanism. Moskovits' group prepared an autonomous plasmonic water splitting material capable of generating H_2 and O_2 (Fig. 13f) and thus serving simultaneously as photocathode and photoanode, respectively [117]. The photocatalyst with remarkable long-term operational stability consisted of a uniform Au nanorod array capped with a TiO_2 layer (Fig. 13d, e). A cobalt-based oxygen evolution catalyst was deposited on the exposed Au sides, whereas Pt NPs were deposited on TiO_2 as reducing sides. Considering high surface areas for enhancing the activity of photoanodes, DeSario et al. reported the preparation of gold-titania aerogels (3D Au/TiO_2) via the sol-gel method and investigated the performance in photoelectrochemical water splitting [173]. The synthesis contained the simultaneous gelation of colloidal alkanethiolate-stabilized Au nanoparticles in the Ti-precursor sol. Supercritical drying followed by calcination resulted in 5 nm sized Au NPs embedded in the TiO_2 network consisting of 12 nm big anatase crystallites. A second preparation route consisted of TiO_2 aerogels with Au NPs simply deposited via a deposition–precipitation method with urea (DP Au/TiO_2). Comparing the 3D Au/TiO_2 and DP Au/TiO_2 showed the superior visible light activity of 3D Au/TiO_2 (55-fold enhancement factor under visible light compared to TiO_2 aerogel) for photoelectrochemical water splitting. The higher performance was not only attributed to the higher surface area and porosity maintained in 3D Au/TiO_2 but also to the increased three-phase boundary sites consisting of Au/TiO_2/solvent and/or to the stronger interaction between Au LSPR and TiO_2 because of the embedded Au NPs. Although the concept of visible light active photoelectrodes via LSPR active metal NPs combined with semiconductors was demonstrated, the enhancement for visible light harvesting remains relatively low compared to conventional dye or quantum dot sensitization [174]. Therefore, further effort is necessary to obtain highly efficient photoelectrodes.

The use of sacrificial electron donors for photocatalytic hydrogen generation (such as methanol [175], ethanol [137], glycerol [176] etc. [177–179]) to replenish the photogenerated holes in the semiconductor or metal NP plays an important role in order to improve the performance of photocatalysts [180]. Moreover, the chemically more facile photocatalytic oxidation of biomass or biomass-derived compounds compared to water might give a sustainable photocatalytic reforming route for hydrogen production. Seh et al. investigated the effect of different morphologies of Au/TiO_2, either core-shell or Janus structure (i.e., the Au NP is non-centrosymmetrically attached to a TiO_2 particle, which results in a particle with different surface properties at opposing sides), on photocatalytic hydrogen generation in aqueous isopropanol solution under visible light irradiation [139]. They found that bare Au NPs (50 nm) and amorphous TiO_2 exhibited no or strongly diminished photocatalytic performance, respectively. On the other hand, core-shell or Janus structures of Au/TiO_2 (Fig. 14a, b) showed visible light activity

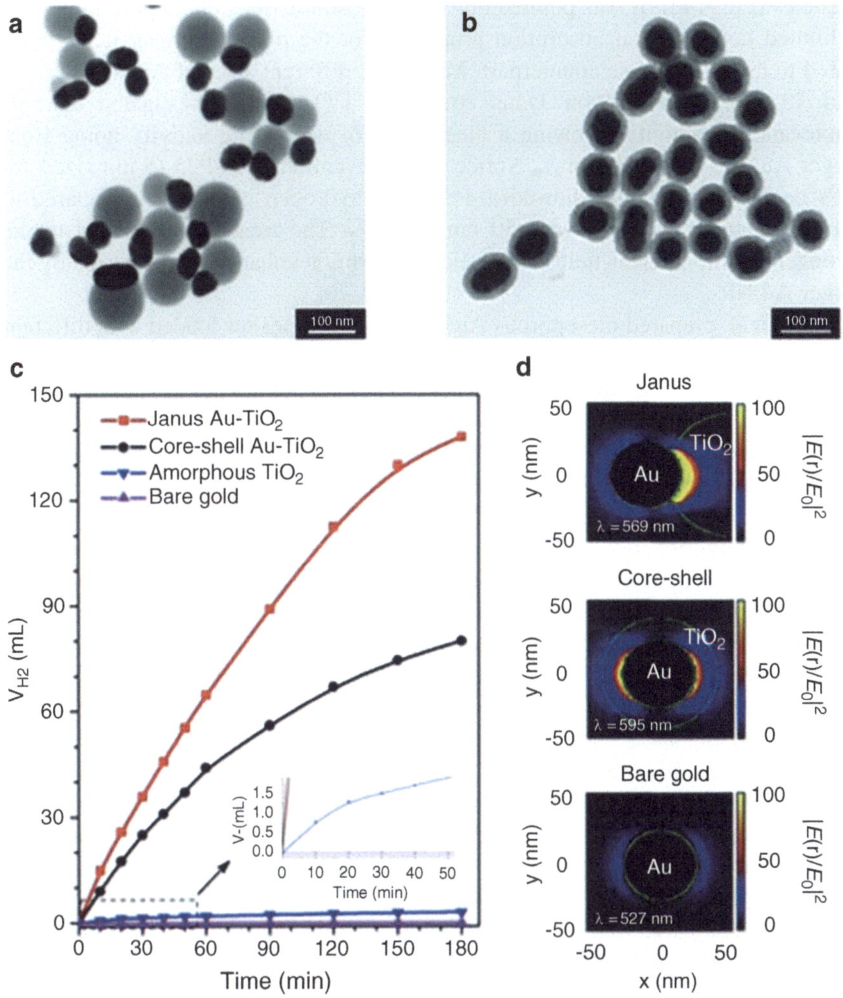

Fig. 14 TEM images of (**a**) Janus-structured and (**b**) core-shell-structured Au/TiO_2 materials. (**c**) Hydrogen evolution over time for different materials. Bare Au NPs and amorphous TiO_2 show negligible hydrogen production. (**d**) DDA simulations of plasmonic near-field maps for Janus, core-shell and bare Au NPs. The electric near-field intensity enhancements $|E(\mathbf{r})/E_0|^2$ are shown at each LSPR wavelength indicated. Reprinted with permission from [139]. Copyright 2012, John Wiley and Sons

with Janus structured Au/TiO_2 being the most active (Fig. 14c). The activity was attributed to the interaction of the generated electromagnetic fields near the Au NPs with the amorphous TiO_2. Discrete dipole approximation (DDA) simulations revealed a 1.75× larger absorbed power of plasmonic near-fields for amorphous TiO_2 with Janus structure compared to the core-shell structured Au/TiO_2. This was ascribed to the stronger localization of the plasmonic near-fields in the Janus particles (Fig. 14d). It was pointed out via DDA simulations that amorphous TiO_2 exhibited better optical absorption properties for the plasmonic near-fields compared to its bulk anatase counterpart. Moreover, different sizes of Au NPs (30, 50, and 70 nm) supported on Janus structured TiO_2 were investigated towards photocatalytic activity, showing a clear trend of decreasing activity going from larger Au NPs to smaller ones. Noticeably, conventional Au/P25 (5 nm sized Au NPs deposited on TiO_2) showed a 4× lower hydrogen generation compared to Janus structured Au/TiO_2 with 50 nm Au NPs. The trend was ascribed to the stronger plasmonic near-fields and optical absorption enhancements caused by the larger Au NPs.

Fang et al. prepared mesoporous Au/TiO_2 nanocomposites loaded with different amounts of Au and reported visible light activity for hydrogen production in aqueous solution with ascorbic acid as the sacrificial electron donor [181]. Au/TiO_2 materials prepared with Pluronic P123 copolymer as structure-directing agent resulted in higher activities compared to a control sample prepared without copolymer. According to Fang et al., this higher activity was not assigned solely to the higher surface area. Moreover, P123-assisted sol–gel preparation resulted in impurity and defect states in TiO_2 (confirmed by UV–vis spectroscopy). This slight visible light activity combined with the electromagnetic field enhancement of Au nanoparticles is claimed as one reason for the higher water reduction activity. The highest photocatalytic activity, observed for 2% Au/TiO_2, was attributed to electromagnetic field enhancement and direct electron transfer. Tanaka et al. reported visible light-driven H_2 generation from various alcohol–water (2-propanol, methanol, ethanol, glycerol, benzyl alcohol; 50 vol.%) and ammonia–water (14 wt%) mixtures with Au/TiO_2 photocatalysts [178]. The materials were prepared by a single-step (SS) or multi-step (MS) photodeposition method, where samples prepared by the MS method showed superior photocatalytic performance compared to the samples prepared by the SS method. Highest H_2 evolution rates were obtained in methanol/water mixtures with 1 wt% MS-Au/TiO_2 showing a 20× higher rate compared to 1 wt% SS-Au/TiO_2. The MS method results in a bimodal Au particle size distribution (1.4 and 13 nm), which is claimed to be one reason for the enhanced performance [36, 178]. Larger Au NPs were responsible for stronger absorbance because of LSPR and hence higher activity, whereas smaller Au NPs were claimed as active cocatalysts for H_2 production.

Higher activities through polydisperse Au NPs on TiO_2 were also observed by Rayalu et al. [36] Furthermore, the group very recently extended the studies on MS-Au/TiO_2 for H_2 generation using 2-propanol and NH_3 as sacrificial electron donors [182]. Different effects, such as amount of deposited Au, number of photodeposition steps, post-calcination treatment, etc., on the photocatalytic performance were investigated. In 2013 Tanaka et al. reported the preparation of 1 wt % Au/TiO_2 via colloidal photodeposition and addition of different cocatalysts via photodeposition (Pt, Pd, Ru, Rh, Ag, Cu, and Ir) and investigated the effect on H_2 generation with 2-propanol as substrate [183]. Remarkably, the highest photocatalytic performance was observed by depositing Pt NPs (0.5 wt%, 4 nm size) on Au/TiO_2 – roughly a fivefold higher H_2 evolution rate compared to 1 wt%

Au/TiO_2 consisting solely of large Au NPs (13 nm). Compared to the generation rate previously reported for Au/TiO_2 consisting of small and large Au NPs [178], the activity is enhanced by a factor of 2.4, indicating that Pt NPs act as effective reduction sites [137]. Pt/TiO_2 showed no activity under visible light irradiation because of the lack of visible light absorption. Accordingly, action spectrum analysis revealed the LSPR effect of Au NPs as a source of visible light activity. It was concluded that hot electrons generated through Au LSPR are injected into the CB of TiO_2 and migrate to the cocatalyst. In this regard, Zhang et al. could show that by using TiO_2 nanofibers decorated with Au and Pt NPs, negligible H_2 generation was observed under 550-nm irradiation, whereas simultaneous irradiation with 420 and 550 nm showed a remarkable increase in activity (2.5 × higher H_2 production rate compared to single 420-nm irradiation) [184]. It was assumed that, because of the higher conductivity of electrons in the CB band of TiO_2 photoexcited by 420-nm irradiation, the electron transfer from Au NPs (electrons generated by LSPR through 550-nm irradiation) to Pt NPs becomes feasible. A synergistic effect of UV and visible light irradiation was also observed by Li's group, who showed that illumination of Au/TiO_2 samples with UV and visible light resulted in higher hydrogen evolution rates than the sum of the evolution rates separately obtained under UV and under visible light irradiation [185]. It was concluded that excited electrons from Au LSPR are transferred to the CB of TiO_2. Furthermore, the local electric field generated enhances the charge separation in TiO_2 under UV light irradiation. Eventually, small Au NPs act as cocatalysts and trap the electrons from the CB of TiO_2, thus resulting in a triple synergistic promotion effect under UV–vis light irradiation. The size effect of the Au NPs on the photocatalytic performance for H_2 generation was further investigated by Wei and co-workers [126]. Two different visible light ranges ($\lambda > 400$ nm and $\lambda > 435$ nm) were investigated for Au/TiO_2, revealing two different mechanisms for H_2 generation. The deposition–precipitation method resulted in Au NPs with sizes of ca. 4 nm (S-Au/TiO_2) and the photodeposition method resulted in particle sizes of ca. 70 nm (L-Au/TiO_2) deposited on TiO_2. Under $\lambda > 400$ nm irradiation, it was shown that S-Au/TiO_2 had a 20× higher H_2 production than L-Au/TiO_2. On the other hand, under $\lambda > 435$ nm irradiation, L-Au/TiO_2 still showed activity, whereas S-Au/TiO_2 did not produce H_2 (Fig. 15a). It was claimed that for S-Au/TiO_2, a more efficient charge separation for small Au NPs (electrons transferred from TiO_2 to Au) resulted in enhanced activities under $\lambda > 400$ nm irradiation. This mechanism is typically observed under UV-light irradiation and requires light absorption from the semiconductor, which in the case of TiO_2 seems to occur to a slight extent even under $\lambda > 400$ nm irradiation. Because of the low LSPR absorption intensity of S-Au/TiO_2, an inefficient charge transfer from Au NPs to TiO_2 was suggested to be the reason for the absent H_2 production compared to L-Au/TiO_2 at $\lambda > 435$ nm irradiation. The stronger LSPR of L-Au/TiO_2 results in a more efficient charge transfer and a higher accumulation of electrons with higher reduction potential in the CB of TiO_2 (Fig. 15b).

Besides the dependency of the amount [181, 186–189] and size [31, 37, 64, 107, 187, 188, 190, 191] of loaded Au NPs on the photocatalytic performance, the exposed facets play an important role in H_2 generation as well. Cui and Lu showed

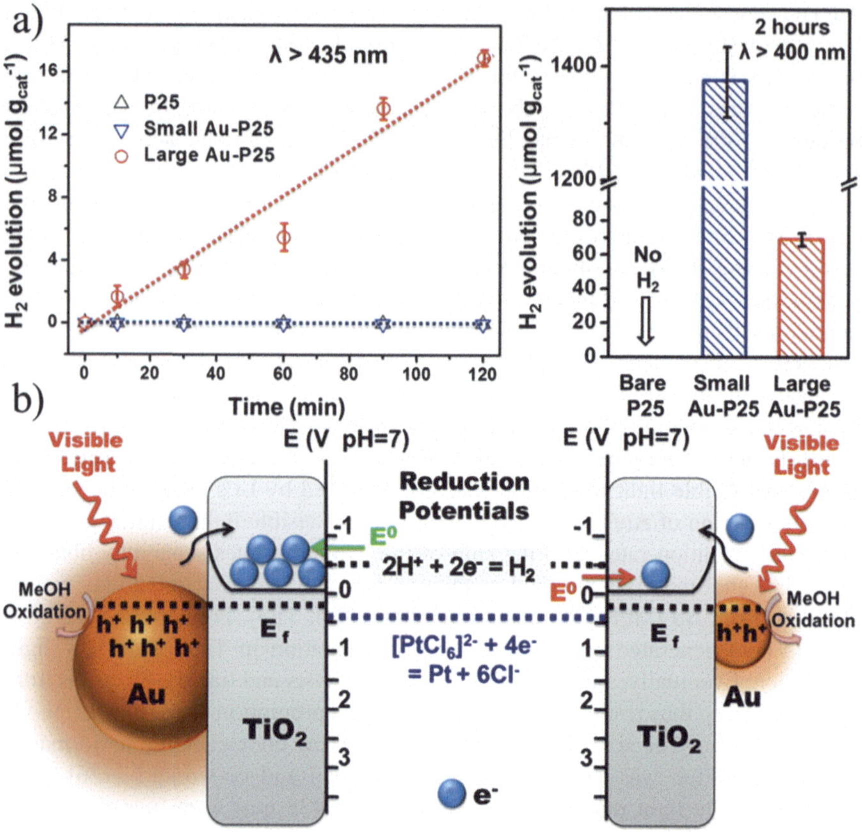

Fig. 15 (**a**) H_2 evolution for Au/TiO_2 materials with different Au particle sizes and pristinte TiO_2 under $\lambda > 435$ nm and $\lambda > 400$ nm irradiation. (**b**) Schematically drawn mechanism for H_2 production under$\lambda > 435$ nm irradiation for S-Au/TiO_2 and L-Au/TiO_2. Note that it is proposed that charge transfer in L-Au/TiO_2 results in electron accumulation in TiO_2 CB with higher reduction potential. Reprinted with permission from [126]. Copyright 2014, American Chemical Society

that Au{111}/TiO_2 (Au octahedral shape) had the highest activity compared to Au {100/111}/TiO_2 (truncated cubic shape) and Au{100}/TiO_2 (cubic shape) under visible light irradiation [192]. They correlated the different photocatalytic activities to the different electron transfer efficiencies of the materials. This, in turn, was attributed to the different Fermi levels of Au{111} facet and Au{100} facet. The results were supported by photoelectrochemical analyses and photoluminescence spectroscopy. Furthermore, it was shown that Au{111}/TiO_2 exhibited the lowest apparent activation energy for hydrogen evolution. Wang et al. prepared graphene-based Au/TiO_2 photocatalysts and investigated photocatalytic hydrogen generation under visible light irradiation [193]. They could show that graphene modified Au/TiO_2 had a 1.7-fold higher activity than the analogous Au/TiO_2 sample under visible light irradiation. It was proposed that electrons generated in TiO_2 by the LSPR of Au (either by DET or LEMF mechanism) are transferred to graphene because of the intermediate redox potential of graphene/graphene$^-$ (−0.08 eV) compared to the CB of TiO_2 (−0.24 eV) and hydrogen generation H^+/H_2 (0 eV). Hence, both the TiO_2 surface and the graphene surface should function as active sites for H_2 generation. The higher activity of graphene-based Au/TiO_2 compared to Au/TiO_2 was attributed to the great electron transport properties of graphene and with that the reduced charge recombination. More recently, Liu et al. observed a twofold enhancement in the H_2 generation rate for Au/graphene/TiO_2 material compared to Au/TiO_2 under UV–vis irradiation in methanol [194]. Singh et al. furthermore reported a 21-fold enhancement of hydrogen production under UV-vis irradiation for Au on graphene/TiO_2 compared to pure TiO_2 [195].

The examples discussed above indicate that the combination of a semiconductor with a metallic nanocrystal that has the ability to absorb visible light is an appealing strategy to boost the efficiency of the materials for solar energy conversion to fuel through water splitting by generating clean hydrogen. The impact of the material design was further pointed out, showing remarkable variations in photocatalytic performances.

4.2 CO_2 Reduction

Inspired by natural photosynthesis, the photocatalytic reduction of CO_2 to hydrocarbon fuels (CH_3OH, CH_4, etc.) is another highly investigated approach for solar energy conversion to fuels. Another advantage of this route – besides solar energy conversion to fuel – is CO_2 capture and its use as a building block for hydrocarbon and fine chemical production to reduce CO_2 emissions, which has a dramatic influence on the climate change. So far, much effort has been expended on using and improving (visible light active) photocatalysts to reduce CO_2 [196–199]. Basically, the process of photocatalytic CO_2 reduction with H_2O as reducing compound is regarded as more challenging in comparison with water splitting because of the more negative redox potentials to various products and multi-electron processes [200, 201].

Hou et al. reported visible light-driven CO_2 conversion to hydrocarbon fuels with Au/TiO_2 [43]. Experiments with different irradiation wavelengths (254, 365, and 532 nm) gave further insight into the reaction mechanism during photocatalytic reduction. As shown in Fig. 16a, visible light irradiation resulted in a 24-fold enhancement in CH_4 formation as the only product for Au/TiO_2 compared to bare TiO_2. Interestingly, irradiation with 254-nm UV light resulted in further products (ethane, formaldehyde, and methanol) for Au/TiO_2 but still only methane for TiO_2 (Fig. 16b). This was ascribed to different excitation mechanisms during irradiation. Visible light irradiation triggered Au LSPR which excited electron-hole pairs in TiO_2 through the LEMF mechanism. The conduction band edge of TiO_2 lies above the reduction potential of CO_2/CH_4 but below the other reduced products (Fig. 16c).

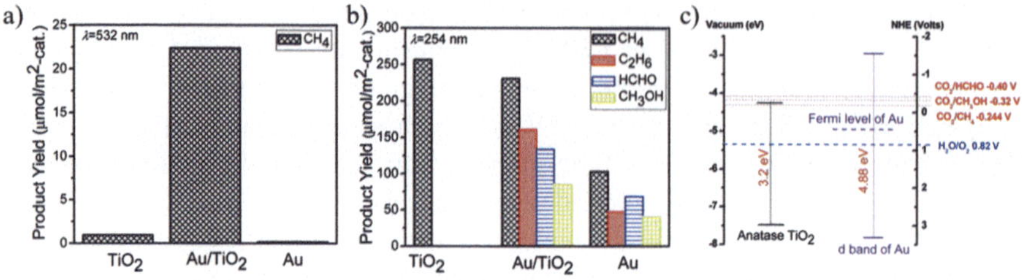

Fig. 16 (**a**) Product yields of methane after 15 h visible light irradiation for bare TiO_2 and Au NPs, and Au/TiO_2. (**b**) Corresponding product yields under $\lambda = 254$ nm irradiation for 15 h. (**c**) Band position of TiO_2, anatase, Fermi level of Au with corresponding excitation from the d band with 254 nm, and redox potentials of CO_2 to various products. Reprinted with permission from [43]. Copyright 2011, American Chemical Society

Hence, only methane formation is favored. The importance of the TiO_2 surface for this catalytic process is evident by the fact that negligible amounts of methane are formed for bare Au NPs. It was proposed that with 254-nm irradiation, interband transitions occur from the d band to unoccupied states in the sp band of Au. As seen in Fig. 16c, the reduction potential is then sufficient to drive the reduction to the other products. Then 365-nm UV irradiation again leads to methane as the only product.

Very recently, Neatu et al. reported the efficient conversion of CO_2 to methane with Au–Cu alloy NPs (Au/Cu 1:2, ca. 5 nm size) on P25 [202]. It was demonstrated that under simulated sunlight irradiation, a selectivity of 97% for methane formation could be achieved. The monometallic Au/TiO_2 and Cu/TiO_2 samples showed a preferred hydrogen evolution or lower activity to methane formation, respectively. It is mentioned that the selectivity towards CH_4 is because of the Cu bonding to CO as intermediate. Remarkably, irradiation with UV light results in higher H_2 generation than under visible light irradiation, whereas CH_4 formation is only observed under visible light. Control samples consisting of pristine TiO_2 and Cu-TiO_2 showed no activity under visible light irradiation. It is proposed that, under visible light irradiation, the Au LSPR transfers electrons to the Cu sites which further reduce CO_2^- to CO. The positive holes oxidize H_2O and generate protons. Consecutive reactions result in CH_4 formation. An et al. reported the effective reduction of CO_2 to methanol under visible light irradiation with AgX:Ag (X=Cl, Br) nanomaterials utilizing the LSPR effect of Ag NPs [44]. Xue and co-workers fabricated TiO_2 nanofibers decorated with Au and Pt NPs and investigated the photoreduction of CO_2 to methane under UV–vis irradiation [48]. It was found that AuPt/TiO_2 exhibited a 1.8× and 1.4× higher methane production compared to Au/TiO_2 and Pt/TiO_2, respectively. The higher activity was attributed to the improved charge separation produced by the Pt NPs as electron sinks and reduction sites and the enhanced visible light response from the Au LSPR effect. Mankidy et al. investigated the effect of TiO_2 decorated with Ag, Pt, bimetallic AgPt, and core-shell Ag@SiO_2 on the photocatalytic performance for CO_2 reduction under UV–vis light irradiation [47]. It was found that the synergistic effect of depositing

$Ag@SiO_2$ as visible light harvester through LSPR and metal cocatalyst (Pt, AgPt) resulted in superior activity. The best photocatalyst regarding activity and selectivity to CH_4 was the combination of $Ag@SiO_2$ and AgPt cocatalyst. It was concluded that $Ag@SiO_2$ deposition increased the electron-hole pair generation in TiO_2, whereas Ag, Pt, and bimetallic Ag-Pt NPs optimized CH_4 and CO product selectivity. Wang et al. assigned the catalytic reduction of CO_2 to methane with Au/ZnO to the plasmonic heating effect [159]. Temperature-calibrated Raman spectra of ZnO phonons revealed that the heat could be controlled up to 600 °C under green laser ($\lambda = 532$ nm) irradiation. The agreement between the light intensity-dependent product selectivity and the temperature-dependent selectivity in the dark led the authors to the conclusion that plasmonic heating was the reason for the activity. Although using plasmonic photocatalysts for CO_2 reduction gained great interest and showed progress during the last few years, further effort has to be made to enable a sufficient conversion on an industrial scale. Nonetheless, plasmonic photocatalysis might give promising opportunities for reaching this goal.

5 Conclusions and Outlook

The combination of plasmonic properties of NPs with semiconductors for enhancing the performance of the photocatalysts has attracted great interest in recent years. In this regard, the major drawbacks of semiconductor photocatalysts, which are high recombination rates of charge carriers and poor visible light harvesting, have to be solved, and plasmonic photocatalysis looks likely to be a promising route, in particular for solar energy conversion to fuels. It was shown that, besides other routes investigated and reported in the literature [2, 203, 204] for enhancing and enabling visible (solar) light activity, plasmonic photocatalysis has great potential for reaching the aim of clean, green energy conversion. However, further effort has to be made to obtain a more detailed insight into the reaction pathway of plasmonic photocatalysis [17, 100].

The controversially discussed working mechanisms of energy transfer from the metal NP to the semiconductor are governed by the metal-semiconductor configurations and arrangements. Thus, an optimal exploitation of these structures to boost photocatalytic performance is essential. For instance, it was shown that polydisperse Au NPs on TiO_2 enabled enhanced visible light harvesting and catalytic activity for H_2 generation [178] and the utilization of the strongly enhanced near-field effect of Ag LSPR from $Ag@SiO_2@TiO_2$was most efficient with thin 2 nm SiO_2 shells [150]. Furthermore, low-cost plasmonic photocatalysts have to be fabricated to overcome high material costs and open the way for industrial applications. Noble metals – despite their desired plasmonic properties and their utilization as cocatalyst – show their economic limitations as rare and expensive materials. Possible alternatives might be Cu- and Al-based plasmonic photocatalysts. Also nonmetal-based plasmonic photocatalysts, as reviewed in [205], might set the path for low-cost materials as effective photocatalysts. It has to be pointed out that photocatalytic stability is considered one main challenge for efficient photocatalysts and is investigated by various research groups besides the targeted high enhancements in photocatalytic activity. A long life time is highly desired for solar fuel production on an industrial scale and economic feasible applications.

As the future perspective, more effort needs to be devoted to solar energy and surface plasmon-assisted fine chemical synthesis. The unique physical properties of the plasmonic catalysts not only enhance the conversion but also change the traditional reaction mechanism and consequently the product selectivity. In this manner, we focus on the design of plasmonic catalysts for glycerol oxidation – which is a large bio-diesel side product – to more useful fine chemicals.

Acknowledgements This work was financially supported by the MAXNET Energy consortium of Max Planck Society and the Cluster of Excellence RESOLV (EXC 1069) funded by the Deutsche Forschungsgemeinschaft (DFG).

References

1. Fujishima A, Honda K (1972) Electrochemical photolysis of water at a semiconductor electrode. Nature 238(5358):37–38
2. Kudo A, Miseki Y (2009) Heterogeneous photocatalyst materials for water splitting. Chem Soc Rev 38(1):253–278
3. Choi WY, Termin A, Hoffmann MR (1994) The role of metal-ion dopants in quantum-sized TiO_2 – correlation between photoreactivity and charge-carrier recombination dynamics. J Phys Chem 98(51):13669–13679
4. Asahi R, Morikawa T, Ohwaki T, Aoki K, Taga Y (2001) Visible-light photocatalysis in nitrogen-doped titanium oxides. Science 293(5528):269–271
5. Youngblood WJ, Lee S-HA, Maeda K, Mallouk TE (2009) Visible light water splitting using dye-sensitized oxide semiconductors. Acc Chem Res 42(12):1966–1973
6. Ni M, Leung MKH, Leung DYC, Sumathy K (2007) A review and recent developments in photocatalytic water-splitting using TiO_2 for hydrogen production. Renew Sustain Energy Rev 11(3):401–425
7. Chen X, Mao SS (2007) Titanium dioxide nanomaterials: synthesis, properties, modifications, and applications. Chem Rev 107(7):2891–2959
8. Linic S, Christopher P, Ingram DB (2011) Plasmonic-metal nanostructures for efficient conversion of solar to chemical energy. Nat Mater 10(12):911–921
9. Awazu K, Fujimaki M, Rockstuhl C, Tominaga J, Murakami H, Ohki Y, Yoshida N, Watanabe T (2008) A plasmonic photocatalyst consisting of sliver nanoparticles embedded in titanium dioxide. J Am Chem Soc 130(5):1676–1680
10. Nie SM, Emery SR (1997) Probing single molecules and single nanoparticles by surface-enhanced Raman scattering. Science 275(5303):1102–1106
11. Moskovits M (2005) Surface-enhanced Raman spectroscopy: a brief retrospective. J Raman Spectrosc 36(6–7):485–496
12. Atwater HA, Polman A (2010) Plasmonics for improved photovoltaic devices. Nat Mater 9 (3):205–213
13. Anker JN, Hall WP, Lyandres O, Shah NC, Zhao J, Van Duyne RP (2008) Biosensing with plasmonic nanosensors. Nat Mater 7(6):442–453

14. Larsson EM, Langhammer C, Zoric I, Kasemo B (2009) Nanoplasmonic probes of catalytic reactions. Science 326(5956):1091–1094
15. Bardhan R, Lal S, Joshi A, Halas NJ (2011) Theranostic nanoshells: from probe design to imaging and treatment of cancer. Acc Chem Res 44(10):936–946
16. Zheng X, Liu Q, Jing C, Li Y, Li D, Luo W, Wen Y, He Y, Huang Q, Long Y-T, Fan C (2011) Catalytic gold nanoparticles for nanoplasmonic detection of DNA hybridization. Angew Chem Int Ed 50(50):11994–11998
17. Zhang X, Chen YL, Liu R-S, Tsai DP (2013) Plasmonic photocatalysis. Rep Prog Phys 76 (4):1–41
18. Liu Z, Hou W, Pavaskar P, Aykol M, Cronin SB (2011) Plasmon resonant enhancement of photocatalytic water splitting under visible illumination. Nano Lett 11(3):1111–1116
19. Thomann I, Pinaud BA, Chen Z, Clemens BM, Jaramillo TF, Brongersma ML (2011) Plasmon enhanced solar-to-fuel energy conversion. Nano Lett 11(8):3440–3446
20. Ingram DB, Linic S (2011) Water splitting on composite plasmonic-metal/semiconductor photoelectrodes: evidence for selective plasmon-induced formation of charge carriers near the semiconductor surface. J Am Chem Soc 133(14):5202–5205
21. Chen J-J, Wu JCS, Wu PC, Tsai DP (2011) Plasmonic photocatalyst for H_2 evolution in photocatalytic water splitting. J Phys Chem C 115(1):210–216
21. Chen HM, Chen CK, Chen CJ, Cheng LC, Wu PC, Cheng BH, Ho YZ, Tseng ML, Hsu YY, Chan TS, Lee JF, Liu RS, Tsai DP (2012) Plasmon inducing effects for enhanced photoelectrochemical water splitting: X-ray absorption approach to electronic structures. ACS Nano 6 (8):7362–7372
22. Warren SC, Thimsen E (2012) Plasmonic solar water splitting. Energy Environ Sci 5 (1):5133–5146
23. Wang H, You T, Shi W, Li J, Guo L (2012) Au/TiO_2/Au as a plasmonic coupling photocatalyst. J Phys Chem C 116(10):6490–6494
24. Gao H, Liu C, Jeong HE, Yang P (2012) Plasmon-enhanced photocatalytic activity of iron oxide on gold nanopillars. ACS Nano 6(1):234–240
25. Lee J, Mubeen S, Ji X, Stucky GD, Moskovits M (2012) Plasmonic photoanodes for solar water splitting with visible light. Nano Lett 12(9):5014–5019
26. Zhang Z, Zhang L, Hedhili MN, Zhang H, Wang P (2013) Plasmonic gold nanocrystals coupled with photonic crystal seamlessly on TiO_2 nanotube photoelectrodes for efficient visible light photoelectrochemical water splitting. Nano Lett 13(1):14–20
27. Pu YC, Wang GM, Chang KD, Ling YC, Lin YK, Fitzmorris BC, Liu CM, Lu XH, Tong YX, Zhang JZ, Hsu YJ, Li Y (2013) Au nanostructure-decorated TiO_2 nanowires exhibiting photoactivity across entire UV-visible region for photoelectrochemical water splitting. Nano Lett 13(8):3817–3823
28. Solarska R, Bienkowski K, Zoladek S, Majcher A, Stefaniuk T, Kulesza PJ, Augustynski J (2014) Enhanced water splitting at thin film tungsten trioxide photoanodes bearing plasmonic gold-polyoxometalate particles. Angew Chem Int Ed Engl 53(51):14196–14200
29. Zhao X, Wang P, Yan Z, Ren N (2014) Ag nanoparticles decorated CuO nanowire arrays for efficient plasmon enhanced photoelectrochemical water splitting. Chem Phys Lett 609:59–64
30. Zhong Y, Ueno K, Mori Y, Shi X, Oshikiri T, Murakoshi K, Inoue H, Misawa H (2014) Plasmon-assisted water splitting using two sides of the same $SrTiO_3$ single-crystal substrate: conversion of visible light to chemical energy. Angew Chem Int Ed 53(39):10350–10354
31. Gomes Silva C, Juarez R, Marino T, Molinari R, Garcia H (2011) Influence of excitation wavelength (UV or visible light) on the photocatalytic activity of titania containing gold nanoparticles for the generation of hydrogen or oxygen from water. J Am Chem Soc 133 (3):595–602
32. Cao SW, Fang J, Shahjamali MM, Boey FYC, Barber J, Loo SCJ, Xue C (2012) Plasmon-enhanced hydrogen evolution on Au-$InVO_4$ hybrid microspheres. RSC Adv 2(13):5513–5515

33. Pany S, Naik B, Martha S, Parida K (2014) Plasmon induced nano Au particle decorated over S, N-modified TiO_2 for exceptional photocatalytic hydrogen evolution under visible light. ACS Appl Mater Interfaces 6(2):839–846
34. Tanaka A, Hashimoto K, Kominami H (2014) Visible-light-induced hydrogen and oxygen formation over Pt/Au/WO_3 photocatalyst utilizing two types of photoabsorption due to surface plasmon resonance and band-gap excitation. J Am Chem Soc 136(2):586–589
35. Verbruggen SW, Keulemans M, Filippousi M, Flahaut D, Van Tendeloo G, Lacombe S, Martens JA, Lenaerts S (2014) Plasmonic gold-silver alloy on TiO_2 photocatalysts with tunable visible light activity. Appl Catal B Environ 156:116–121
36. Rayalu SS, Jose D, Joshi MV, Mangrulkar PA, Shrestha K, Klabunde K (2013) Photocatalytic water splitting on Au/TiO_2 nanocomposites synthesized through various routes: enhancement in photocatalytic activity due to SPR effect. Appl Catal B Environ 142:684–693
37. Zhou C, Shang L, Yu H, Bian T, Wu L-Z, Tung C-H, Zhang T (2014) Mesoporous plasmonic Au-loaded Ta_2O_5 nanocomposites for efficient visible light photocatalysis. Catal Today 225:158–163
38. Long J, Chang H, Gu Q, Xu J, Fan L, Wang S, Zhou Y, Wei W, Huang L, Wang X, Liu P, Huang W (2014) Gold-plasmon enhanced solar-to-hydrogen conversion on the {001} facets of anatase TiO_2 nanosheets. Energy Environ Sci 7(3):973–977
39. Qu Y, Zhou W, Ren Z, Tian C, Li J, Fu H (2014) Heterojunction Ag-TiO_2 nanopillars for visible-light-driven photocatalytic H_2 production. Chempluschem 79(7):995–1000
40. Luo Y, Liu X, Tang X, Luo Y, Zeng Q, Deng X, Ding S, Sun Y (2014) Gold nanoparticles embedded in Ta_2O_5/Ta_3N_5 as active visible-light plasmonic photocatalysts for solar hydrogen evolution. J Mater Chem A 2(36):14927–14939
41. Rayalu SS, Jose D, Mangrulkar PA, Joshi M, Hippargi G, Shrestha K, Klabunde K (2014) Photodeposition of AuNPs on metal oxides: study of SPR effect and photocatalytic activity. Int J Hydrog Energy 39(8):3617–3624
42. Ma X, Zhao K, Tang H, Chen Y, Lu C, Liu W, Gao Y, Zhao H, Tang Z (2014) New insight into the role of gold nanoparticles in Au@ CdS core-shell nanostructures for hydrogen evolution. Small 10(22):4664–4670
43. Hou W, Hung WH, Pavaskar P, Goeppert A, Aykol M, Cronin SB (2011) Photocatalytic conversion of CO_2 to hydrocarbon fuels via plasmon-enhanced absorption and metallic interband transitions. ACS Catal 1(8):929–936
44. An C, Wang J, Jiang W, Zhang M, Ming X, Wang S, Zhang Q (2012) Strongly visible-light responsive plasmonic shaped AgX:Ag (X=Cl, Br) nanoparticles for reduction of CO_2 to methanol. Nanoscale 4(18):5646–5650
45. Tan JZY, Fernandez Y, Liu D, Maroto-Valer M, Bian JC, Zhang XW (2012) Photoreduction of CO_2 using copper-decorated TiO_2 nanorod films with localized surface plasmon behavior. Chem Phys Lett 531:149–154
46. Li X, Zhuang Z, Li W, Pan H (2012) Photocatalytic reduction of CO_2 over noble metal-loaded and nitrogen-doped mesoporous TiO_2. Appl Catal A Gen 429:31–38
47. Mankidy BD, Joseph B, Gupta VK (2013) Photo-conversion of CO_2 using titanium dioxide: enhancements by plasmonic and co-catalytic nanoparticles. Nanotechnology 24(40)
48. Zhang Z, Wang Z, Cao S-W, Xue C (2013) Au/Pt nanoparticle-decorated TiO_2 nanofibers with plasmon-enhanced photocatalytic activities for solar-to-fuel conversion. J Phys Chem C 117(49):25939–25947
49. Liu EZ, Kang LM, Wu F, Sun T, Hu XY, Yang YH, Liu HC, Fan J (2014) Photocatalytic reduction of CO_2 into methanol over Ag/TiO_2 nanocomposites enhanced by surface plasmon resonance. Plasmonics 9(1):61–70
50. Yu JG, Dai GP, Huang BB (2009) Fabrication and characterization of visible-light-driven plasmonic photocatalyst Ag/AgCl/TiO_2 TiO_2 nanotube arrays. J Phys Chem C 113 (37):16394–16401

51. Chen X, Zheng ZF, Ke XB, Jaatinen E, Xie TF, Wang DJ, Guo C, Zhao JC, Zhu HY (2010) Supported silver nanoparticles as photocatalysts under ultraviolet and visible light irradiation. Green Chem 12(3):414–419
52. Alvaro M, Cojocaru B, Ismail AA, Petrea N, Ferrer B, Harraz FA, Parvulescu VI, Garcia H (2010) Visible-light photocatalytic activity of gold nanoparticles supported on template-synthesized mesoporous titania for the decontamination of the chemical warfare agent Soman. Appl Catal B Environ 99(1–2):191–197
53. Christopher P, Ingram DB, Linic S (2010) Enhancing photochemical activity of semiconductor nanoparticles with optically active Ag nanostructures: photochemistry mediated by Ag surface plasmons. J Phys Chem C 114(19):9173–9177
54. Wang P, Huang BB, Lou ZZ, Zhang XY, Qin XY, Dai Y, Zheng ZK, Wang XN (2010) Synthesis of highly efficient Ag@AgCl plasmonic photocatalysts with various structures. Chem Eur J 16(2):538–544
55. Hu C, Peng T, Hu X, Nie Y, Zhou X, Qu J, He H (2010) Plasmon-induced photodegradation of toxic pollutants with Ag-AgI/Al_2O_3 under visible-light irradiation. J Am Chem Soc 132 (2):857–862
56. Zielinska-Jurek A, Kowalska E, Sobczak JW, Lisowski W, Ohtani B, Zaleska A (2011) Preparation and characterization of monometallic (Au) and bimetallic (Ag/Au) modified-titania photocatalysts activated by visible light. Appl Catal B Environ 101(3-4):504–514
57. Zhang Q, Lima DQ, Lee I, Zaera F, Chi M, Yin Y (2011) A highly active titanium dioxide based visible-light photocatalyst with nonmetal doping and plasmonic metal decoration. Angew Chem 50(31):7088–7092
58. Hou W, Liu Z, Pavaskar P, Hung WH, Cronin SB (2011) Plasmonic enhancement of photocatalytic decomposition of methyl orange under visible light. J Catal 277(2):149–153
59. Wen B, Ma J, Chen C, Ma W, Zhu H, Zhao J (2011) Supported noble metal nanoparticles as photo/sono-catalysts for synthesis of chemicals and degradation of pollutants. Sci China Chem 54(6):887–897
60. Qu Y, Cheng R, Su Q, Duan X (2011) Plasmonic enhancements of photocatalytic activity of Pt/n-Si/Ag photodiodes using Au/Ag core/shell nanorods. J Am Chem Soc 133 (42):16730–16733
61. Anandan S, Pugazhenthiran N, Selvamani T, Hsieh SH, Lee GJ, Wu JJ (2012) Investigation on photocatalytic potential of Au-Ta_2O_5 semiconductor nanoparticle by degrading methyl orange in aqueous solution by illuminating with visible light. Catal Sci Technol 2 (12):2502–2507
62. Jiang L, Zhou G, Mi J, Wu Z (2012) Fabrication of visible-light-driven one-dimensional anatase TiO_2/Ag heterojunction plasmonic photocatalyst. Catal Commun 24:48–51
63. Zhou MH, Zhang J, Cheng B, Yu HG (2012) Enhancement of visible-light photocatalytic activity of mesoporous Au-TiO_2 nanocomposites by surface plasmon resonance. Int J Photoenergy 10
64. Kochuveedu ST, Kim DP, Kim DH (2012) Surface-plasmon-induced visible light photocatalytic activity of TiO_2 nanospheres decorated by Au nanoparticles with controlled configuration. J Phys Chem C 116(3):2500–2506
65. Gomez L, Sebastian V, Arruebo M, Santamaria J, Cronin SB (2014) Plasmon-enhanced photocatalytic water purification. Phys Chem Chem Phys 16(29):15111–15116
66. Golabiewska A, Lisowski W, Jarek M, Nowaczyk G, Zielinska-Jurek A, Zaleska A (2014) Visible light photoactivity of TiO_2 loaded with monometallic (Au or Pt) and bimetallic (Au/Pt) nanoparticles. Appl Surf Sci 317:1131–1142
67. Ayati A, Ahmadpour A, Bamoharram FF, Tanhaei B, Manttari M, Sillanpaa M (2014) A review on catalytic applications of Au/TiO_2 nanoparticles in the removal of water pollutant. Chemosphere 107:163–174
68. Tanaka A, Hashimoto K, Kominami H (2012) Preparation of Au/CeO_2 exhibiting strong surface plasmon resonance effective for selective or chemoselective oxidation of alcohols to

aldehydes or ketones in aqueous suspensions under irradiation by green light. J Am Chem Soc 134(35):14526–14533
69. Wang F, Li C, Chen H, Jiang R, Sun L-D, Li Q, Wang J, Yu JC, Yan C-H (2013) Plasmonic harvesting of light energy for Suzuki coupling reactions. J Am Chem Soc 135(15):5588–5601
70. Huang X, Li Y, Chen Y, Zhou H, Duan X, Huang Y (2013) Plasmonic and catalytic AuPd nanowheels for the efficient conversion of light into chemical energy. Angew Chem Int Ed 52 (23):6063–6067
71. Lang X, Chen X, Zhao J (2014) Heterogeneous visible light photocatalysis for selective organic transformations. Chem Soc Rev 43(1):473–486
72. Garcia MA (2011) Surface plasmons in metallic nanoparticles: fundamentals and applications. J Phys D Appl Phys 44(28):283001
73. Mayer KM, Hafner JH (2011) Localized surface plasmon resonance sensors. Chem Rev 111 (6):3828–3857
74. Hartland GV (2011) Optical studies of dynamics in noble metal nanostructures. Chem Rev 111(6):3858–3887
75. Kelly KL, Coronado E, Zhao LL, Schatz GC (2003) The optical properties of metal nanoparticles: the influence of size, shape, and dielectric environment. J Phys Chem B 107 (3):668–677
76. Ruach-Nir I, Bendikov TA, Doron-Mor I, Barkay Z, Vaskevich A, Rubinstein I (2007) Silica-stabilized gold island films for transmission localized surface plasmon sensing. J Am Chem Soc 129(1):84–92
77. Larsson EM, Alegret J, Kall M, Sutherland DS (2007) Sensing characteristics of NIR localized surface plasmon resonances in gold nanorings for application as ultrasensitive biosensors. Nano Lett 7(5):1256–1263
78. Haes AJ, Van Duyne RP (2002) A nanoscale optical biosensor: sensitivity and selectivity of an approach based on the localized surface plasmon resonance spectroscopy of triangular silver nanoparticles. J Am Chem Soc 124(35):10596–10604
79. Miller MM, Lazarides AA (2005) Sensitivity of metal nanoparticle surface plasmon resonance to the dielectric environment. J Phys Chem B 109(46):21556–21565
80. Pena O, Pal U, Rodriguez-Fernandez L, Crespo-Sosa A (2008) Linear optical response of metallic nanoshells in different dielectric media. J Optical Soc Am B Optical Phys 25 (8):1371–1379
81. Zhao J, Pinchuk AO, McMahon JM, Li S, Ausman LK, Atkinson AL, Schatz GC (2008) Methods for describing the electromagnetic properties of silver and gold nanoparticles. Acc Chem Res 41(12):1710–1720
82. Chen H, Kou X, Yang Z, Ni W, Wang J (2008) Shape- and size-dependent refractive index sensitivity of gold nanoparticles. Langmuir 24(10):5233–5237
83. Haes AJ, Zou SL, Schatz GC, Van Duyne RP (2004) A nanoscale optical biosensor: the long range distance dependence of the localized surface plasmon resonance of noble metal nanoparticles. J Phys Chem B 108(1):109–116
84. Homola J, Yee SS, Gauglitz G (1999) Surface plasmon resonance sensors: review. Sens Actuators B Chem 54(1–2):3–15
85. Homola J (2008) Surface plasmon resonance sensors for detection of chemical and biological species. Chem Rev 108(2):462–493
86. Zheng Y, Zhong X, Li Z, Xia Y (2014) Successive, seed-mediated growth for the synthesis of single-crystal gold nanospheres with uniform diameters controlled in the range of 5–150 nm. Part Part Syst Charact 31(2):266–273
87. Choi KW, Kim DY, Zhong X-L, Li Z-Y, Im SH, Park OO (2013) Robust synthesis of gold rhombic dodecahedra with well-controlled sizes and their optical properties. Crystengcomm 15(2):252–258
88. Zhou X, Liu G, Yu J, Fan W (2012) Surface plasmon resonance-mediated photocatalysis by noble metal-based composites under visible light. J Mater Chem 22(40):21337–21354

89. Jain PK, Lee KS, El-Sayed IH, El-Sayed MA (2006) Calculated absorption and scattering properties of gold nanoparticles of different size, shape, and composition: applications in biological imaging and biomedicine. J Phys Chem B 110(14):7238–7248
90. Link S, El-Sayed MA (1999) Size and temperature dependence of the plasmon absorption of colloidal gold nanoparticles. J Phys Chem B 103(21):4212–4217
91. Link S, El-Sayed MA (2000) Shape and size dependence of radiative, non-radiative and photothermal properties of gold nanocrystals. Int Rev Phys Chem 19(3):409–453
92. Tsukamoto D, Shiraishi Y, Sugano Y, Ichikawa S, Tanaka S, Hirai T (2012) Gold nanoparticles located at the interface of anatase/rutile TiO_2 particles as active plasmonic photocatalysts for aerobic oxidation. J Am Chem Soc 134(14):6309–6315
93. Chen C-J, Chen M-G, Chen CK, Wu PC, Chen P-T, Basu M, Hu S-F, Tsai DP, Liu R-S (2015) Ag-Si artificial microflowers for plasmon-enhanced solar water splitting. Chem Commun 51 (3):549–552
94. Hou X (2014) Nonaqueous fabrication of ZnO/Au nanohybrids with enhanced photocatalytic activity. Mater Lett 137:319–322
95. Primo A, Corma A, Garcia H (2011) Titania supported gold nanoparticles as photocatalyst. Phys Chem Chem Phys 13(3):886–910
96. Cushing SK, Li J, Meng F, Senty TR, Suri S, Zhi M, Li M, Bristow AD, Wu N (2012) Photocatalytic activity enhanced by plasmonic resonant energy transfer from metal to semiconductor. J Am Chem Soc 134(36):15033–15041
97. Zhang Z, Yates JT (2012) Band bending in semiconductors: chemical and physical consequences at surfaces and interfaces. Chem Rev 112(10):5520–5551
98. Hou WB, Cronin SB (2013) A review of surface plasmon resonance-enhanced photocatalysis. Adv Funct Mater 23(13):1612–1619
99. Zhang XM, Chen YL, Liu RS, Tsai DP (2013) Plasmonic photocatalysis. Rep Prog Phys 76 (4):41
100. Kochuveedu ST, Jang YH, Kim DH (2013) A study on the mechanism for the interaction of light with noble metal-metal oxide semiconductor nanostructures for various photophysical applications. Chem Soc Rev 42(21):8467–8493
101. Bumajdad A, Madkour M (2014) Understanding the superior photocatalytic activity of noble metals modified titania under UV and visible light irradiation. Phys Chem Chem Phys 16 (16):7146–7158
102. Watanabe K, Menzel D, Nilius N, Freund H-J (2006) Photochemistry on metal nanoparticles. Chem Rev 106(10):4301–4320
103. Kale MJ, Avanesian T, Christopher P (2014) Direct photocatalysis by plasmonic nanostructures. ACS Catal 4(1):116–128
104. Xiao Q, Jaatinen E, Zhu HY (2014) Direct photocatalysis for organic synthesis by using plasmonic-metal nanoparticles irradiated with visible light. Chem Asian J 9(11):3046–3064
105. Yu K, Tian Y, Tatsuma T (2006) Size effects of gold nanaoparticles on plasmon-induced photocurrents of gold-TiO_2 nanocomposites. Phys Chem Chem Phys 8(46):5417–5420
106. Sakai N, Fujiwara Y, Takahashi Y, Tatsuma T (2009) Plasmon-resonance-based generation of cathodic photocurrent at electrodeposited gold nanoparticles coated with TiO_2 films. ChemPhysChem 10(5):766–769
107. Kowalska E, Abe R, Ohtani B (2009) Visible light-induced photocatalytic reaction of gold-modified titanium(IV) oxide particles: action spectrum analysis. Chem Commun 2:241–243
108. Nishijima Y, Ueno K, Yokota Y, Murakoshi K, Misawa H (2010) Plasmon-assisted photocurrent generation from visible to near-infrared wavelength using a Au-nanorods/TiO_2 electrode. J Phys Chem Lett 1(13):2031–2036
109. Li RH, Chen WX, Kobayashi H, Ma CX (2010) Platinum-nanoparticle-loaded bismuth oxide: an efficient plasmonic photocatalyst active under visible light. Green Chem 12(2):212–215
110. Mubeen S, Hernandez-Sosa G, Moses D, Lee J, Moskovits M (2011) Plasmonic photosensitization of a wide band gap semiconductor: converting plasmons to charge carriers. Nano Lett 11(12):5548–5552

111. Lee YK, Jung CH, Park J, Seo H, Somorjai GA, Park JY (2011) Surface plasmon-driven hot electron flow probed with metal-semiconductor nanodiodes. Nano Lett 11(10):4251–4255
112. Kimura K, Naya S-I, Jin-nouchi Y, Tada H (2012) TiO_2 crystal form-dependence of the Au/TiO_2 plasmon photocatalyst's activity. J Phys Chem C 116(12):7111–7117
113. Lu Y, Yu H, Chen S, Quan X, Zhao H (2012) Integrating plasmonic nanoparticles with TiO_2 photonic crystal for enhancement of visible-light-driven photocatalysis. Environ Sci Technol 46(3):1724–1730
114. Hu XF, Burgi T (2012) Photoinduced electron transfer and photodegradation of malonic acid at Au/TiO_2 investigated by in situ ATR-IR spectroscopy. Appl Catal A Gen 449:139–144
115. Sa J, Tagliabue G, Friedli P, Szlachetko J, Rittmann-Frank MH, Santomauro FG, Milne CJ, Sigg H (2013) Direct observation of charge separation on Au localized surface plasmons. Energy Environ Sci 6(12):3584–3588
116. Wu K, Rodriguez-Cordoba WE, Yang Y, Lian T (2013) Plasmon-induced hot electron transfer from the Au Tip to CdS rod in CdS-Au nanoheterostructures. Nano Lett 13 (11):5255–5263
117. Mubeen S, Lee J, Singh N, Kraemer S, Stucky GD, Moskovits M (2013) An autonomous photosynthetic device in which all charge carriers derive from surface plasmons. Nat Nanotechnol 8(4):247–251
118. Naya S-I, Niwa T, Kume T, Tada, H (2014) Visible-light-induced electron transport from small to large nanoparticles in bimodal gold nanoparticle-loaded titanium(IV) oxide. Angew Chem Int Ed 53(28):7305–7309
119. Bian ZF, Tachikawa T, Zhang P, Fujitsuka M, Majima T (2014) Au/TiO_2 superstructure-based plasmonic photocatalysts exhibiting efficient charge separation and unprecedented activity. J Am Chem Soc 136(1):458–465
120. DuChene JS, Sweeny BC, Johnston-Peck AC, Su D, Stach EA, Wei WD (2014) Prolonged hot electron dynamics in plasmonic-metal/semiconductor heterostructures with implications for solar photocatalysis. Angew Chem Int Ed 53(30):7887–7891
121. Marchuk K, Willets KA (2014) Localized surface plasmons and hot electrons. Chem Phys 445:95–104
122. Takahiro K, Naya S-I, Tada H (2014) Highly active supported plasmonic photocatalyst consisting of gold nanoparticle-loaded mesoporous titanium(IV) oxide over layer and conducting substrate. J Phys Chem C 118(46):26887–26893
123. Kazuma E, Tatsuma T (2014) In situ nanoimaging of photoinduced charge separation at the plasmonic au nanoparticle-TiO_2 interface. Adv Mater Interfaces 1(3):1400066
124. Li JT, Cushing SK, Zheng P, Senty T, Meng FK, Bristow AD, Manivannan A, Wu NQ (2014) Solar hydrogen generation by a CdS-Au-TiO_2 sandwich nanorod array enhanced with Au nanoparticle as electron relay and plasmonic photosensitizer. J Am Chem Soc 136 (23):8438–8449
125. Wu L, Li F, Xu Y, Zhang JW, Zhang D, Li G, Li H (2015) Plasmon-induced photoelectrocatalytic activity of Au nanoparticles enhanced TiO_2 nanotube arrays electrodes for environmental remediation. Appl Catal B Environ 164:217–224
126. Qian K, Sweeny BC, Johnston-Peck AC, Niu W, Graham JO, DuChene JS, Qiu J, Wang Y-C, Engelhard MH, Su D, Stach EA, Wei WD (2014) Surface plasmon-driven water reduction: gold nanoparticle size matters. J Am Chem Soc 136(28):9842–9845
127. Tian Y, Tatsuma T (2005) Mechanisms and applications of plasmon-induced charge separation at TiO_2 films loaded with gold nanoparticles. J Am Chem Soc 127(20):7632–7637
128. Zhai WY, Xue SJ, Zhu AW, Luo YP, Tian Y (2011) Plasmon-driven selective oxidation of aromatic alcohols to aldehydes in water with recyclable Pt/TiO_2 nanocomposites. ChemCatChem 3(1):127–130
129. Zheng Z, Huang B, Qin X, Zhang X, Dai Y, Whangbo M-H (2011) Facile in situ synthesis of visible-light plasmonic photocatalysts M@TiO_2 (M = Au, Pt, Ag) and evaluation of their photocatalytic oxidation of benzene to phenol. J Mater Chem 21(25):9079–9087

130. Furube A, Du L, Hara K, Katoh R, Tachiya M (2007) Ultrafast plasmon-induced electron transfer from gold nanodots into TiO_2 nanoparticles. J Am Chem Soc 129(48):14852–14853
131. Du LC, Furube A, Yamamoto K, Hara K, Katoh R, Tachiya M (2009) Plasmon-induced charge separation and recombination dynamics in gold-TiO_2 nanoparticle systems: dependence on TiO_2 particle size. J Phys Chem C 113(16):6454–6462
132. Du LC, Furube A, Hara K, Katoh R, Tachiya M (2013) Ultrafast plasmon induced electron injection mechanism in gold-TiO_2 nanoparticle system. J Photochem Photobiol C Photochem Rev 15:21–30
133. Tian Y, Tatsuma T (2004) Plasmon-induced photoelectrochemistry at metal nanoparticles supported on nanoporous TiO_2. Chem Commun 16:1810–1811
134. Priebe JB, Karnahl M, Junge H, Beller M, Hollmann D, Brueckner A (2013) Water reduction with visible light: synergy between optical transitions and electron transfer in Au-TiO_2 catalysts visualized by in situ EPR spectroscopy. Angew Chem Int Ed 52(43):11420–11424
135. Hirakawa T, Kamat PV (2005) Charge separation and catalytic activity of Ag@TiO_2 core-shell composite clusters under UV-irradiation. J Am Chem Soc 127(11):3928–3934
136. Rosseler O, Shankar MV, Du MK-L, Schmidlin L, Keller N, Keller V (2010) Solar light photocatalytic hydrogen production from water over Pt and Au/TiO_2(anatase/rutile) photocatalysts: influence of noble metal and porogen promotion. J Catal 269(1):179–190
137. Bamwenda GR, Tsubota S, Nakamura T, Haruta M (1995) Photoassisted hydrogen production from a water-ethanol solution: a comparison of activities of Au·TiO_2 and Pt·TiO_2. J Photochem Photobiol A Chem 89(2):177–189
138. Kumar MK, Krishnamoorthy S, Tan LK, Chiam SY, Tripathy S, Gao H (2011) Field effects in plasmonic photocatalyst by precise SiO_2 thickness control using atomic layer deposition. ACS Catal 1(4):300–308
139. Seh ZW, Liu S, Low M, Zhang S-Y, Liu Z, Mlayah A, Han M-Y (2012) Janus Au-TiO_2 photocatalysts with strong localization of plasmonic near-fields for efficient visible-light hydrogen generation. Adv Mater 24(17):2310–2314
140. Li H, Lu W, Tian J, Luo Y, Asiri AM, Al-Youbi AO, Sun X (2012) Synthesis and study of plasmon-induced carrier behavior at Ag/TiO_2 nanowires. Chem Eur J 18(27):8508–8514
141. Park JY, Kim SM, Lee H, Naik B (2014) Hot electron and surface plasmon-driven catalytic reaction in metal-semiconductor nanostructures. Catal Lett 144(12):1996–2004
142. Jiang DH, Zhou W, Zhong XH, Zhang YG, Li XH (2014) Distinguishing localized surface plasmon resonance and Schottky junction of Au-Cu_2O composites by their molecular spacer dependence. ACS Appl Mater Interfaces 6(14):10958–10962
143. Qiu J, Zeng G, Pavaskar P, Li Z, Cronin SB (2014) Plasmon-enhanced water splitting on TiO_2-passivated GaP photocatalysts. Phys Chem Chem Phys 16(7):3115–3121
144. Amrollahi R, Hamdy MS, Mul G (2014) Understanding promotion of photocatalytic activity of TiO_2 by Au nanoparticles. J Catal 319:194–199
145. Maier SA, Atwater HA (2005) Plasmonics: localization and guiding of electromagnetic energy in metal/dielectric structures. J Appl Phys 98(1):011101
146. Ingram DB, Christopher P, Bauer JL, Linic S (2011) Predictive model for the design of plasmonic metal/semiconductor composite photocatalysts. ACS Catal 1(10):1441–1447
147. Hao E, Schatz GC (2004) Electromagnetic fields around silver nanoparticles and dimers. J Chem Phys 120(1):357–366
148. Esteban R, Borisov AG, Nordlander P, Aizpurua J (2012) Bridging quantum and classical plasmonics with a quantum-corrected model. Nat Commun 2012:3
149. Chung T, Lee S-Y, Song EY, Chun H, Lee B (2011) Plasmonic nanostructures for nano-scale bio-sensing. Sensors 11(11):10907–10929
150. Zhang X, Zhu Y, Yang X, Wang S, Shen J, Lin B, Li C (2013) Enhanced visible light photocatalytic activity of interlayer-isolated triplex Ag@SiO_2@TiO_2 core-shell nanoparticles. Nanoscale 5(8):3359–3366

151. Zhang X, Zhao J, Wang S, Dai H, Sun X (2014) Shape-dependent localized surface plasmon enhanced photocatalytic effect of ZnO nanorods decorated with Ag. Int J Hydrog Energy 39 (16):8238–8245
152. Li J, Cushing SK, Bright J, Meng F, Senty TR, Zheng P, Bristow AD, Wu N (2013) Ag@Cu2O core-shell nanoparticles as visible-light plasmonic photocatalysts. ACS Catal 3 (1):47–51
153. Melancon MP, Lu W, Yang Z, Zhang R, Cheng Z, Elliot AM, Stafford J, Olson T, Zhang JZ, Li C (2008) In vitro and in vivo targeting of hollow gold nanoshells directed at epidermal growth factor receptor for photothermal ablation therapy. Mol Cancer Ther 7(6):1730–1739
154. Baffou G, Quidant R (2013) Thermo-plasmonics: using metallic nanostructures as nano-sources of heat. Laser Photonics Rev 7(2):171–187
155. Jain PK, Huang X, El-Sayed IH, El-Sayed MA (2008) Noble metals on the nanoscale: optical and photothermal properties and some applications in imaging, sensing, biology, and medicine. Acc Chem Res 41(12):1578–1586
156. Baffou G, Quidant R, Girard C (2009) Heat generation in plasmonic nanostructures: Influence of morphology. Appl Phys Lett 94(15):153109
157. Herzog JB, Knight MW, Natelson D (2014) Thermoplasmonics: quantifying plasmonic heating in single nanowires. Nano Lett 14(2):499–503
158. Baffou G, Polleux J, Rigneault H, Monneret S (2014) Super-heating and micro-bubble generation around plasmonic nanoparticles under cw illumination. J Phys Chem C 118 (9):4890–4898
159. Wang C, Ranasingha O, Natesakhawat S, Ohodnicki PR Jr, Andio M, Lewis JP, Matranga C (2013) Visible light plasmonic heating of Au-ZnO for the catalytic reduction of CO_2. Nanoscale 5(15):6968–6974
160. Christopher P, Xin H, Linic S (2011) Visible-light-enhanced catalytic oxidation reactions on plasmonic silver nanostructures. Nat Chem 3(6):467–472
161. Adleman JR, Boyd DA, Goodwin DG, Psaltis D (2009) Heterogenous catalysis mediated by plasmon heating. Nano Lett 9(12):4417–4423
162. Evanoff DD, Chumanov G (2005) Synthesis and optical properties of silver nanoparticles and arrays. ChemPhysChem 6(7):1221–1231
163. Wiley BJ, Chen Y, McLellan JM, Xiong Y, Li Z-Y, Ginger D, Xia Y (2007) Synthesis and optical properties of silver nanobars and nanorice. Nano Lett 7(4):1032–1036
164. Sankar M, Nowicka E, Carter E, Murphy DM, Knight DW, Bethell D, Hutchings GJ (2014) The benzaldehyde oxidation paradox explained by the interception of peroxy radical by benzyl alcohol. Nat Commun 2014:5
165. Zhou J, Ren F, Zhang S, Wu W, Xiao X, Liu Y, Jiang C (2013) SiO_2-Ag-SiO_2-TiO_2 multi-shell structures: plasmon enhanced photocatalysts with wide-spectral-response. J Mater Chem A 1(42):13128–13138
166. Lin S-J, Lee K-C, Wu J-L, Wu J-Y (2012) Plasmon-enhanced photocurrent in dye-sensitized solar cells. Sol Energy 86(9):2600–2605
167. Abdi FF, Dabirian A, Dam B, van de Krol R (2014) Plasmonic enhancement of the optical absorption and catalytic efficiency of $BiVO_4$ photoanodes decorated with Ag@SiO_2 core-shell nanoparticles. Phys Chem Chem Phys 16(29):15272–15277
168. Li JT, Cushing SK, Zheng P, Meng FK, Chu D, Wu NQ (2013) Plasmon-induced photonic and energy-transfer enhancement of solar water splitting by a hematite nanorod array. Nat Commun 4:1–8
169. Kong B, Tang J, Selomulya C, Li W, Wei J, Fang Y, Wang Y, Zheng G, Zhao D (2014) Oriented mesoporous nanopyramids as versatile plasmon-enhanced interfaces. J Am Chem Soc 136(19):6822–6825
170. Zhang X, Liu Y, Lee S-T, Yang S, Kang Z (2014) Coupling surface plasmon resonance of gold nanoparticles with slow-photon-effect of TiO_2 photonic crystals for synergistically enhanced photoelectrochemical water splitting. Energy Environ Sci 7(4):1409–1419

171. Zhan ZY, An JN, Zhang HC, Hansen RV, Zheng LX (2014) Three-dimensional plasmonic photoanodes based on Au-embedded TiO_2 structures for enhanced visible-light water splitting. ACS Appl Mater Interfaces 6(2):1139–1144
172. Erwin WR, Coppola A, Zarick HF, Arora P, Miller KJ, Bardhan R (2014) Plasmon enhanced water splitting mediated by hybrid bimetallic Au–Ag core-shell nanostructures. Nanoscale 6 (21):12626–12634
173. DeSario PA, Pietron JJ, DeVantier DE, Brintlinger TH, Stroud RM, Rolison DR (2013) Plasmonic enhancement of visible-light water splitting with Au-TiO_2 composite aerogels. Nanoscale 5(17):8073–8083
174. Wang G, Ling Y, Wang H, Lu X, Li Y (2014) Chemically modified nanostructures for photoelectrochemical water splitting. J Photochem Photobiol C Photochem Rev 19:35–51
175. Bowker M, Millard L, Greaves J, James D, Soares J (2004) Photocatalysis by au nanoparticles: reforming of methanol. Gold Bull 37(3-4):170–173
176. Daskalaki VM, Kondarides DI (2009) Efficient production of hydrogen by photo-induced reforming of glycerol at ambient conditions. Catal Today 144(1–2):75–80
177. Su R, Tiruvalam R, Logsdail AJ, He Q, Downing CA, Jensen MT, Dimitratos N, Kesavan L, Wells PP, Bechstein R, Jensen HH, Wendt S, Catlow CRA, Kiely CJ, Hutchings GJ, Besenbacher F (2014) Designer titania-supported Au-Pd nanoparticles for efficient photocatalytic hydrogen production. ACS Nano 8(4):3490–3497
178. Tanaka A, Sakaguchi S, Hashimoto K, Kominami H (2012) Preparation of Au/TiO_2 exhibiting strong surface plasmon resonance effective for photoinduced hydrogen formation from organic and inorganic compounds under irradiation of visible light. Catal Sci Technol 2 (5):907–909
179. Fu X, Long J, Wang X, Leung DYC, Ding Z, Wu L, Zhang Z, Li Z, Fu X (2008) Photocatalytic reforming of biomass: a systematic study of hydrogen evolution from glucose solution. Int J Hydrog Energy 33(22):6484–6491
180. Hisatomi T, Kubota J, Domen K (2014) Recent advances in semiconductors for photocatalytic and photoelectrochemical water splitting. Chem Soc Rev 43(22):7520–7535
181. Fang J, Cao SW, Wang Z, Shahjamali MM, Loo SCJ, Barber J, Xue C (2012) Mesoporous plasmonic Au-TiO_2 nanocomposites for efficient visible-light-driven photocatalytic water reduction. Int J Hydrog Energy 37(23):17853–17861
182. Tanaka A, Sakaguchi S, Hashimoto K, Kominami H (2014) Photocatalytic reactions under irradiation of visible light over gold nanoparticles supported on titanium(IV) oxide powder prepared by using a multi-step photodeposition method. Catal Sci Technol 4(7):1931–1938
183. Tanaka A, Sakaguchi S, Hashimoto K, Kominami H (2013) Preparation of Au/TiO_2 with metal cocatalysts exhibiting strong surface plasmon resonance effective for photoinduced hydrogen formation under irradiation of visible light. ACS Catal 3(1):79–85
184. Zhang ZY, Li AR, Cao SW, Bosman M, Li SZ, Xue C (2014) Direct evidence of plasmon enhancement on photocatalytic hydrogen generation over Au/Pt-decorated TiO_2 nanofibers. Nanoscale 6(10):5217–5222
185. Yan J, Wu G, Guan N, Li L (2013) Synergetic promotion of the photocatalytic activity of TiO_2 by gold deposition under UV-visible light irradiation. Chem Commun 49 (100):11767–11769
186. Sreethawong T, Yoshikawa S (2005) Comparative investigation on photocatalytic hydrogen evolution over Cu-, Pd-, and Au-loaded mesoporous TiO_2 photocatalysts. Catal Commun 6 (10):661–668
187. Murdoch M, Waterhouse GIN, Nadeem MA, Metson JB, Keane MA, Howe RF, Llorca J, Idriss H (2011) The effect of gold loading and particle size on photocatalytic hydrogen production from ethanol over Au/TiO_2 nanoparticles. Nat Chem 3(6):489–492
188. Yuzawa H, Yoshida T, Yoshida H (2012) Gold nanoparticles on titanium oxide effective for photocatalytic hydrogen formation under visible light. Appl Catal B Environ 115:294–302

189. Jovic V, Chen WT, Sun-Waterhouse D, Blackford MG, Idriss H, Waterhouse GIN (2013) Effect of gold loading and TiO_2 support composition on the activity of Au/TiO_2 photocatalysts for H_2 production from ethanol-water mixtures. J Catal 305:307–317
190. Kowalska E, Mahaney OOP, Abe R, Ohtani B (2010) Visible-light-induced photocatalysis through surface plasmon excitation of gold on titania surfaces. Phys Chem Chem Phys 12 (10):2344–2355
191. Yan JQ, Wu GJ, Dai WL, Guan NJ, Li LD (2014) Synthetic design of gold nanoparticles on anatase TiO_2 {001} for enhanced visible light harvesting. ACS Sustain Chem Eng 2 (8):1940–1946
192. Cui ET, Lu GX (2014) New evidence for the regulation of photogenerated electron transfer on surface potential energy controlled co-catalyst on TiO_2 – the investigation of hydrogen production over selectively exposed Au facet on Au/TiO_2. Int J Hydrog Energy 39 (15):7672–7685
193. Wang Y, Yu JG, Xiao W, Li Q (2014) Microwave-assisted hydrothermal synthesis of graphene based Au-TiO_2 photocatalysts for efficient visible-light hydrogen production. J Mater Chem A 2(11):3847–3855
194. Liu Y, Yu H, Wang H, Chen S, Quan X (2014) Efficient H_2 production over Au/graphene/ TiO_2 induced by surface plasmon resonance of Au and band-gap excitation of TiO_2. Mater Res Bull 59:111–116
195. Singh GP, Shrestha KM, Nepal A, Klabunde KJ, Sorensen CM (2014) Graphene supported plasmonic photocatalyst for hydrogen evolution in photocatalytic water splitting. Nanotechnology 25(26):1–11
196. Roy SC, Varghese OK, Paulose M, Grimes CA (2010) Toward solar fuels: photocatalytic conversion of carbon dioxide to hydrocarbons. ACS Nano 4(3):1259–1278
197. Dhakshinamoorthy A, Navalon S, Corma A, Garcia H (2012) Photocatalytic CO_2 reduction by TiO_2 and related titanium containing solids. Energy Environ Sci 5(11):9217–9233
198. Tu W, Zhou Y, Zou Z (2014) Photocatalytic conversion of CO_2 into renewable hydrocarbon fuels: state-of-the-art accomplishment, challenges, and prospects. Adv Mater 26 (27):4607–4626
199. Das S, Daud WMAW (2014) A review on advances in photocatalysts towards CO_2 conversion. RSC Adv 4(40):20856–20893
200. Habisreutinger SN, Schmidt-Mende L, Stolarczyk JK (2013) Photocatalytic reduction of CO_2 on TiO_2 and other semiconductors. Angew Chem Int Ed 52(29):7372–7408
201. Ma Y, Wang X, Jia Y, Chen X, Han H, Li C (2014) Titanium dioxide-based nanomaterials for photocatalytic fuel generations. Chem Rev 114(19):9987–10043
202. Neatu S, Macia-Agullo JA, Concepcion P, Garcia H (2014) Gold-copper nanoalloys supported on TiO_2 as photocatalysts for CO_2 reduction by water. J Am Chem Soc 136 (45):15969–15976
203. Qu Y, Duan X (2013) Progress, challenge and perspective of heterogeneous photocatalysts. Chem Soc Rev 42(7):2568–2580
204. Wang Z, Liu Y, Huang B, Dai Y, Lou Z, Wang G, Zhang X, Qin X (2014) Progress on extending the light absorption spectra of photocatalysts. Phys Chem Chem Phys 16 (7):2758–2774
205. Lou Z, Wang Z, Huang B, Dai Y (2014) Synthesis and activity of plasmonic photocatalysts. ChemCatChem 6(9):2456–2476

Top Curr Chem (2016) 371: 253–324
DOI: 10.1007/128_2015_650

Published online: 13 August 2015

Advanced and In Situ Analytical Methods for Solar Fuel Materials

Candace K. Chan, Harun Tüysüz, Artur Braun, Chinmoy Ranjan, Fabio La Mantia, Benjamin K. Miller, Liuxian Zhang, Peter A. Crozier, Joel A. Haber, John M. Gregoire, Hyun S. Park, Adam S. Batchellor, Lena Trotochaud, and Shannon W. Boettcher

Abstract In situ and operando techniques can play important roles in the development of better performing photoelectrodes, photocatalysts, and electrocatalysts by helping to elucidate crucial intermediates and mechanistic steps. The development of high throughput screening methods has also accelerated the evaluation of relevant photoelectrochemical and electrochemical properties for new solar fuel materials. In this chapter, several in situ and high throughput characterization tools are discussed in detail along with their impact on our understanding of solar fuel materials.

Keywords Electrocatalysis · In operando · In situ · Photocatalysis · Solar fuels

Contents

1 Overview of Recent In Situ Studies for Characterizing Solar Fuel Materials

Candace K. Chan
School for Engineering of Matter, Transport and Energy,
Arizona State University, Tempe, AZ 85287, USA
candace.chan@asu.edu

Harun Tüysüz
Max-Planck-Institut für Kohlenforschung, Kaiser-Wilhelm-Platz 1, 45470
Mülheim an der Ruhr, Germany
tueysuez@kofo.mpg.de

The revisiting of the field of solar fuels, now almost 45 years after the seminal work by Fujishima and Honda [1], has been aided by the development of improved measurement techniques and synthesis methods which were previously unavailable. The establishment of standard methods, efficiency definitions, and reporting protocols for photoelectrochemical systems for solar hydrogen production is considerable progress towards obtaining accurate and consistent evaluation of the performance of photoelectrode and photocatalyst materials. Readers interested in these protocols are encouraged to consult [2, 3] for methods pertaining to photoelectrochemical systems and the chapter by Takanabe for slurry photocatalyst configurations.

Equally important is the development of improved characterization techniques to better understand the mechanisms of photocatalytic or electrocatalytic reactions. Because of the nature of photoelectrochemistry, in which the reaction conditions comprise light excitation, aqueous electrolyte, and often electrical biasing, the materials properties in operando or in situ can be difficult to study. In situ characterization is particularly useful for enabling improved correlation between observations, such as quantities of generated hydrogen or measured photocurrents, with the multitude of processes that work together to make the end results possible (Fig. 1).

This has, for example, drawn attention to important considerations related to side reactions that were previously not identified or quantified. For instance, simultaneous acquisition of current-voltage curves with O_2 detection on photoanodes (either using electrochemical mass spectroscopy [4] or fluorescence-based sensor probes [5, 6]) has revealed important potential-dependent photocorrosion processes in n-type (In,Ga)N nanowires [4] and electrolyte salt decomposition observations in

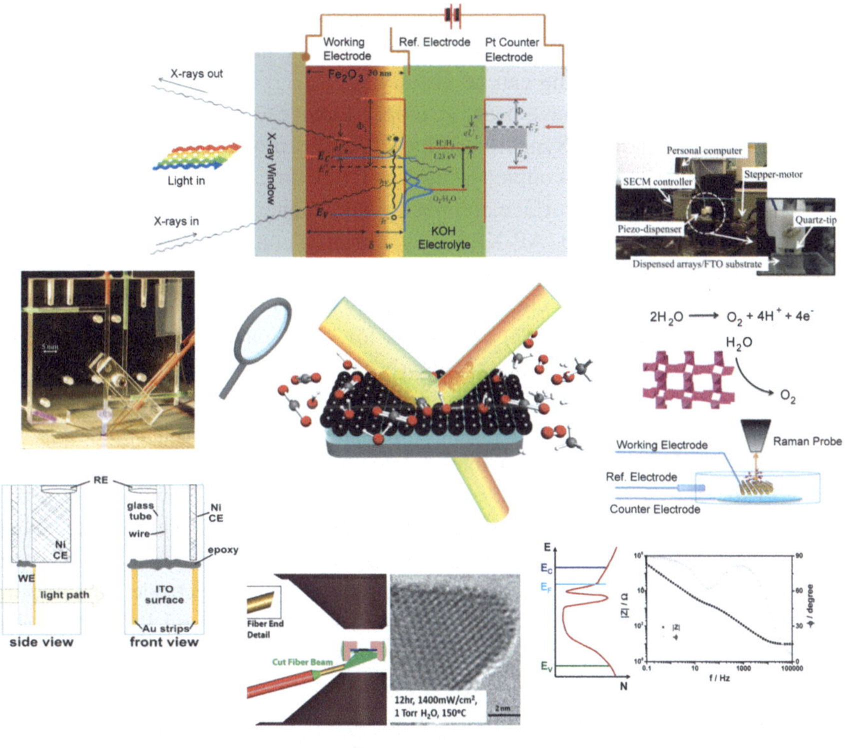

Fig. 1 Characterization methods for studying solar fuel absorbers and catalysts reviewed in this chapter including, starting at the top and going clock-wise: in situ X-ray spectroscopies, high throughput scanning electrochemical microscopy, in situ Raman spectroscopy, impedance spectroscopy techniques, operando TEM, in situ optical characterization, and high throughput screening platforms. This figure is partially adopted from other figures in this chapter

WO_3 [5, 6]. Although the photocatalytic decomposition of electrolyte salt anions (e.g., Cl^-, HSO_4^-, ClO_4^-) by WO_3 is certainly an undesired side-reaction for solar-driven water splitting, the concept was exploited in a regenerative photoelectrochemical cell using a reversible anionic redox couple that demonstrated a very large open-circuit photovoltage [7]. This illustrates just one example of how in situ characterization can not only improve fundamental understanding but also lead to completely new insights, directions, and opportunities.

In terms of understanding mechanistic aspects related to the photocatalyst or electrocatalyst materials, in situ spectroscopic methods have long been used in the field of heterogeneous catalysis. A review of common methods used by surface scientists to study materials under reaction conditions, such as ambient pressure X-ray photoelectron spectroscopy (AP-XPS), high-pressure scanning tunneling microscopy (HP-STM), sum frequency generation (SFG) vibrational spectroscopy, and time-resolved Fourier transform infrared (FTIR) spectroscopy can be found in the review by Somorjai et al. [8]. These techniques are particularly useful for obtaining surface composition and oxidation state information (AP-XPS), elucidating the surface structure (HP-STM), and identifying reaction intermediates (SFG, FTIR). In situ experiments are particularly advantageous when using surface sensitive techniques, as artifacts from contaminants introduced during sample transfers or ion beam cleaning can be avoided [9].

Many of these methods have been developed further to probe materials relevant for solar fuel applications. Synchrotron-based X-rays provide high intensity excitation sources which, when coupled with in situ reactors, allow for the monitoring of electronic structure and surface compositional changes as a function of temperature and/or applied voltage. AP-XPS at synchrotron sources has been adopted in various configurations for investigating materials used in energy applications. Schlögl et al. described the chemical reactivity of a Pt electrode surface under oxygen evolution reaction (OER) conditions for low temperature gas phase electrolysis using in situ near-ambient-pressure (NAP)-XPS with a water pressure in the region of millibars [10]. The combination of this technique with online product analysis allowed for detection of the intermediate species during gas phase water electrolysis. An AP-XPS setup capable of detecting high kinetic energy photoelectrons (up to 7 keV) was constructed by Liu and co-workers which could reach pressures up to 110 Torr [11]. A three-electrode electrochemical cell was achieved by employing a "dip and pull" technique to create a stable, thin electrolyte layer tens of nanometers thick on a Pt working electrode. The Pt electrode was investigated under OER conditions and the formation of both Pt^{2+} and Pt^{4+} interfacial species was identified, with the divalent state proposed to be the result of a kinetics-driven product. In another in situ AP-XPS study, Jaramillo and Nilsson showed that the surface sites on amorphous molybdenum sulfides experienced a gradual chemical composition change from MoS_3 to MoS_2 under hydrogen evolution conditions [12].

In Sect. 2 of this chapter, Braun discusses several other characterization techniques commonly performed at synchrotron sources, such as X-ray absorption spectroscopy (XAS) and extended X-ray absorption fine structure (EXAFS).

These methods have long been used to investigate the structures of natural photosynthesis components and have also been extended to solar fuel materials, such as α-Fe_2O_3 (hematite) photoanodes under water oxidation conditions [13] and the Co–Pi water oxidation catalyst developed by Nocera's group under active biasing conditions [14, 15]. In the latter case, the determination of the catalyst structure was particularly challenging because of the amorphous nature of Co–Pi, making X-ray diffraction and transmission electron microscopy (TEM) methods less useful. However, by performing XAS on Co–Pi inside an electrolysis cell during activity catalysis, Kanan et al. were able to infer that the active material was composed of bis-μ-oxo/hydroxo-linked Co ions and developed a molecular cobaltate cluster model to describe the local structure responsible for the catalytic activity [15]. In situ XAS has also recently been used to probe the transfer of carriers between the light absorber and the co-catalyst where the chemical reactions take place. For example, Yoshida et al. studied the photoexcited hole transfer from $SrTiO_3$ photoelectrodes to MnO_x co-catalysts and found the migration of holes was greater under applied biases and stronger light irradiation, as expected based on the photoelectrochemical activity measurements [16].

Electron paramagnetic resonance (EPR) or electron spin resonance (ESR) is a powerful spectroscopy technique used to explore paramagnetic compounds, species, or radicals to provide information on their geometric and electronic structure and chemical environments. EPR probes paramagnetic species by means of magnetic dipole transitions and a spectrum is obtained by sweeping the externally applied magnetic field. This spectroscopy technique is commonly used for the detection and identification of free radicals and paramagnetic centers. For instance, Cox et al. used ex situ EPR spectroscopy to resolve the electronic structure of the oxygen-evolving complex in photosystem II prior to O–O bond formation [17]. Gray and co-workers studied the charge separation of the titania photocatalyst Degussa P25 (which contains small rutile crystallites mixed with anatase crystallites) under light irradiation using EPR and attributed the enhanced photocatalytic activity to the junction areas between the two phases, which allowed rapid electron transfer from rutile to anatase [18]. Britt and Nocera applied the EPR method in operando electrochemical water oxidation at neutral pH using Co–Pi as catalyst [19]. Their EPR study showed paramagnetic signals corresponding to populations of both Co(II) and Co(IV) species. They found that when increasing the voltage, the population of Co(IV) rises whereas that of Co(II) decreases. This phenomenon confirmed the formation of Co(IV) species during water oxidation which strongly supported the mechanism of OER over Co–Pi. Combination of this spectroscopy technique with operando conditions can provide information about intermediate species and enable better mechanistic understanding of various chemical reactions.

Absorption and vibrational spectroscopies have long been important characterization tools for researchers working on light absorbing and catalytic materials, so it is not unexpected that there has been much research activity in developing in situ methods for characterizing solar fuel materials. Transient absorption (TA) methods have been quite useful for probing charge carrier dynamics and lifetimes. For example, TA spectra acquired (from the visible to near IR wavelengths at femtosecond to microsecond timescales) under biasing that promoted electron or hole accumulation conditions allowed for the spectral signatures of electrons and holes in hematite photoanodes to be determined [20]. TA kinetics studies showed that 98% of the photogenerated holes decayed within 6 μs, indicating that electron-hole recombination is a major limiting factor in the efficiency of hematite-based electrodes. In another example, ultrafast TA was used to probe the kinetics of the first hole transfer (namely the oxidation of hydroxide ion to form hydroxyl radical) at the $SrTiO_3$/electrolyte interface as a function of surface hole potential in an in situ electrochemical cell, demonstrating that TA is an effective method for separating out the kinetics of a single step from subsequent steps under relevant, i.e., high quantum efficiency, reaction conditions [21]. In situ photoelectrochemical ultrafast TA can also be used to examine the role of defects and surface treatments on the charge carrier dynamics and, hence, generated photocurrents, as was recently performed on ZnO nanowires [22].

In situ Raman spectroscopy is also being increasingly used to study spectral features of solar fuel materials under reaction conditions, particularly as Raman spectroscopy does not have interference from the water signal as in Fourier transform infrared (FTIR) spectroscopy, its complementary technique. In Sect. 3, Ranjan presents the fundamentals of Raman spectroscopy and recent developments in in situ and surface enhanced Raman methods for probing the water oxidation and CO_2 reduction reactions, particularly for investigating the structures of metal oxide electrocatalysts. Various methods used to enhance Raman signals and improve sensitivity of the technique are also described.

Time-resolved FTIR spectroscopy can be used to obtain information regarding the molecular nature of active sites, with in situ FT-IR able to provide real time monitoring of reaction intermediates. Frei et al. were able to detect a surface hydroperoxide intermediate on iridium oxide water oxidation catalysts during operation with a visible light excited sensitizer [23]. On cobalt oxide water oxidation catalysts, time resolved FT-IR spectroscopy showed formation of intermediates, namely surface superoxides (the three-electron oxidation intermediate absorbing at 840 cm^{-1}) and oxo Co(IV) sites (the one-electron oxidation intermediate absorbing at 1,013 cm^{-1}) [24]. In situ FT-IR has also been employed to monitor the intermediates and products of CO_2 reduction. For instance, the formation of formate and methoxy adsorbates could be detected over Cu based catalysts [25]. From the combination of the aforementioned in situ methods, it is clear that a great deal of information regarding the nature of the electronic, structural, and surface properties of the catalyst, as well as the important intermediates in the chemical reactions and catalyst active sites, can be obtained.

Electrochemical methods, long used for evaluation of photocurrent generation, can also reveal a great amount of fundamental materials properties. In Sect. 4, La Mantia provides an overview of electrochemical impedance methods used to study the semiconductor/electrolyte interface. These techniques, commonly used in semiconductor photoelectrochemical characterization, were expanded to obtain a better understanding of mechanistic aspects in dye-sensitized solar cells and now for photoelectrochemical water-splitting systems [26].

For visualization of catalyst structures in operando conditions, in situ microscopy methods can provide valuable information. Scanning probe techniques such as atomic force microscopy (AFM), scanning tunneling microscopy (STM), and scanning electrochemical microscopy (SECM) have been well adapted to accomodate liquid cells, biasing, and light irradiation functionalities. In STM, a fine conducting probe is held close to the sample surface and an electrical signal is produced from the tunneling of electrons between the surface and the probe tip. The sharp probe tip can be slowly scanned across the surface at a distance on the order of the diameter of one atom. This kind of scan can provide the finest details of the surface and can enable the study of surface structure atom by atom. By using low temperature STM, Hou and co-workers were able to observe tbe photocatalytic dissociation of water on terminal Ti sites of a TiO_2 surface under UV light irradiation [27]. An STM study on the adsorption of water on a reduced rutile TiO_2 (110) surface showed that, at a temperature of 50 K, an isolated water monomer adsorbs on top of a Ti(5f) atom [28]. When the coverage is increased, water molecules form one-dimensional chain structures along the Ti row direction. This study provided insights into the nature of hydrogen bonding in the initial stages of wetting of the TiO_2 photocatalyst. In another study using STM, the dissociation of CO_2 molecules was investigated on a TiO_2 (110) surface [29]. The observed dependence of the dissociation rate in the tunneling current suggested that the reduction of CO_2 was induced by a single electron attachment. Kelvin probe force microscopy (KPFM) has been widely applied to solar cells to study the surface potential of materials. Liu et al. adopted this technique to a photochemically relevant nanowire nanostructure and examined the photoresponse of the surface potential at the semiconductor/electrolyte interface and semiconductor/semiconductor (p-Si/n-TiO_2) heterojunctions [30]. Under UV illumination, a higher surface potential was observed on the n-TiO_2 side relative to the potential of the p-Si side, which was attributed to the recombination of majority carriers at the Si/TiO_2 interface. These findings show that this approach can be used to investigate charge separation and transport in heterojunctions to understand better the fundamental processes occuring and enable better design of photoeletrodes.

The development of in situ TEM methods has the potential for substantially improving our understanding of the atomic-level structure-property relationships of photocatalysts and electrocatalysts under reaction conditions. The investigation of heterogenous catalyst particles under dynamic reaction conditions (e.g., in the presence of gas mixtures) using in situ environmental TEM has been going on for some time [31, 32]. Recently, the field of in situ TEM in general has undergone an explosive evolution, particularly with regard to (or perhaps, as a result of) sophisticated sample holders, with once highly novel liquid cells and electrical/electrochemical biasing setups now available at many institutions. The liquid cell configuration reported in 2012 based on solutions encapsulated between two graphene layers and used to observe colloidal platinum nanocrystal growth [33] is impressive considering that graphene was only isolated and properly characterized for the first time in 2004 [34]. A recent review contains more than 30 citations since 2011 describing in situ TEM studies on lithium battery materials [35]; it is not unreasonable to expect a similar, if not greater, output for solar fuel materials, which includes many more potential candidates for investigation, considering all the photoabsorber and electrocatalysts possible for water splitting and CO_2 reduction.

In Sect. 5, the contribution by Miller, Zhang, and Crozier describes the state-of-the-art for using in situ TEM to characterize solar water-splitting-related materials. Similar to the simultaneous photocurrent and in situ O_2 detection experiments described earlier, in situ TEM studies have led to some unexpected observations, such as the formation of an amorphous, Ti^{3+} rich layer one to two monolayers thick on anatase TiO_2 under exposure to light and water vapor [36] (which is described in more detail in Sect. 5), and photocorrosion and morphology changes in Ni/NiO core-shell co-catalysts decorated on Ta_2O_5 [37] and TiO_2 light absorbers [38]. In situ observations of electrolytic H_2 production using gold electrodes have also revealed new insights into the nucleation processes of H_2 bubbles, namely that they initially dissolve in the solution and nucleate about 6–8 nm away, not directly on the electrodes [39].

The development of aberration-corrected electron microscopes allows for spatial resolution down to 0.5 Å [40], with 1 Å routine. However, the energy resolution of electron energy loss spectroscopy (EELS) performed in an electron microscope is typically around 1 μm or more. Recently, the electron probe has been optimized to an energy width of ~10 meV and size <1 nm using a monochromated scanning transmission electron microscope (STEM), which has enabled the access of vibrational signals never possible before, such as the EELS detection of hydrogen in TiH_2 [41]. EELS has also been used to quantitatively analyze the composition of gas (mixtures of CO, CO_2, CH_4, O_2, and H_2, air) in an environmental TEM with accuracies of 15% or better [42, 43], which, along with mass spectrometry (e.g., using a residual gas analyzer interfaced to the reaction cell [44]), undoubtedly proves to be useful for characterizing products of water splitting and CO_2 reduction catalyst materials with simultaneous atomic structural characterization of the particles during the reaction.

It did not take long for the community to realize that the simple binary oxides such as TiO_2, WO_3, and Fe_2O_3 would be inadequate as light absorbers for solar fuel applications because of the sheer number of materials requirements (e.g., stability, band gaps, band edge energies, charge carrier lifetime, etc.). Without even considering different crystal structures and complex composition ratios, there are about 20,000 possible ternary oxide compositions and >220,000 possible combinations for quaternary oxides [45]. For this reason, combinatorial and high throughput methods have been investigated to aid with materials discovery. Although high throughput analysis of photoelectrochemical properties is more involved than, say, screening the binding affinity of potential drugs to biological targets [46], there has been significant success in establishing the methods to perform three-electrode electrochemical measurements to obtain a variety of relevant properties, such as the dark open circuit potential (OCP), OCP shift under illumination, peak-photocurrent, steady state photocurrent, external quantum efficiency (EQE), and incident-photon-conversion-to-electron efficiency (IPCE). With this information, one can quickly evaluate several critical properties about the semiconductor, for example whether it is an *n*- or *p*-type material, the Fermi level or flat-band potential, the steady-state photocurrent onset voltage, the potential-dependent photocurrent, surface recombination rate, and presence of possible light or voltage-induced corrosion reactions [47]. In Sect. 6, Haber and Gregoire provide an overview of the high throughput screening methods employed for characterizing solar fuels materials. The contribution from Park in Sect. 7 focuses on the scanning electrochemical microscopy (SECM) methods utilized for screening photocatalysts, and in Sect. 8 Batchellor et al. describe in more detail the high throughput UV–Vis absorption screening methods.

Acknowledgements We would like to thank the MAXNET Energy Consortium of the Max Planck Society and the RESOLV Cluster of Excellence (EXC 1069) funded by the Deutsche Forschungsgemeinschaft (DFG) for support. We thank Dr. Kun Chen for his help with preparing Fig. 1.

2 X-Ray and Neutron Spectroscopy Methods for Photosynthesis and Photoelectrochemistry

Artur Braun
Empa, Swiss Federal Laboratories for Materials Science and Technology,
8600 Dübendorf, Switzerland
Artur.Braun@alumni.ethz.ch

2.1 Introduction

This section presents an overview of the X-ray and neutron spectroscopy methods used in the field of photosynthesis and photoelectrochemistry. A conceptual difference arises from these related but different fields. The analytical methods used in photosynthesis are mainly derived from molecular structure and similar chemists' concepts, whereas photoelectrochemistry is typically semiconductor photoelectrochemistry and thus derived from condensed matter physics and surface science. Looking at these two dissimilar fields, I consider photoelectrochemistry on organic materials to range in between them.

2.2 Characterization of Hydrogenases

Photoelectrochemical water splitting plus CO_2 reduction to produce solar hydrocarbon fuels is basically the reproduction of the photosynthesis cycle found in nature. Hydrogenases are enzymes that catalyze the reversible oxidation of molecular hydrogen. The center of the hydrogenase macromolecule is a unit of two metal ions (Fe and Ni) which are bridged by two sulfur ions (Fig. 2).

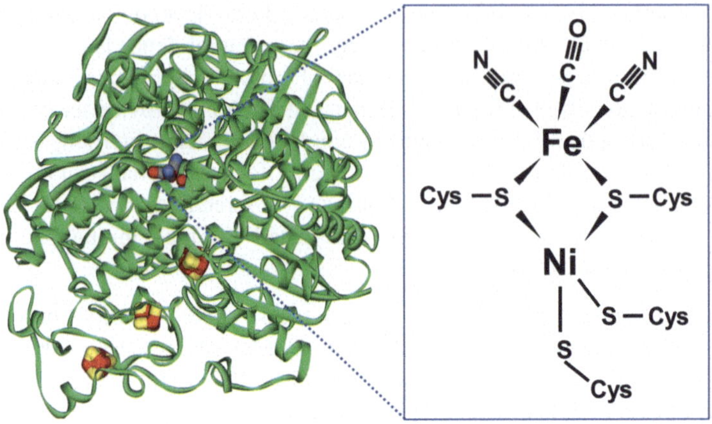

Fig. 2 Structural representation of the hydrogenase macro-molecule from *Desulfovibrio gigas* bacteria, plus the structural formula of the [NiFe] active site

The crystal structures of such enzymes are known largely from multiwavelength anomalous X-ray diffraction (MAD phasing) [48, 49] obtained from single crystals. To understand the function of hydrogenases, information on the chemical bond lengths has been a topic of interest for a long time. X-Ray absorption spectroscopy (XAS) has been of great service for this task. EXAFS [50] spectroscopy allows for the determination of the first and second molecular coordination shells between Ni–S or Ni–Fe, for example. For this, the EXAFS oscillations are Fourier transformed into real space information. In contrast to X-ray diffraction, EXAFS is a structure determination method which requires no crystallographic long range order. Short range order such as in a molecule is sufficient. The X-ray spectroscopy methods including EXAFS are element specific. It is therefore possible to discriminate different bond lengths in a molecule with metal centers. The hydrogenases contain an iron–sulfur cluster, which spectroscopically cannot be distinguished from the aforementioned Fe–S bond. Therefore, determination of the Fe–S bond length has always been problematic, whereas the Ni–S bond can be identified with ease by EXAFS because there is no other Ni-containing component in hydrogenases which could cause interference with the Ni–S co-factor.

Figure 3 shows a Ni EXAFS spectrum of the $Ni[(S_2C_2)Ph_2]_2$ model compound with absorption oscillations in the range 8,350–9,000 eV, indicating near range order in the molecular coordination shells around the Ni central atom. This X-ray absorption spectrum is then transformed from the X-ray energy scale (eV) to the electron wave momentum reciprocalspace, i.e., the k-vector, whereas the absorbed intensity is multiplied with the third power of the k-vector (Fig. 3, top right). The Fourier transform of this signal then yields the radial distribution of statistically significant molecular distances in the coordination shell, as shown in Fig. 3 (lower right).

The majority of Ni hydrogenase EXAFS studies report the presence of trace amounts of Cu [52] in the sample or in the measurement apparatus. Because Cu is next to Ni in the periodic table of elements, it shows up in the high energy tail of the

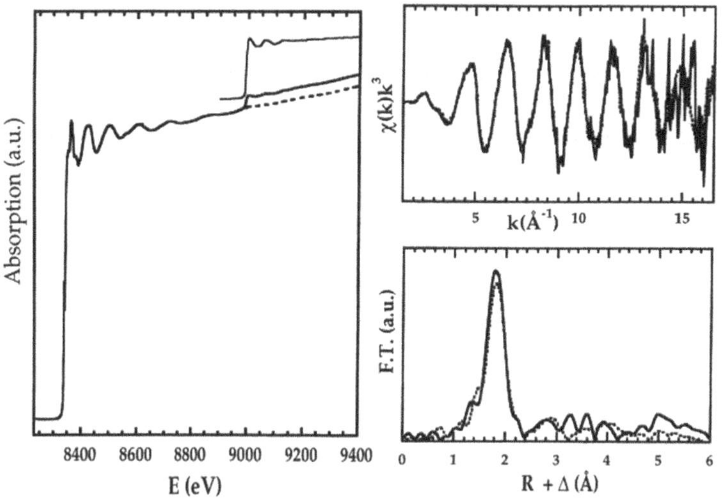

Fig. 3 (*Left*) Demonstration of the dual energy window subtraction procedure. Ni spectrum for a Cu-doped sample of $Ni[(S_2C_2)Ph_2]_2$ (*solid line*); Cu spectrum for the same (*solid line*) starting at 9,000 eV; and result of subtracting a weighted portion of the Cu spectrum from the Ni spectrum (*dashed line*). (*Right*) Comparison of k space (*top*) and Fourier-transformed (*bottom*) EXAFS spectra for a $Ni[(S_2C_2)Ph_2]_2$ sample without Cu contamination (*solid line*) vs corrected spectra for a Cu-doped sample (*dashed line*). Reprinted from Gu et al. [51] with kind permission from Dr. Weiwei Gu. Copyright 2003, Elsevier

EXAFS spectrum of Ni and thus impairs the accuracy of Ni EXAFS data. This is clearly visible in Fig. 3 (left) where at 9,000 eV a new absorption edge from Cu shows up. Hence, the EXAFS oscillations from Ni which extend beyond 9,000 eV are overshadowed from the Cu absorption, and therefore the determined Ni–S distances are not accurate. With a dual energy window subtraction procedure, as presented by Dr. Weiwei Gu [52], the EXAFS spectrum can be corrected for the Cu contamination and thus yields Ni-S distances with improved accuracy [51]. The difference in results when using this subtraction procedure is illustrated in Fig. 4 (left), where the Ni EXAFS spectra from the *Desulfovibrio gigas* hydrogenase were treated in the conventional way (dashed spectrum) or with the range-extended procedure (solid line spectrum).

Because of the double energy window correction, two Ni–S bonds in as-isolated hydrogenase are found with distance of 2.2 Å, as well as Ni–S bonds in the 2.35 Å range. The Ni–O distance was evident at 1.91 Å. The as-isolated Ni–Fe distance could not be unambiguously determined. Upon reduction in hydrogen, two Ni–S distances persist at 2.2 Å, but the aforementioned other Ni–S distances increase to 2.47 Å. The Ni–Fe distances are determined to be 2.52 Å, supported by good simulations and in agreement with crystal structures of reduced hydrogenase. From good fits to structural models, it appears that a Ni–O distance with 2.03 Å is present – information which was not evident from the crystal structure is obtained from XRD. Based on these data it was, for example, possible to discuss the spin state of the reduced hydrogenase.

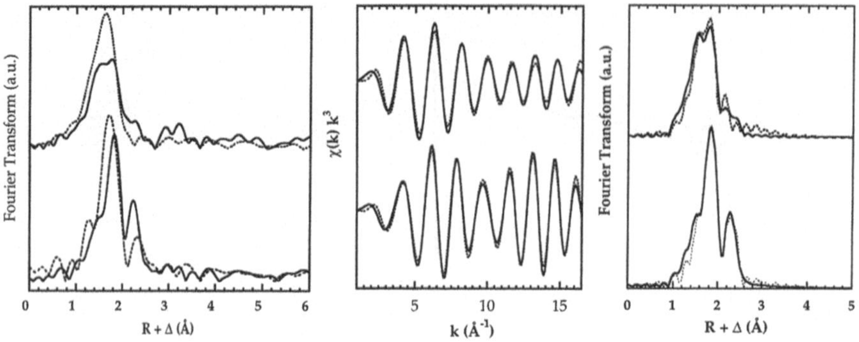

Fig. 4 (*Left*) Conventional (*dashed curves*) and range-extended (*solid curves*) EXAFS Fourier transforms for (*top line*) as-isolated and (*bottom line*) for H reduced hydrogenase from *desulfovibrio gigas*. (*Middle*) Fourier-filtered EXAFS k-space spectra (*solid line*) and best fits (*dotted line*) for (*top line*) as-isolated and (*bottom line*) H reduced hydrogenase. (*Right*) EXAFS Fourier transforms for the spectra (*solid line*) and the best fits (*dashed line*) reported in the *left panel*. Reprinted with permission from Gu et al. [51]. Copyright 2003, Elsevier

X-Ray spectroscopy methods are not only element specific. Their advantage is that they can reveal the molecular structure and electronic structure, which includes, for example, the determination of the oxidation states of cations or ligands in compounds and also spin states. Changes in the oxidation state manifest typically in a shift of the X-ray absorption or X-ray emission spectrum vs the X-ray energy scale; this is consequently called a "chemical shift". In many cases, the chemical shift can be used for a valuable quantitative determination of the oxidation state, more so and also with more ease than, for example, by chemical titration [53, 54].

The oxygenases are proteins which contain the oxygen evolving the complex in Photosystem II (PS II), which is a Mn–Ca–O cluster with two manganese ions bridged by two oxygen ions. It is understood that virtually all oxygen in the atmosphere is created by the PS II of the plankton in the oceans. Manganese undergoes a complete redox cycle during photosynthesis, which can be monitored with XAS at the Mn K-shell. Of interest is, for example, whether the Mn^{3+} species exists in the high spin or low spin configuration. The electronic spin leaves no easy visible signature in the XAS or X-ray absorption near edge structure (XANES) spectra at the Mn K-edge.

An alternative method is X-ray emission spectroscopy, which probes the occupied states. This method is very sensitive for the spin state. In this particular case, we applied K-β spectroscopy, where we excited the sample with 11,000 eV X-rays and then probed the K-β emission profile. It should be noted that the Mn redox cycles observed with X-ray spectroscopy, in this particular case K-β spectroscopy, uncovers the analogous chemical processes around Mn in PS II and in Mn oxide-based lithium battery cathodes [54, 55].

The aforementioned methods are typical for X-rays with energies that probe the 3d metal K-shell absorption edges, such as at 3,500 eV photon energy and above. Because of their high attenuation, they are suitable for in situ experiments with spectro-electrochemical cells. One recent example is shown by Kanan et al. [15] where XANES and EXAFS spectra at the cobalt K-edge were recorded for their cobalt–phosphate water oxidation catalysts. It was thus possible to monitor changes of the Co oxidation state and bonding lengths upon electrocatalyst operation for water splitting.

The hydrogenase also undergoes a catalytic cycle which includes "unready,""ready," and "active" states, during which the electronic structures of Ni and Fe change (Fig. 5, left). As already mentioned in the beginning, the Fe electronic structure is very elusive because it escapes spectroscopic detection because of the proximity of Fe in the [NiFe] active sites to the Fe–S cluster nearby. A functional description of the involved Ni-states is provided in Sect. 7.3.1

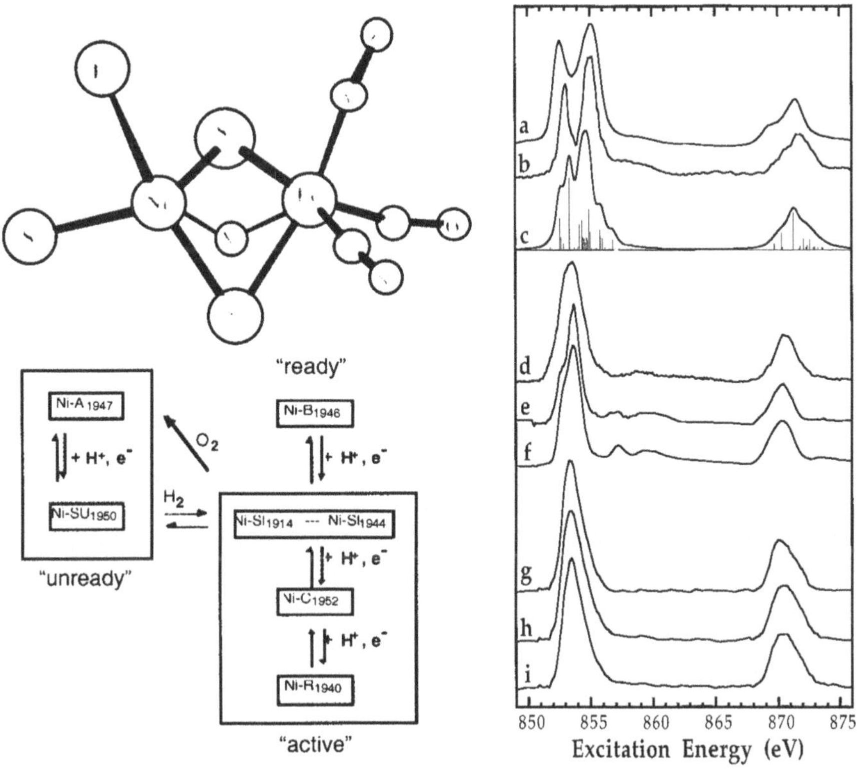

Fig. 5 (*Left*) Relationship between different forms of hydrogenase throughout the catalytic cycle of unready, ready, and active states. *Four-digit numerals* refer to the CO infrared stretching vibration frequencies (in wavenumbers, 1/cm) of the corresponding Ni– bonds. (*Right*) Comparison of oxidized *Desulfovibrio gigas* spectra with those of Ni(III) model compounds. *Top to bottom*: ("ionic Ni(III)") (*a*) K_3Ni-(III)F_6, (*b*) Ni^{III}(cyclam)$Cl_2(ClO_4)$, and (*c*) simulation of a Ni(III) L-edge spectrum. Covalent complexes: (*d*) $[Ni^{III}pdtc_2]^{2-}$, (*e*) (*n*-Bu_4)Ni^{III} $[S_2C_2(CF_3)_2]_2$, and (*f*) (Et_4N) $[Ni^{III}(S_2C_2Ph_2)]_2$. Spectra from as-isolated *Desulfovibrio gigas* hydrogenase using open (*g*), capped (*h*), and sealed (*i*) containers. Reprinted (adapted) with permission from Wang et al. [56] with kind permission from Dr. Hongxin Wang. Copyright 2000, American Chemical Society

"Overview of the catalytic cycle" in [57]. These states have been investigated with infrared spectroscopy, electroparamagnetic resonance spectroscopy, and X-ray spectroscopy. Figure 5 (right) [56] shows a set of Ni L-edge spectra (Ni2p) which show a pronounced spectroscopic doublet with L_3 and L_2 absorption peaks at around 853 and 871 eV ($2p_{3/2}$ and $2p_{1/2}$ states). Specifically, the spectra of hydrogenase are compared with Ni(III) containing model compounds. The three top spectra are obtained from model Ni compounds with preferential ionic bonding, showing distinct fine structures, particularly in the $2p_{3/2}$ state at around 850–857 eV, with the crystal field peak splitting into t_{2g} and e_g orbital symmetry. Spectrum "c" shows, for example, a simulated multiplet spectrum. Covalent compounds, such as the complexes in spectra d, e, and f show typically more diffuse spectra with less peculiar sharp features. However, additional features at around 857 eV which originate from charge transfer processes are visible as satellite peaks. The bottom three spectra, g, h, and i, were recorded from *Desulfovibrio gigas* hydrogenase which was kept under different protective environments. Spectrum g was measured in the ultrahigh vacuum chamber after exposure to ambient environment. Spectrum h was recorded with the sample being transported from a nitrogen glove box to the ultrahigh vacuum chamber at the synchrotron end station in a capped sample contained, and spectrum i was obtained from the hydrogenase which was transported in an airtight or sealed cell to the synchrotron beamline. Hydrogenases are sensitive to oxygen exposure and therefore subject to degradation, which can to some extent be detected in spectral changes.

Hence, although all aforementioned compounds including the hydrogenases were nominally in the Ni(III) state, peculiar differences in the spectra suggest that L-edge spectroscopy gains more information than just the chemical valence of the cation. Multiplet simulations allow for elaborate electronic structure determination such as charge transfer, spin states, and crystal field splitting [58].

The high oxidation states of cations often go along with 3d electron holes, the determination of which can be particularly well accomplished for the case of Ni with L-edge near edge X-ray absorption fine structure (NEXAFS) spectroscopy. This was demonstrated by Dr. Hongxin Wang in Fig. 6, which shows the systematic variation of Ni L-edge multiplets of various Ni model compounds with oxidation states from Ni(I) to Ni(IV). A first observation is that the $2p_{3/2}$ peak intensity, the "white line," is increasing vs the normalization threshold of the spectra with increasing oxidation state. Quantitative "study" of the normalized and integrated intensity of the L_3 and L_2 peaks turns out to scale linearly with the suggested nominal formal Ni oxidation states. Moreover, the same linear behavior turns out to hold for the number of Ni 3d holes with surprisingly good accuracy.

2.3 *Characterization of Metal Oxides*

There may be situations where we are interested in the molecular structure of the ligand molecules in electrode compounds, for example in metal oxides or metal

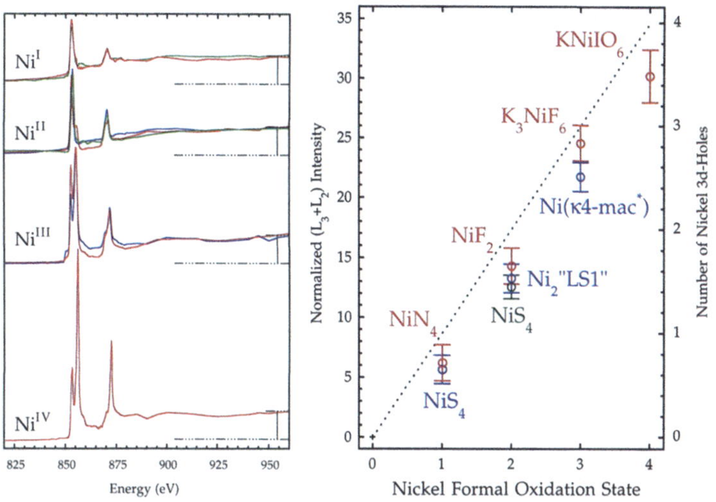

Fig. 6 (*Left*) Comparison of normalized L edge absorption spectra of Ni complexes. *From top to bottom*: (1) Na[Ni^{I}"S_4"] (*green*) and Na[Ni^{I}"N_4"] (*red*); (2) [Ni^{II}_2"LS1"] (*blue*), $Ni^{II}F_2$ (*red*), and Ni^{II}"S_4" (*green*); (3) Li[Ni^{III}(κ^4-mac*)] (*blue*) and $K_3Ni^{III}F_6$ (*red*); (4) $KNi^{IV}IO_6$ (*red*). (*Right*) Integrated intensity of the L edge resonance, normalized to the non-resonance absorption background, as a function of formal oxidation state (I, II, III, IV). The *dotted green line* represents the best estimate for the purely ionic limit, using an 8.66:1 intensity-to-3d holes conversion ratio (where Ni(κ^4-mac*) is Li[Ni(κ^4-mac*)]. Reprinted (adapted) from Wang et al. [59] with kind permission from Dr. Hongxin Wang. Copyright 1998, American Chemical Society

chalcogenides. The relevance of the anion ligands is given by the fact that they are highly functional as far as the electronic transport properties of the compounds are concerned, such as in electrodes in batteries, ceramic fuel cells, and photoelectrochemical cells. For example, the oxygen ions in metal oxides provide charge transport via the oxidation state of the cations and their spins. The conducted electrons interact with these quantities via the exchange interaction, which is rationalized by the Goodenough–Kanamori rules. In addition to this, the O2p orbital can provide electron holes, either by doping or by external transient excitation. We begin this discussion with an example from solid oxide fuel cell cathodes, which are typically built from metal oxides with ABO_3-type perovskite structure such as LaSrMn oxide. An illustrative example is LaSrFe oxide. The parent compound $LaFeO_3$ is an electronic insulator with a large band gap and is thus not conducting. The iron oxidation state is Fe^{3+}. Substitution on the A-site of the La^{3+} with Sr^{2+} forces the Fe^{3+} partially into Fe^{4+}. To a chemist, the suggestion of Fe^{4+} species sounds exotic. Yet, the substitution forces the Fe into an extreme oxidation state which makes the LaSrFe-oxide conducting because of the formation of electron holes, which actually originate from the O2p orbitals of the adjacent oxygen. The electronic structure of Fe is $3d^64s^2$. When we remove four electrons by oxidation, the Fe^{4+} then becomes $3d^5\underline{L}$, where $\underline{L}$ stands for the O2p ligand hole. The chemist can live with this notation because the half full 3d shell is maintained, and the Fe is spectroscopically still a $3d^5$-type species.

Experimental spectroscopic evidence is then provided in the oxygen K-shell X-ray absorption spectra, which are recorded with X-ray energy of around 520–560 eV. Whereas the $LaFeO_3$ with Fe^{3+} has in the pre-edge a nice e_g–t_{2g} doublet, the substituted compound with a portion of Fe^{4+} shows an additional peak in the NEXAFS spectra, representative of the electron hole, $\underline{L}$. It has been shown that the relative spectral weight of this hole peak scales exponentially with the electronic conductivity of the compound [60, 61].

We have applied oxygen NEXAFS spectroscopy to iron oxide photoelectrodes after they had been operated under DC bias in KOH electrolyte for different durations. It turned out that photoelectrochemical (PEC) operation forms a new structure in the valence band which can be associated with an oxyhydroxy-type surface functionality. The new peak is quite elusive because it is positioned in the intensity minimum between the well pronounced e_g–t_{2g} doublet and the Fe^4sp-O2p hybridized states. Experiments with films of different thickness show that the new state is formed at the surface [13]. Its proximity to the readily known FeOOH peak suggests that its structure is similar to that of –OOH, but it is not identical to FeOOH because of the slight energy difference.

Carrying out such experiments in situ is very challenging. To allow for sufficient X-ray photon flux to the photoelectrode, ultrathin X-ray windows made from 100 nm thick Si_3N_4 are used. These are generally rigid enough to sustain the pressure difference between ambient pressure and the ultrahigh vacuum chamber at the synchrotron end station. Figure 7 illustrates how the photoelectrode assembly is mounted on a PEEK plastic sample holder which contains liquid electrolyte in a three-electrode arrangement developed at Lawrence Berkeley National Laboratory.

With this experimental setup, it was for the first time possible to look into the role of the oxygen atoms of iron oxide during PEC water oxidation. From Fig. 7 (top, left) it becomes clear that the X-rays are passing through the vacuum chamber to the Si_3N_4 window, the metal current collector, the iron oxide photoelectrode, and the KOH electrolyte. Only the latter two lend their spectroscopic signature to the oxygen NEXAFS spectrum because they contain oxygen. The attenuation length for X-rays of 520 eV is around 0.4 μm. The transmission of the electrode assembly is below 50%.

The green solid line spectrum in Fig. 8 was obtained from the liquid cell shown in Fig. 7 in operation under light and 500 mV bias potential. Comparison of the spectra recorded under bias potentials from 100 to 900 mV under dark and light conditions shows the striking emergence of two new spectral signatures in the oxygen pre-edge which could be associated with O2p-type and Fe3d-type electron holes. These can be switched on and off with a bias ranging from 400 to 800 mV and visible light [65]. The shaded area peaks between 525 and 527 eV, as obtained by deconvolution and least square fitting of the spectrum, have been associated with transitions into the charge transfer band (CTB) and upper Hubbard band (UHB) [65]. It turns out that the higher energy hole state can be associated with a capacitive surface defect state as determined from impedance spectroscopy [66, 67]. Closer investigation of the green "in situ"spectrum in Fig. 8 shows that it is actually a convolution from hematite and water (KOH electrolyte) spectra. A "dry" hematite

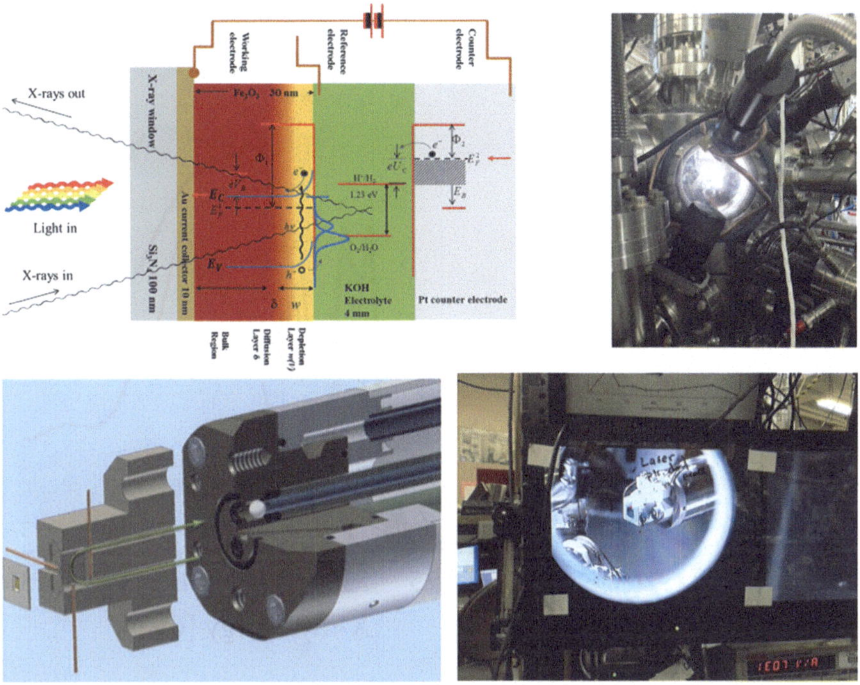

Fig. 7 (*Top left*) Schematic of hematite film coated on metal current collector on Si_3N_4 X-ray window, facing the electrolyte at the right side. The *red–yellow color* gradient illustrates the parabolic profile of the charge carrier depletion layer produced by the applied DC bias. Energy band levels and charge transport directions are shown along with two emerging hole states in the valence band density of states. (*Bottom left*) Design of the PEEK® plastic liquid cell with Si_3N_4 frame, with three-electrode wires, electrolyte compartment, channels, and cell holder for insertion in UHV recipient. Reprinted with permission from Jiang et al. [62]. Copyright 2010, Elsevier. (*Top right*) UHV recipient with attached solar simulator, shining light through vacuum window on liquid cell during PEC operation and NEXAFS data acquisition. (*Bottom right*) Video camera image of the mounted PEEK® cell in operation during NEXAFS data recording. Beamline 7 at the Advanced Light Source

spectrum (orange solid line) with the typical t_{2g}–e_g doublet at 528.6 and 530 eV from hybridized O2p – Fe3d states is shown for reference. We have deconvoluted this part of the "in situ" spectrum for the well-known t_{2g}–e_g doublet plus a smaller peak at 531 eV, which is known to originate from FeOOH (three dotted peaks with orange color).

At 533.25 eV, the in situ spectrum shows a clear peak which is not known from hematite, but from liquid water [63, 64, 68], which we have plotted along with two experimental NEXAFS spectra obtained from liquid water (solid dark blue spectrum) and water in the gas phase (light blue solid spectrum) [63]. When comparing the in situ spectra recorded at 100–900 mV under light (not shown here), we notice an increase of the intensity in the spectral minimum at 532 eV when the potential is approaching 500 mV. At about this X-ray energy, one expects a peak which can be

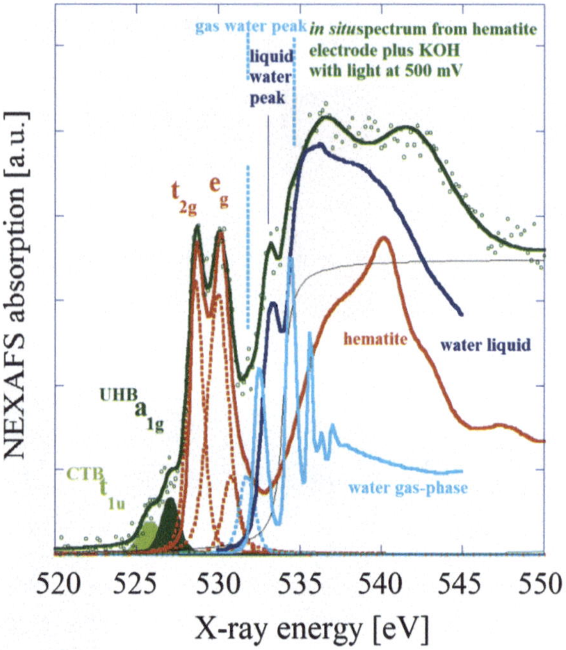

Fig. 8 Oxygen 1 s NEXAFS spectra of hematite film in contact with KOH electrolyte in the in situ liquid cell under light and 500 mV DC (*green spectrum*), spectrum from dry hematite nanoparticles in UHV (*red spectrum*), spectrum of liquid water and gas phase water from Myneni et al. [63] and Hetényi et al. [64]

attributed to the gas phase of water. The relative energy position of this peak depends on the available hydrogen bonds which constitute ice (solid water), liquid water, and water vapor [63, 64]. According to Cavalleri et al., "breaking an accepting H-bond has very little effect on the absorption spectrum. A broken donating H-bond, however, is identified through a strong pre-edge feature in XAS" [68]. I believe that in photoelectrochemical water splitting approaching 500 mV bias, the increase in the intensity minimum plus a very slight redistribution of intensity towards lower energy by 0.5 eV is an experimental manifestation of hydrogen bond breaking.

2.4 Proton Pump Bacteriorhodopsin and Neutron Spectroscopy Methods

In the previous section, we saw how X-ray (NEXAFS) spectra allow for indirect conclusions on the protons in the aqueous KOH electrolyte during PEC water splitting. Because X-rays interact with electrons, X-ray-based methods are not best suited for direct detection and investigation of light elements with very few electrons, particularly hydrogen or protons. Rather, methods with a large scattering cross section for protons are suitable for this purpose, such as neutron scattering and neutron spectroscopy.

Bacteriorhodopsin is a membrane protein which, by absorption of light, can build up a proton concentration gradient and thus produce energy for the ATP-synthase. Bacteriorhodopsin contains chromophores and can be considered to be a genuine proton pump [69, 70]. Prior to light absorption, the chromophores are present as isomer mixtures of *cis*- and *trans*-configuration. Upon light absorption, the *trans*-configurations become converted into *cis*-configurations, which cause morphological changes in the protein with an energetic imbalance of protons, which in turn has influence on the protonation of functional Schiff bases in the protein. The proton is moved in the extracellular direction to regain energetic balance. Further proton shifts follow before the protein regains its initial configuration (=morphological) and energetic state, ready for another such proton pump action. These steps are characterized by protein states which can be determined and distinguished by optical spectroscopy, and is thus called a "photocycle" (Fig. 9).

The aforementioned conformational changes in the proteins occur during light activation and cause changes in the proton transport properties. Such conformational changes can be detected better by neutron diffraction, rather than by X-ray diffraction, because of the large scattering contrast between the protein and associated water. Using deuterated water (D_2O) allows for enhanced scattering contrast in neutron diffraction. Figure 10 shows neutron diffractograms of bacteriorhodopsin recorded at low temperature (93 K) in two different photo-excited

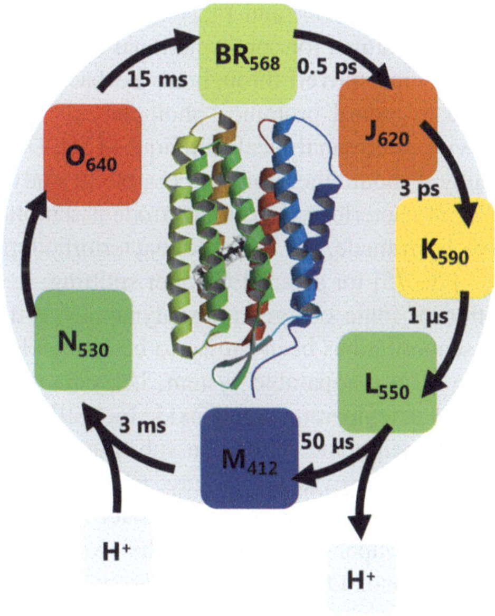

Fig. 9 Schematic for the photocycle of the membrane protein bacteriorhodopsin with characteristic spectroscopy states and proton motion activity. Adapted from Pieper et al. [71] and Patzelt et al. [72]

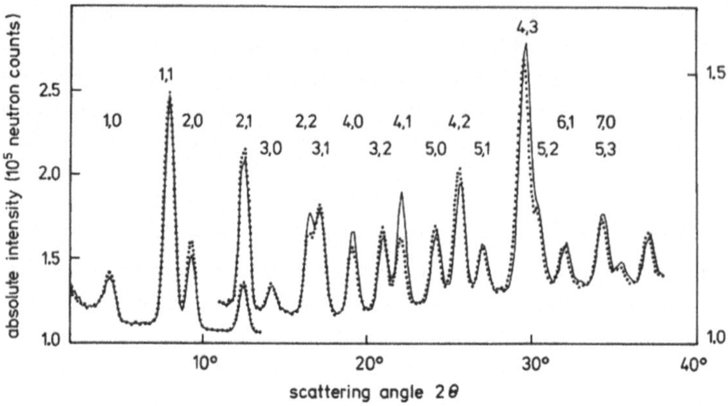

Fig. 10 Neutron diffraction intensities of purple membrane stacks in D_2O of the light-adapted ground state BR-568(−) and of the M intermediate of the photocycle (. . .) at 93 K. Neutron counts as a function of the detector angle 2θ are shown. Reflections are indexed according to a hexagonal lattice. Reproduced from Dencher et al. [73] with kind permission from the author. Copyright 1989, National Academy of Sciences

states, i.e., BR-568 (solid line) and at an intermediate position in the photocycle, referred to as state M. The Bragg reflections are indexed vs a hexagonal lattice. We notice minute but clearly distinguishable changes in the peak heights for the (2,0), (2,1), (2,2), (3,1), (4,0), (4,1), and (4,2) indexed Bragg peaks, which are interpreted as structural signatures for conformational changes [73].

Because protons are being moved throughout the bacteriorhodopsin transport channels, it is possible to record protonic "photocurrents" in bacteriorhodopsin electrode assemblies with electroanalytical techniques [74]. Using electrochemical impedance spectroscopy it should therefore be possible to study the proton dynamics and charge transfer in bacteriorhodopsin electrode assemblies [74]. Such electrode assemblies have been made, for example, bacteriorhodopsin/TiO_2 nanotube arrays hybrid systems [70, 75] for solar PEC water splitting.

For the investigation of ionic charge carrier dynamics at the molecular scale, quasi-elastic neutron scattering has been shown to be a suitable method [76, 77]. I show here an example on a comparable system, i.e., ceramic proton-conducting electrolytes for intermediate temperature solid oxide fuel cells [78]. In this particular case, we have a ceramic membrane of yttrium substituted barium cerate, such as $BaCe_{0.9}Y_{0.1}O_{3-\delta}$ with ABO_3 perovskite structure, which contains engineered oxygen vacancies to be filled with oxygen ions from water molecules when the membrane is exposed to water vapor or humidity. The two protons from the water molecule forms hydroxyl bonds O–H with the adjacent oxygen ions in the crystal lattice. Upon temperature activation, the O–H bonds break and the H^+ protons become mobile charge carriers [79]. The neutron scattering spectrum typically shows a peak with a finite width, which originates from the elastically scattered neutrons from the atoms in the crystalline structure of the probed sample. The temperature-activated mobility of the protons causes a diffusive proton motion

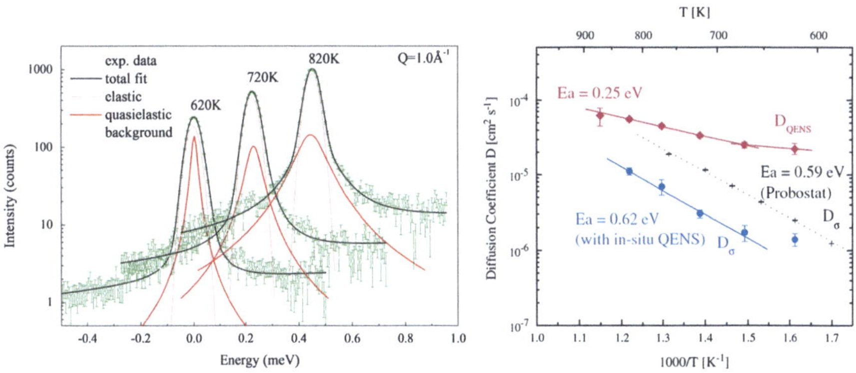

Fig. 11 Neutron scattering spectra from ceramic proton conductor recorded at 620, 720, and 820 K, deconvoluted into elastic and quasi-elastic contributions. The quasi-elastic lines broaden significantly with increasing temperature. Reproduced with permission from Chen et al. [77]. Copyright 2013, Elsevier

visible in the neutron spectrum as a broadening of the peaks, the quasi-elastic scattering, as demonstrated in Fig. 11. The width of this diffuse scattered intensity is given by the diffusivity of the protons. By deconvolution of the neutron spectrum into elastic and quasi-elastic contributions we are able to determine the diffusivity and, via the Nernst–Einstein relation, the proton conductivity at the molecular scale.

To study the proton dynamics in bacteriorhodopsin, such as with quasi-elastic neutron scattering (QENS), we need to be sure about the particular spectroscopic state in the photocycle. Because these photoactivated states have limited lifetimes, it is necessary to synchronize the photoactivation steps with the QENS probing steps. Practically, this amounts to a typical combined spectroscopic pump probe experiment. This has been done recently with success (Fig. 12) [71].

At this point, it should be stated that the right sample preparation becomes more important the more sophisticated an experiment becomes. For example, in the aforementioned X-ray absorption studies, particularly for quantitative EXAFS studies in a transmission sampling geometry (Fig. 3), it is necessary to optimize the sample thickness to "one absorption length," which is the sample thickness weighted with its mass density, to avoid loss of photons by extensive absorption and avoid spectroscopic artifacts by multiple scattering. This holds particularly for operando and in situ experiments where several methods are applied to the very same sample. In a heterogeneous sample with different elements, the optimum sample thickness also depends on which particular element is being investigated [50]. Soft X-ray spectroscopy experiments (NEXAFS) are typically done in reflection mode, where the requirements for optimum sample thickness are more relaxed. In the in situ NEXAFS studies on photoelectrochemical water splitting (Fig. 7), however, the sample thickness needs to be chosen carefully when specific spectral information is required on the electrode bulk, depletion layer range, electrode/electrolyte interface, and electrolyte region. With a 100 nm thick iron oxide film,

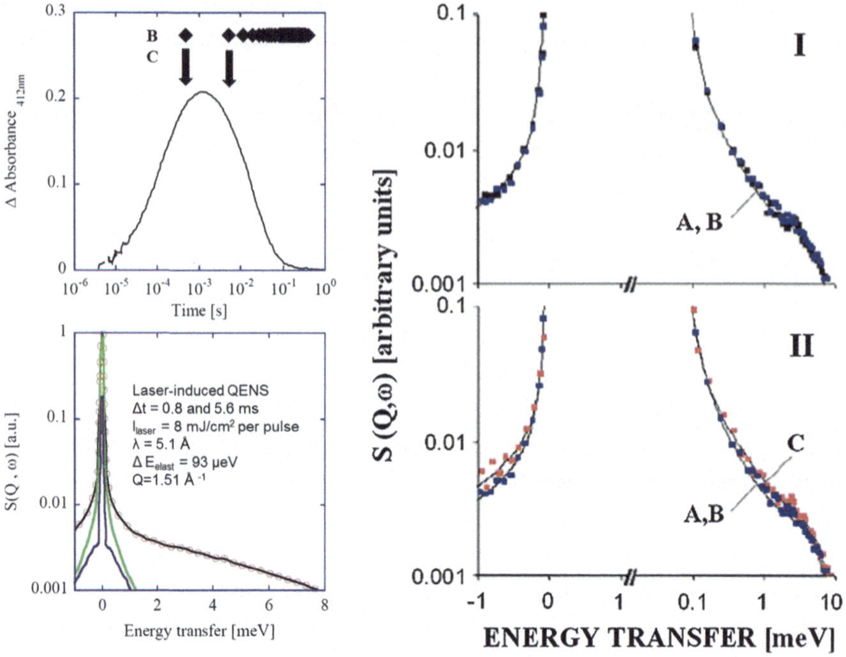

Fig. 12 Transient absorbance changes (*top left*) in bacteriorhodopsin at 412 nm at room temperature after laser excitation at 532 nm and $t = 0$ obtained using a laser pulse energy of 8 mJ cm^{-2} and a laser pulse repetition time of 400 ms, i.e., exactly those conditions used for the light-excited QENS experiments. *Black diamonds* and *arrows* indicate the arrival times of the neutron pulses at the sample relative to the laser pulse in two different measurement modes, i.e., experiments without (B, *black diamonds*) and with time selection (C, *arrows*), respectively. Here, the *letters B and C* correspond to the subscripts of the fit functions to the respective QENS spectra (*right*). Reprinted with permission from Pieper et al. [71]. Copyright 2009, John Wiley and Sons

no information from the bulk and depletion layer can be extracted and no signature from the electrolyte is visible in the measurement geometry shown in Fig. 7.

It should be noted that exposure to X-rays can cause radiation damage to the sample, although X-rays are typically not as damaging to the sample as, for example, the electron beam in an electron microscope. Neutrons, in contrast, hardly cause any radiation damage to the samples shown here. One shortcoming of neutron experiments is the need for a large amount of sample, which is typically a result of the small neutron flux. A QENS experiment needs around 300 mg of protein materials, which is a very large amount in the protein community. The optical absorption of this amount of material is so huge that virtually all light from excitation is absorbed. Then there arises another problem. The fraction of bacteriorhodopsin reaching the intermediate state “M” depends on the absorbance of the protein and the quantum efficiency of the entire photocycle. To allow for combined optical and QENS experiments it was necessary to lower the sample amount to 30 mg, which required a correspondingly longer data acquisition time [80, 81].

3 In Situ Raman Spectroscopy for Probing Solar Fuel Reactions

Chinmoy Ranjan
Max Planck Institute for Chemical Energy Conversion, Stiftstrasse 34-36, 45470 Muelheim an der Ruhr, Germany
chinmoy.ranjan@cec.mpg.de

3.1 Introduction

The phenomenon of Raman scattering was first reported by the Indian physicist C.V. Raman in 1928 [82]. Raman scattering is very low in probability compared to elastic light scattering (Rayleigh scattering). To measure a Raman spectrum, one needs a high intensity monochromatic light source. Initially, such facilities were cumbersome affairs with mercury lamps, dark rooms, and filters. Only after the advent of lasers could Raman spectroscopy become an effective tool in a chemist's laboratory and increase its usage for molecular spectroscopy. Cheap and easily maintained lasers and improved optical detection via charge coupled detection (CCD) devices have now made Raman spectroscopy a routine technique. Figure 13 shows a schematic drawing of a typical Raman spectrometer.

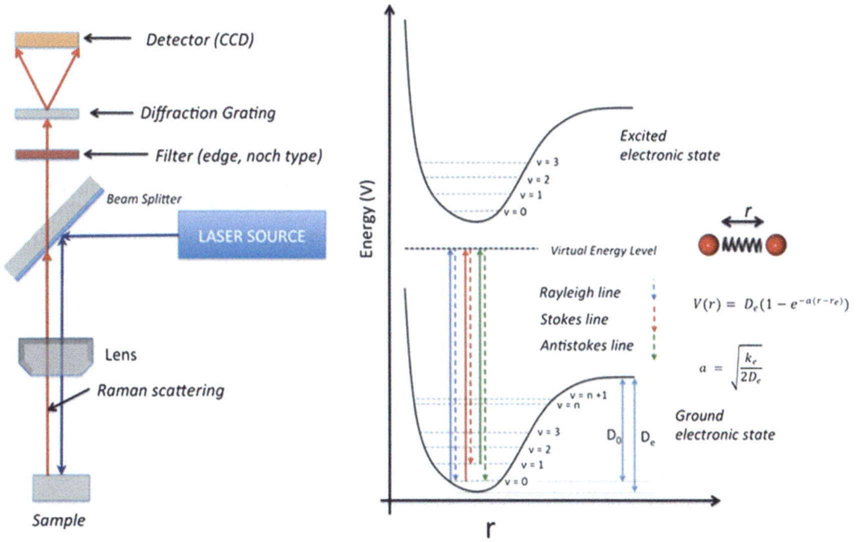

Fig. 13 (*Left*) Schematic diagram of a typical Raman spectrometer. (*Right*) The phenomena of light scattering in a vibrating diatomic molecule

3.2 *Overview of Light Scattering*

3.2.1 Raman Effect

When monochromatic electromagnetic radiation is shines on a vibrating molecule, three types of scattering events can occur (Fig. 13).

1. The photon is elastically scattered (Rayleigh scattering); molecules are excited from a vibrational level $\nu = 0$ to a virtual level and then return back to the $\nu = 0$ level. In such a case, the scattered light is of identical wavelength to the exciting light. This event has the highest probability.
2. The photon is inelastically scattered (Raman scattering); molecules are excited from vibrational level $\nu = 0$ to a virtual level and come back to $\nu \geq 1$ level. Thus the scattered light is of higher wavelength than the exciting light. Such a line in the spectrum is called a stokes line and is the most common type of scattering probed within Raman spectroscopy.
3. Inelastic scattering may also result from molecules being excited from level $\nu = n$ (where $n \geq 1$) to the virtual level and then return to the $\nu < n$ level with the scattered light having a lower wavelength than the exciting light. Such lines are called Raman antistokes lines. For any molecular system elastic (type 1) scattering is most common (Rayleigh). These lines are very high in intensity and can easily overpower type 2 (stokes) and type 3 (antistokes) lines unless filtered out from the spectrum. A typical Raman spectrum consists of inelastically scattered stokes lines (type 2) which are much higher in intensity compared to antistokes lines because of the much lower population of $\nu > 0$ levels compared to the lowest $\nu = 0$ level at room temperature.

3.2.2 Molecular Vibrations

Under harmonic approximation, a vibrating diatomic molecule can be essentially thought of as two atoms linked to each other by a spring of force constant k. The potential energy of such a spring can be expressed as a parabolic potential $V(\Delta r) = 1/2k\Delta r^2$ where Δr is the displacement from equilibrium position $\mathbf{r_e}$. Real systems are more complex and better represented by a Morse-type potential $V(\Delta r) = D_e(1 - e^{-a\Delta r})$ as shown in Fig. 13. The strength of the bond is reflected in the force constant k_e (steepness of the $V(\Delta r)$) around the equilibrium position. D_e is the bond dissociation energy.

A vibrating molecule can be thought of as an oscillating dipole interacting with the electrical component of the electromagnetic spectrum. In absorption spectroscopy, a polychromatic light source shines on a molecule and the quanta of light absorbed, hν, corresponds to the difference in the energy levels within the molecule. When the photon energy matches the energy difference between two vibrational levels, then vibrations inside the molecules get excited. In general, infrared (IR) and Raman spectroscopy are both used to probe vibrations within molecules. However, the two phenomena are physically very different. IR spectroscopy is physically based on absorption and Raman spectroscopy is based on scattering as explained earlier. Being physically very different, the two spectroscopies have very different selection rules. For a vibrational mode to be IR active, the vibration must result in a change in the overall dipole moment (μ) of the molecule ($\frac{\partial \mu}{\partial r} \neq 0$). In a Raman active mode, the vibration should change the polarisability (α) of the molecule ($\frac{\partial \alpha}{\partial r} \neq 0$). For a simple molecule this usually means that antisymmetric vibrations are IR-active and symmetric vibrations are Raman-active. In complex molecules where the symmetry is not straightforward, vibrations usually contribute either more or less towards Raman and IR intensities. The two techniques essentially provide complementary information on vibrations. Arrival of quantum chemical calculations has further developed the technique with the possibility to model and assign complex vibrational modes present in large molecules, solids, and even proteins.

3.3 Recent Developments

Over the years, Raman spectroscopy has become a very powerful technique to collect molecular data because of the high chemical information content and its ease of application. Raman spectroscopy can be performed under ambient or specialized atmospheres (high pressure, high vacuum) and this allows studies of catalytic systems under in situ conditions. Coupling of Raman spectrometers to optical and scanning probe microscopes have resulted in coupling chemical information to highly spatially resolved information provided by these microscopes. Because Raman scattering is a weak phenomenon, significant enhancement in the Raman signal is desirable when studying molecules in low concentrations. Various methods that enhance Raman signals have been developed over the years, i.e., surface enhanced Raman spectroscopy (SERS) [83], resonance Raman spectroscopy (RRS) [84, 85], SERRS (combination of SERS and RS), tip-enhanced Raman spectroscopy (TERS) [86], etc. Raman spectroscopy remains a developing field with new additions to a well-known technical repertoire. In the following section we cover only a small part of recent applications of Raman spectroscopy in the context of challenges relevant to production of solar fuels.

3.3.1 Plasmonic Enhancement Effect

Surface Enhanced Raman Spectroscopy (SERS) allows for the study of chemical species in low concentrations. Signal enhancement results from electromagnetic enhancement through localized surface plasmons [83, 87]. Surface plasmons are collective oscillations of conduction electrons in a metal. For electromagnetic enhancement, a substrate made of Ag, Au, or Cu has to be used although surface plasmons are not limited to only these metals. Nanoscale dimensions significantly enhance the SERS effect and this makes surface roughening useful. As described in the later sections, SERS has been used to good effect to study electrode materials. The same effect when achieved through scanning probe microscopy (SPM) is referred to as Tip Enhanced Raman Spectroscopy (TERS) where a tip made of Au/Ag/Cu is brought close to the sample to observe the electromagnetic enhancement effect on the Raman spectra of the molecules [86]. Sometimes both substrate and tip are made of Au, allowing for electromagnetic enhancement within the cavity formed between the tip and substrate [88]. TERS thus allows for the use of planar (no roughness) and well-defined single crystalline surfaces as substrate materials. A tip curvature of several nanometers results in high spatial resolution in the nanometer range as only molecules in the cavity region are detected. Raman laser spots typically a few micrometers in size are used. The tip is placed within this region. Such an arrangement usually provides very high spatial resolution in the nanometer range. In combination with SPM, the TERS allows for mapping of chemical heterogeneities in the nanometer scale. Recently, Zhang and coworkers managed to map the submoleculer details of a porphyrin-type molecule (**H_2-TBPP**) supported on Ag(111) surface using TERS, opening up the possibility of chemically studying a single catalytic site in situ [89].

3.3.2 Resonance Enhancement Effect

Resonance enhancement happens when the intermediate virtual level involved in Raman scattering (Fig. 13) is close to a vibrational level of the molecule in the first electronically exited state [84, 85]. Such a situation provides significant enhancement in Raman scattering (~10^4 to 10^6). This allows for studying molecules in extremely low concentrations (10^{-8} M). Resonance Raman (RR) remains ideal for studying chromophores present within proteins which can be prepared and studied in situ in extremely low concentrations. Lately, new catalytic systems have been designed to mimic biologically active systems in the context of oxygen activation chemistry (oxygen reduction reaction). When adsorbed on a typical SERS active surface, the enhancement effect can be derived from both RR and SERS effects (SERRS). SERRS has been used to investigate electron transfer (redox) processes in cytochrome C immobilized on electrodes [90]. In a similar fashion, Chatterjee et al. have used SERRS to study ligand effects on Fe-porphyrin-based catalysts in the oxygen reduction reaction [91]. Their studies indicate that a high spin Fe (II) center is likely to prefer an inner sphere model of electron transfer to reduce oxygen compared to a low spin Fe(II) center which prefers an outer sphere electron transfer to reduce O_2 to O_2^-.

Raman spectroscopy continues to develop with additions of new techniques and applications, making it increasingly useful and accessible. In the next section we describe some of the recent applications of in situ Raman spectroscopy in the context of solar fuel production.

3.4 Applications

Renewable hydrogen remains central to a sustainable energy economy [92]. A large number of methods such as hydroelectricity, wind turbines, solar panels, etc., generate electricity that needs to be utilized or stored. Water splitting remains one of the primary ways to convert electricity to chemical energy and store it long term. Platinum remains as a close to ideal catalyst for the hydrogen evolution reaction (HER) but the OER still has significant challenges. Although a large number of transition and noble metal oxides show some activity for oxygen evolution, the inherent slowness of oxygen evolution and catalyst corrosion effectively mean such systems are far from efficient. Because most catalysts change when they are active, it becomes very important to study them under in situ conditions. SERS remains one of the most accessible techniques for studying electrochemical water oxidation on metal oxides because there is no interference from water signal in the range for typical M–O vibrations for the oxides. Au is relatively inert for oxygen evolution compared to transition metal catalysts such as Co, Ni, and Mn oxides. This allows for such oxides to be studied for the OER using Au as a substrate.

Yeo et al. have studied cobalt oxides as OER catalysts using in situ SERS [93]. The Co oxides were deposited on Au electrodes in 0.4 monolayer (ML) to 87 ML quantities. The sub-monolayer CoO_x (with Raman peak at 609 cm^{-1}) showed high activity with turnover frequencies ~40 times that of bulk or thick layers of electrodeposited Co. The peak red shifted to 579 cm^{-1} with OER. Authors argued that such a red shift is associated with oxidation of Co^{3+} species all the way to Co^{4+} species with analogies to similar oxidation of $LiCoO_2$ to form Co^{4+} species in Li-ion cathodes. For thick Co oxide layers, the usual deposition occurs in the form of Co_3O_4, which is the thermodynamically stable oxide under deposition conditions. Under conditions of OER, the Co_3O_4 is oxidized to CoO(OH) with its characteristic peak at 505 cm^{-1}. Eventually, CoO_x species with Co–O stretching frequency at 579 cm^{-1} could be detected. It is believed that it is the CoO_x species that are really active for the OER. Studies on Ni oxides [94] by the same authors revealed that upon OER, the clean Ni surface forms a layer of γ-NiOOH which upon potential cycling-based aging results in formation of the more active β-NiOOH. Based on ratios of peak intensities of peaks at 479 and 560 cm^{-1}, the authors proposed formation of an active yet different oxide that formed above 0.52 V (vs Hg/HgO reference) compared to the already known forms of Ni oxides and hydroxides.

Ranjan et al. have studied electrodeposited manganese oxides as catalysts for water oxidation using the in situ cell shown in Fig. 14 [95]. SERS results show a catalyst whose structure is potential dependent (Fig. 14). Although the deposited catalyst has an α-MnO_2 type structure at open circuit potential, the catalyst changes to a different MnO_x form upon water oxidation reactions at potentials above 1.4 V vs RHE (reversible hydrogen electrode).

The catalyst/electrolyte interface is complex and is influenced by both potential and electrolyte composition. As the data has all the implicit information of potential

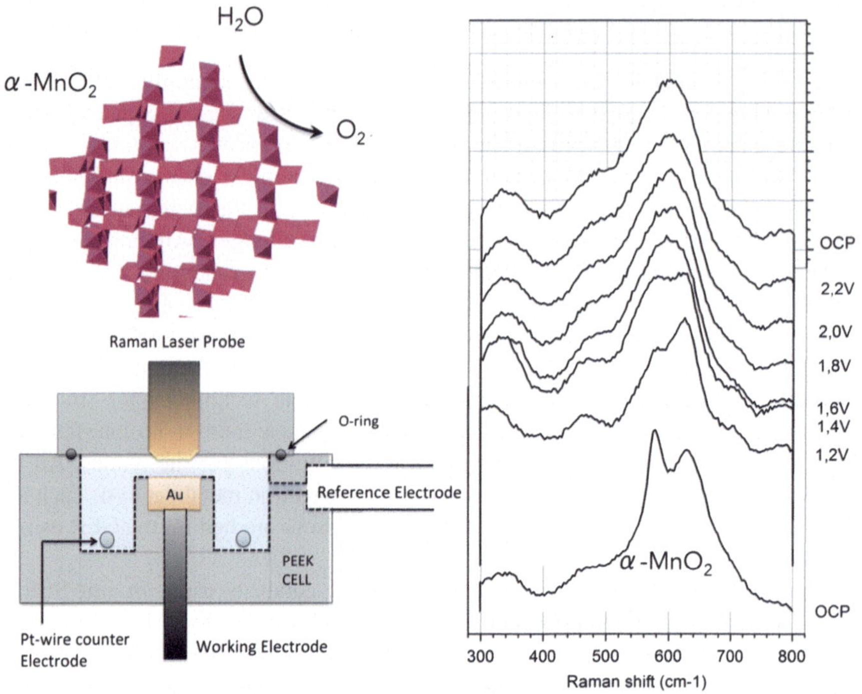

Fig. 14 (*Left*) Cross section of an in situ SERS cell for probing water oxidation reaction on metal oxides. A typical cell consists of a Teflon or polyether ether ketone (PEEK) body which provides good chemical resistance at room temperature in aqueous electrolytes. Available Raman probe, reference (e.g., Ag/AgCl), and counter electrodes (platinum wire) are usually accommodated through the cell design. A roughened Au pellet is a typical substrate because of its relative inertness at water oxidation potentials. In the cell shown above, the Au pellet is contacted through a glassy carbon rod. A platinized Pt wire acts as the counter electrode. (*Right*) In situ Raman spectra showing structural changes occurring in electrodeposited manganese oxide during water oxidation. Potentials shown are with respect to the RHE

and electrolyte composition, its can be systematically used to understand and single out the dominant effects (of potential, pH, and ions in electrolytes) on the nature of catalyst structure [95]. Such structural information can also be correlated with electrochemical activity with appropriate product yield measurements. All the in situ structural information of the catalyst can be visualized using contour plots as shown in Fig. 15. This shows two very distinct structural profiles of the catalyst–electrolyte system in 0.1 M NaOH and 0.1 M H_2SO_4 mostly arising out of the difference in pH in this case. The method is effective even for identifying subtle differences.

Pure hydrogen comes with its own challenges of storage. Liquid hydrocarbon fuels have the advantage of being commonly used within the existing infrastructure and provide very high energy densities compared to gaseous H_2. Natural photosynthesis combines water oxidation to oxygen with CO_2 reduction to biomass and

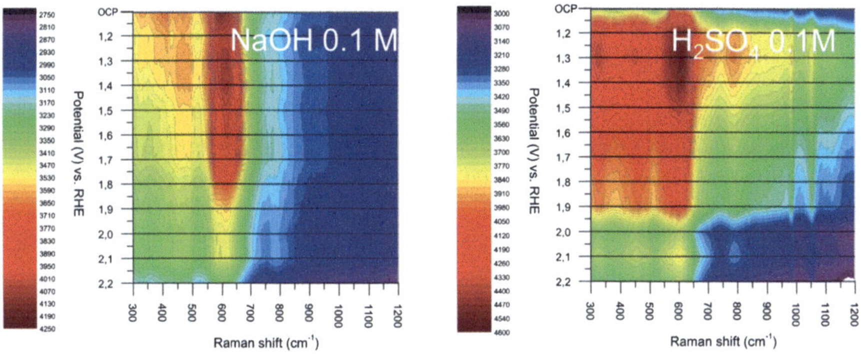

Fig. 15 Contour plots of Raman spectra vs applied potential (vs RHE) of electrodeposited manganese oxides on Au in alkaline and acidic media

oxygen. If atmospheric CO_2 could be used to create the hydrocarbons, the overall scheme would remain carbon neutral, even with the use of hydrocarbons as fuels. This makes CO_2 reduction a key component of producing solar fuels. CO_2 reduction has been attempted using both chemical and electrochemical techniques but we focus on the electrochemical part in the following text.

As in electrochemical water splitting, electrochemical CO_2 reduction poses great challenges. The nature of carbonaceous species derived from CO_2 depends heavily on the pH of the electrolytes and additives. Carbonates and bicarbonates dominate at alkaline pH. Because CO_2 reduction happens at very negative potentials (~1.4 V vs Ag/AgCl), losses through the HER are common on various catalysts. For increased efficiency of CO_2 reduction, suppression of H_2 evolution is desirable. Schmitt and coworkers studied Ag catalysts for CO_2 reduction with DAT (3,5-diamino-1,2,4-triazole) as an additive using in situ SERS [96]. In 1 M KOH and $Ca(OH)_2$ solution, addition of 10 mM DAT caused the H_2 evolution onset to shift negative by 200 mV. In the potential regime for CO evolution from CO_2 (−1.4 V vs Ag/AgCl), the currents were higher in the presence of DAT. In situ SERS studies showed that during a cathodic potential sweep, carbonate species are removed from the electrode at −0.5 V. Ag in the presence of DAT forms an Ag–DAT complex which suppresses H_2 evolution. On the Ag surface, CO can bind to various surface sites which show a very strong potential-dependent Stark tuning of the C–O stretch frequency, indicative of the strong electronic influence of the Ag surface on adsorbed CO. In the presence of DAT, the C–O stretching frequency for on-top coordinated CO shifted to higher values with reduced Stark tuning, indicating reduced coordination of CO to the Ag surface. This effect is believed to result in easier removal of CO from the catalyst surface, thus reducing the overpotential of CO_2 reduction. During the CO_2 reduction experiment on Ag, CH_2 stretching and scissoring peaks were also formed alongside CO; such CH_2 signals were indicative of the formation of hydrocarbon chains. Formation of liquid hydrocarbon fuels from water and CO_2 using renewable energy remains the holy grail of solar fuel production.

4 Electrochemical Impedance Spectroscopy Applied to Photoelectrochemical Cells

Fabio La Mantia
Semiconductor and Energy Conversion – Center for Electrochemical Sciences, Ruhr-Universität Bochum, Universitätsstr. 150, 44780 Bochum, Germany
fabio.lamantia@rub.de

4.1 Introduction

Electrochemical impedance spectroscopy is a method widely used to investigate electrochemical cells. It involves perturbing the system with a sinusoidal wave and recording its response. Under small perturbations, the response of the system is also a sinusoidal wave with the same frequency as the perturbation. It does, however, show a time delay. The impedance spectrum is then defined by the ratio of the Fourier transform of the potential, u, and the Fourier transform of the current:

$$Z(\omega) = \frac{F[u](\omega)}{F[i](\omega)} \tag{1}$$

where the Fourier transform of a function g is given by

$$F[g](\omega) = \int_0^T g(t)\mathrm{e}^{-j\omega t} dt \tag{2}$$

where t is the time domain, T is the total recording time of the signal, and j is the imaginary unit. Given the finiteness of the data set, the Fourier transform is defined with the integral between 0 and T, instead of the usual mathematical definition with the integral between $-\infty$ and $+\infty$. As with the impedance spectrum, it is possible to define an admittance spectrum:

$$Y(\omega) = \frac{F[i](\omega)}{F[u](\omega)} = \frac{1}{Z(\omega)} \tag{3}$$

The impedance is a complex number and is composed of a real part and an imaginary part. The real part represents the resistance for the charged species to move through the system, whereas the imaginary part represents the difficulty for the charged species to be separated in the system.

One of the most intriguing advantages of electrochemical impedance spectroscopy is that it can be used to explore a wide range of time domains, from 10^{-6} down to 10^5 s. Therefore, it can be used to observe a variety of phenomena, such as charge accumulation and transfer at an interface, transport of electroactive species, adsorption/desorption phenomena, and chemical reactions related to electroactive species.

The shape of the impedance spectra is a result of the convolution of the different phenomena in the time domain, each having a well-defined shape and dependence in the frequency domain. At the same time, this is also one of its disadvantages: because of the complexity of phenomena that can occur in an electrochemical system, impedance spectra can be very difficult to interpret without a proper physico-chemical model of the system.

It also has to be taken into account that, when a two-electrode cell configuration is used, the impedance of the whole electrochemical cell is measured, i.e., the positive and negative electrode and also the electrolyte. In contrast, if only the impedance of the working electrode is to be measured, it is important that a three-electrode cell configuration is used. However, the use of a reference electrode can cause severe distortions in the impedance spectra, depending on the position of the reference electrode, the relative dimensions of the counter and working electrodes, and the ratio between the volume of the electrolyte and the area of the solid/liquid interface [97, 98].

4.2 *Representation of Impedance Spectra*

The impedance is a complex number, and therefore its representation in a graphical form is more complicated than the graphical representation of real data. The most common representations are the Nyquist plot and the Bode plot. In the Nyquist plot, the opposite of the imaginary part of the impedance, $-\mathrm{Im}(Z)$, is represented as a function of the real component of the impedance, $\mathrm{Re}(Z)$, as shown in Fig. 16.

It is important that the graph is plotted with equal scales for the x and y axes. By using this representation, different graphical methods exist to extrapolate the physical constants from the data. However, the explicit information on the frequency domain is lost, and therefore often the most important and/or characteristic frequencies are

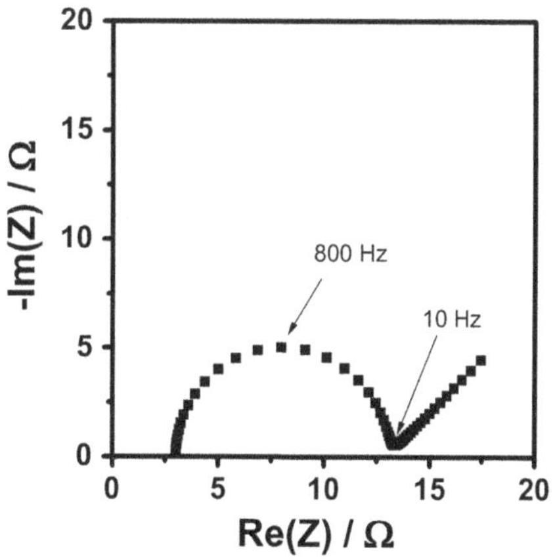

Fig. 16 Nyquist plot of an impedance spectrum, which reports the negative of the imaginary part vs the real part of the impedance using equal scales for the axes. Sometimes, characteristic frequencies are indicated

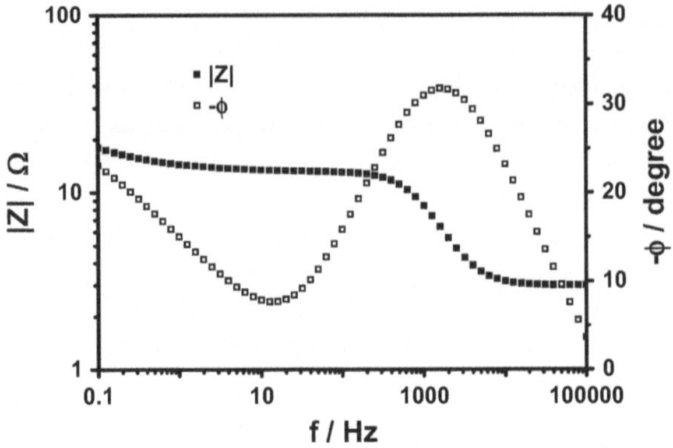

Fig. 17 Bode plot of an impedance spectrum. Frequency and modulus of the impedance are reported in logarithmic scale

indicated in the plot. In the Bode plot, the modulus and phase of the impedance are reported in two separate axes or graphs as a function of the frequency (Fig. 17).

Typically, the frequency and modulus of the impedance are reported in logarithmic scales. The modulus, Z, and the phase, ϕ of the impedance are defined through the real and imaginary components by (4) and (5), respectively:

$$|Z(\omega)| = \sqrt{\mathrm{Re}(Z)^2 + \mathrm{Im}(Z)^2} \tag{4}$$

$$\phi(\omega) = a\tan\left(\frac{\mathrm{Im}(Z)}{\mathrm{Re}(Z)}\right) \tag{5}$$

The Bode plot is particularly useful when the value of the modulus of the impedance changes by several orders of magnitudes in the frequency range explored. For example, this occurs in blocking electrodes, in which the imaginary part of the impedance scales linearly with the inverse of the frequency. Under these conditions, the Nyquist plot appears as a straight line that tends towards infinity as the frequency decreases, and the features at high frequency are not clearly visible. There is a third possible representation of impedance data, which is less common because it uses a 3-D plot (Fig. 18). In this case, the x axis is logarithmic and reports the value of the frequency, whereas the y and z axes are linear and report the value of real and imaginary part of the impedance, respectively. This last graph has the same advantages of the Nyquist plot, which is the projection in the y–z plane of the 3-D curve, but maintains the frequency information of the data set.

4.3 *Validation of the Data: Kramer–Kronig Transformations*

The real and the imaginary parts of the impedance spectra of an electrochemical system are not two sets of independent data. A correlation exists between these two sets of data which is called the Kramer–Kronig transformation. For a system that

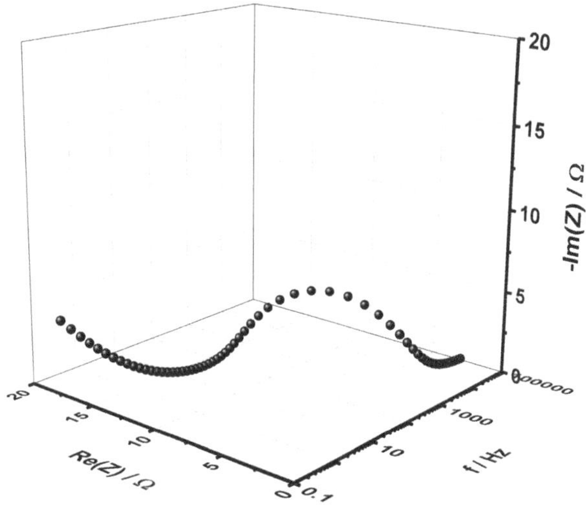

Fig. 18 Three-dimensional plot of an impedance spectrum. The frequency axis is in logarithmic scale. The other two axes report the real and the negative of the imaginary part of the impedance

responds to the principles of linearity, causality, and stability, the following equations hold:

$$\mathrm{Re}[Z(\omega)] = R_\infty - \frac{2}{\pi}\int_0^\infty \frac{x \cdot \mathrm{Im}[Z(x)] - \omega \cdot \mathrm{Im}[Z(\omega)]}{x^2 - \omega^2} dx \tag{6}$$

$$\mathrm{Im}[Z(\omega)] = \frac{2\omega}{\pi} \cdot \int_0^\infty \frac{\mathrm{Re}[Z(x)] - \mathrm{Re}[Z(\omega)]}{x^2 - \omega^2} dx \tag{7}$$

From (6) and (7) it becomes clear that the use of the Kramer–Kronig transformations is limited by the impossibility of measuring the impedance spectrum between 0 and ∞ frequency. Several techniques have been developed for solving the Kramer–Kronig transformation by using a finite set of impedance data. The most commonly used is an algorithm based on a wide distribution of Voigt elements (a capacitance and a resistance in parallel) in series with different time constants [99, 100], or a general equivalent circuit which considers a cascade of time constants [101]. Another very powerful technique is based on the logarithmic Hilbert transform. The logarithmic Hilbert transform uses the principle of the Kramer–Kronig transformation to determine (8) and (9), which connect the modulus and the phase of the set of impedance data:

$$\ln[|Z(\omega)|] = \ln[R_\infty] - \frac{2}{\pi}\int_0^\infty \frac{x\phi(x) - \omega\phi(\omega)}{x^2 - \omega^2} dx \tag{8}$$

$$\phi(\omega) = \frac{2\omega}{\pi} \cdot \int_0^\infty \frac{\ln[|Z(x)|] - \ln[|Z(\omega)|]}{x^2 - \omega^2} dx \tag{9}$$

Although a proper transformation requires the full set of data between 0 and ∞, (8) can be approximated by (10) [102]:

$$\ln[|Z(\omega)|] \approx K + \frac{2}{\pi}\int_{\omega_0}^{\omega} \frac{\phi(x)}{x}dx - \frac{\pi}{6}\frac{d[\phi(x)]}{d\ln\omega} \tag{10}$$

where K is a constant and ω_0 is the highest frequency investigated. If the phase of the impedance data is known with sufficient precision, it is possible to use the Hilbert transform to confirm the stability, causality, and linearity of the system, thus removing eventual instabilities in time, which are the main artifacts in electrochemical systems. Especially in photoelectrochemical systems, when the impedance spectrum is acquired under illumination, instabilities can be observed because the impedance itself depends on the intensity of the light; in fact most sources are unstable over time. In these circumstances, the Hilbert transform is a powerful tool for improving the quality of the data at low frequencies. Several programs for EIS analysis contain algorithms that check the impedance data for consistency through the Kramer–Kronig transformation.

4.4 Physical Modeling and Equivalent Circuit

Extracting physical information from the impedance data is a complicated task and requires the development of appropriate models that describe, in a simplified way, the electrode/electrolyte interface and bulk phenomena related to transport and recombination of charge carriers or chemical reactions. The model should take into consideration that the electrode/electrolyte interface may be non-planar, as in the case of the mesoporous electrodes often used in dye-sensitized solar cells and photoelectrochemical cells. The model should also consider whether the impedance spectra were acquired only for the working electrode, or in a two-electrode cell configuration. In this section the principles of physical modeling and fitting are presented.

4.4.1 Equivalent Circuit of a Photoelectrochemical Cell in the Dark

Equivalent circuits are based on the mathematical description of the physicochemical phenomena in the investigated system. The description of the electrode/electrolyte interface has to include the charge accumulation and the charge transfer. The charge accumulation represents the formation of the double layer at the liquid/solid interface, and may include adsorption phenomena (e.g., H^+ or CO_2 adsorption); the charge transfer is the passage of the electron from the electrode to the reactive species. In the equivalent circuit, these two processes appear in parallel and typically can be represented by a capacitor and a resistor, respectively (Fig. 19a). However, when mass transport through diffusion also has to be taken into account, a so-called Warburg element has to be added to the previous equivalent circuit. For example, in photoelectrochemical cells used for water splitting, diffusion of protons

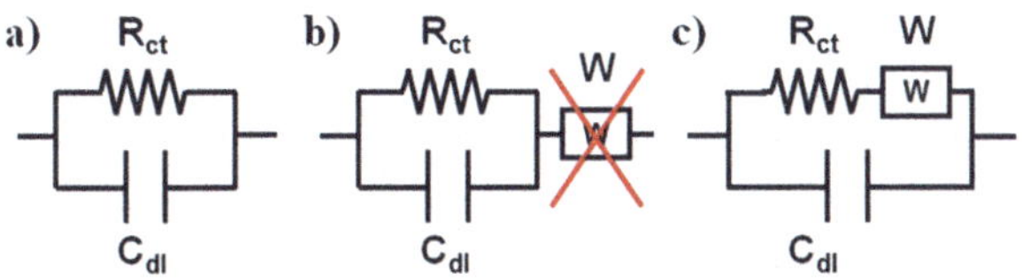

Fig. 19 (**a**) Equivalent circuit of an electrochemical interface. (**b**, **c**) The model includes the diffusion process, although (**b**) is wrong from the physico-chemical point of view, and (**c**) is the right representation for the diffusion. R_{ct} is the charge transfer resistance; C_H is the capacitance of the interface; W is the Warburg element, representing the diffusion

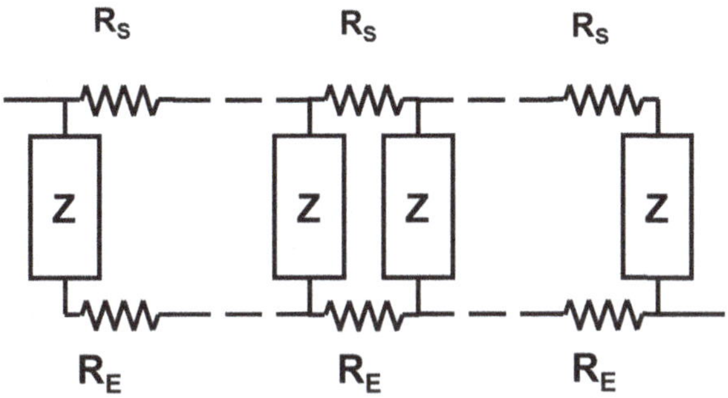

Fig. 20 Equivalent circuit of a porous electrode. R_S and R_E are the ionic and electronic resistances in the pores, respectively; Z is the interfacial impedance, which could be represented by the equivalent circuit in Fig. 19a or c

to the photocathode must occur for production of hydrogen. Similarly, in the case of CO_2 reduction, the Warburg element is related to the phenomenon of diffusion of CO_2 from the bulk of the solution to the surface of the electrode. Although the diffusion process is located in the bulk of the solution, the Warburg element takes into account not the diffusion itself but the variation in concentration of reactants and products at the electrode interface, which is controlled by the transport mechanism. Therefore, the Warburg element is located in series with the charge transfer process, and not in series with the whole interface (Fig. 19b, c).

As anticipated, a porous electrode composed of nanostructured or mesoporous semiconductors (e.g., titania, tungsten oxide, iron oxide) has a further complication in the modeling of the *ac* response: the interface between the electrode and electrolyte is distributed in space. Under these circumstances, the electrode cannot be considered as composed of a single interface, but rather the different interfaces are distributed and separated by the ohmic drop in solution and in the solid phase. In general, this class of equivalent circuits are called transmission line models. Rather sophisticated ones can be found in [101, 103–106]. Using the transmission line model, and the description of the electrode/electrolyte interface previously developed, one obtains an equivalent circuit as in Fig. 20. Other examples of equivalent circuits used in the literature for describing the semiconductor in dye-sensitized solar cells can be found in [107]. Such models are rarely considered in the description of the impedance spectra in the case of the photogeneration cells.

4.4.2 AC Response of a Semiconductor in the Dark

Although the charge accumulation at the interface is typically described by a capacitive element, it is experimentally observed that, even for strongly polarizable electrodes, the interface rarely behaves as a pure capacitor. In the case of the semiconductor/electrolyte interface, the situation is further complicated by the fact that the lower concentration of charge carriers in the electrolyte generates a considerably extended space charge region in the semiconductor. In the case of an ideal semiconductor/electrolyte interface, the charge accumulation in the semiconductor can be described by a pure capacitance, which is dependent on the polarization potential, U_{E}, through the Mott–Schottky equation [108]:

$$C_{\mathrm{SC}}^{-2}(U_{\mathrm{E}}) = \pm\left(\frac{2}{\varepsilon\varepsilon_0|e|N_{\mathrm{D,A}}}\right)\left(U_{\mathrm{E}} - U_{\mathrm{fb}} \mp \frac{k_{\mathrm{B}}T}{|e|}\right) \tag{11}$$

where U_{fb} is the flat band potential, ε the relative dielectric constant, ε_0 the permittivity of vacuum, $N_{\mathrm{D,A}}$ the number of donors or acceptors per unit of volume, and $|e|$ the absolute charge of the electron. The + or − is dependent on the type of semiconductor, with + for *n*-type and − for *p*-type. The Mott–Schottky equation can be used only for crystalline semiconductors with a single donor or acceptor level completely ionized. This is typically a rare case, especially in nanocrystalline materials, where the large presence of defects necessitates a proper treatment, which is closer to the description of an amorphous semiconductor. In particular, it has to be taken into consideration that the high defectiveness of some semiconductors, such as TiO_2, can influence their electronic structure and create electronic states in the band gap (Fig. 21a). Such electronic states are localized, and therefore they can only exchange electrons with the conduction or valence band. However,

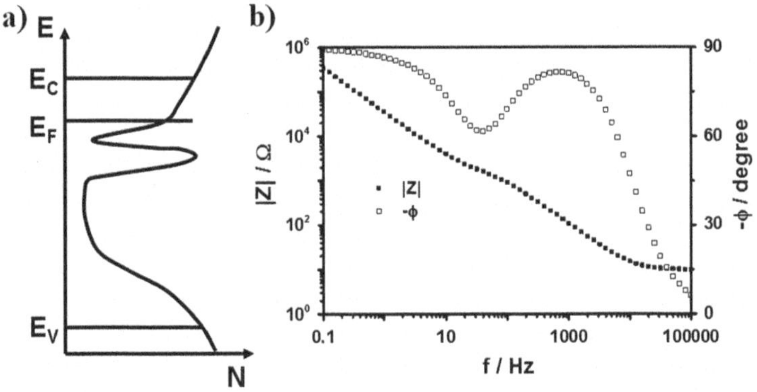

Fig. 21 (**a**) Electronic structure of a highly defective semiconductor. E_{C} is the bottom of the conduction, E_{V} is the top of the valence band, and E_{F} is the Fermi energy level. (**b**) Bode plot of the impedance of the semiconductor in (**a**). The minimum in the phase at 30 Hz shows the frequency dispersion caused by the presence of electronic states in the mobility gap

their mobility is too low to contribute to the conductivity of the charge carriers in the extended bands. The exchange of electrons between the states in the mobility gap and the extended bands is typically slow, and can be detected by impedance spectroscopy, in the form of a time constant dispersion, as shown in Fig. 21b. A time constant dispersion represents a non-ideal capacitance, i.e., a capacitance the value of which is dependent on the frequency. The frequency dependence of the capacitance depends on the energy losses during the accumulation of charge carriers, caused by the electron capture/emission process in the mobility gap. In fact, the capacitance of the semiconductor can be approximated by (12) [109]:

$$C_{SC}^{-1}(U_E,\omega) \approx \pm\left(\frac{1}{\sqrt{\varepsilon\varepsilon_0|e|N}}\right)\left[1+\ln\left(\frac{U_E-U_{fb}}{\psi_C}\right)\right] \quad (12)$$

$$\psi_C = -k_B T \ln(\omega\tau_0) - \Delta E_F \quad (13)$$

where N is the density of states in the mobility gap, τ_0 the time constant of electron exchange between extended bands and localized states, and ΔE_F the difference between the edges of the extended bands and the Fermi level (conduction band for an n-type semiconductor, valence band for a p-type semiconductor). More information is given in [110–115].

4.4.3 Equivalent Circuit of a Photoelectrochemical Cell Under Illumination

To understand the limitations, behavior, and charge transfer/recombination mechanism, electrochemical solar cells are often investigated in dark conditions and under illumination. In the latter case, one often refers to photoimpedance. The illumination generates charge carriers in the semiconductors, which are separated by means of the electric field. The minority charge carrier is then transferred to the reactive species in the electrolyte, whereas the majority carrier travels along the semiconductor and is transferred to the other electrode. The minority carrier can therefore be used for the electrochemical reaction. As a consequence, the photoimpedance depends on the intensity of the light source and the wavelength of light through the amount of minority photogenerated charge carriers. Because the light is absorbed following an exponential law (the Beer–Lambert law), the charge carriers are not generated homogeneously. The same phenomenon appears in a dye-sensitized solar cell, in which the dye is distributed along the thickness of the mesoporous TiO_2. The spatial distribution of photogenerated carriers influences the photoimpedance spectra by introducing a further source of time constant distribution. Under these conditions, the porous electrode model should take into consideration not only the potential drop in the solid phase and liquid phase but also the inhomogeneity of the interfacial impedance, Z, which now depends on the position inside the electrode (see Fig. 20). The equivalent circuit is no longer given by the classic transmission line model, which can be found in the literature. The exact solution in this case is unclear, and typically a constant phase element is used instead of a capacitor to describe the time dispersion at the distributed interface. The constant phase element takes into account both 2-D and 3-D time constant distributions [116, 117], and its interpretation is rather arbitrary. Another possibility is to consider an exponentially changing charge transfer resistance, R_{ct}, and to solve the transmission line model numerically for the different frequencies.

4.5 Final Remarks

Electrochemical impedance spectroscopy is widely used to obtain relevant information about the mechanism of reaction and/or limitations in electrochemical systems, as, for example, in the analysis of the solar water-splitting mechanism in hematite photoanodes [26]. Especially when photoimpedance (impedance under illumination) is recorded, it is important that the light source is stable and that the impedance data are validated through Kramer–Kronig transform algorithms. The recovery of information on the phenomena occurring in the photoelectrochemical cells from impedance spectra requires the application of a physical model, which can be expressed though an equivalent circuit. Here, it is important to describe the system with proper complexity, also taking into account the experimental data. The model should take into consideration that the electrode/electrolyte interface may be distributed in the space, and that the light absorption and charge carrier photogeneration may not be constant throughout the whole system. The use of a constant phase element to describe the capacitance of the interface could introduce an uncertainty in the physical meaning of the parameters.

5 In Situ Characterization of Photocatalytic Materials in an Environmental Transmission Electron Microscope

Benjamin K. Miller, Liuxian Zhang, and Peter A. Crozier (✉)
School for Engineering of Matter, Transport and Energy, Arizona State University, Tempe, AZ 85287, USA
crozier@asu.edu

5.1 Introduction

Heterogeneous catalysis is a key process for converting reactants into products and plays a critical role in chemical and energy transformations [118, 119]. During catalysis, the structure of the catalyst may vary in a complex way with changes in the pressure of reactants and products as well as temperature. To build a structure–reactivity–performance correlation and achieve a fundamental understanding of catalytic mechanisms, the electronic and geometric structure of the catalyst must be determined at the atomic level. For many catalysts, the active form of the material may exist only under reaction conditions and must be characterized to develop a fundamental understanding of structure–reactivity relations. Environmental transmission electron microscopy (ETEM) is at present the only technique available to perform in situ characterization of a high surface area catalyst with atomic resolution. In this approach, reactants are allowed to flow over the TEM sample (often during in situ heating) and the changes in the nanocatalysts are recorded using TEM imaging, diffraction, or spectroscopy techniques [120]. For in situ observation of photocatalysts, the materials must be observed not only in the presence of reactant and product species but also during in situ light illumination. In addition to allowing the active form of the catalyst to be observed, the in situ approach can also provide important information on deactivation mechanisms such as photocorrosion of co-catalyst components [121, 122].

5.2 *In Situ TEM for Photocatalytic Observations*

Only a couple of groups have successfully incorporated light sources into environmental transmission electron microscopes, permitting gas–solid–light interactions to be observed at atomic resolution on nanoparticle systems. So far, no groups have developed in situ capabilities that combine liquid cells with light illumination or liquid, light, and electrical biasing for photoelectrocatalysis. The first work to attempt to study photocatalytic processes was performed on a system in which an optical fiber was attached to a high vacuum TEM (i.e., no in situ gas or liquid exposure). The system was employed to observe the photodecomposition of polyhydrocarbon compounds on a TiO_2 film [123]. The same group was also able to observe photocatalytic nucleation of Au nanoparticles on a TiO_2 nanoparticle under UV irradiation [124]. Two groups have developed in situ light illumination systems for ETEM. The Danish Technical University group has developed a light illumination holder compatible with an FEI Titan ETEM [125]. Lenses are used to focus a laser beam onto the TEM sample and their system was employed to investigate the photodecomposition of a Cu_2O photocatalyst to Cu metal under UV irradiation in an H_2O environment [126]. They also showed in situ photodeposition of Pt nanoparticles on a support [125].

The Arizona State University (ASU) group has also developed a fiber optic-based system that allows light of variable wavelengths to illuminate a photocatalyst in the environmental TEM [127]. In their design, the fiber is introduced through a port on the side of the TEM which provides greater flexibility in the choice of sample holder. One advantage of this approach is that thermal processing of the catalyst (such as reduction or oxidation of co-catalyst components) can be performed in situ in the ETEM. This allows the system to be put into a well-defined state, making fundamental structure–reactivity measurements possible. The

ASU system is also compatible with liquid cell holders, allowing, for example, water-splitting catalysts to be studied in the presence of liquid or vapor phase reactants. A brief description of the ASU system is given below. More details can be found in the paper by Miller and Crozier [127].

5.3 Design Considerations for In Situ Illumination Under Reactive Gas Conditions

A number of constraints are imposed on the design of the in situ illumination system so that the electron optical performance of the TEM is not degraded. The TEM is designed to allow the local structure of a material to be imaged with atomic resolution. It is important to be able to tilt samples so that low-order crystallographic zone-axis orientations are parallel to the incident electron beam. The optical fiber must be introduced to the ETEM without negatively affecting sample tilting. The location of the maximum light intensity on the sample must coincide with the electron optical axis. Moreover, it would be ideal if substantial light intensity can strike the sample when it is tilted by at least ±20° to allow the crystal orientation to be varied. A further constraint of the design was the necessity to be able to employ hot stages in experiments. Many hot stages have a somewhat bulky furnace around the TEM sample which can block light from an oblique angle striking the center of the sample.

A schematic diagram of a design which addresses these constraints is shown in Fig. 22 [36]. In this case, the fiber was inserted into an available port (often used for installing an energy dispersive X-ray spectrometer) on a FEI Tecnai F20 environmental transmission electron microscope. In this instrument, the TEM sample sits between the upper and lower pole pieces of the objective lens and the pole piece gap is 5.4 mm. The EDX port is located perpendicular to the TEM sample rod port and the fiber approaches the sample region at close to a horizontal orientation. A 600-μm silica core, silica clad, solarization resistant fiber was used to carry the light into the microscope. This fiber is compatible with visible and ultraviolet light,

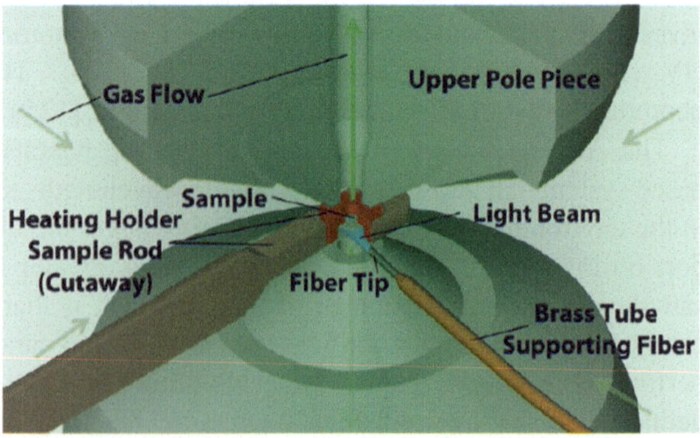

Fig. 22 Schematic diagram of the overall geometry of in situ fiber illumination system. Reprinted with permission from Zhang et al. [36]. Copyright 2013, American Chemical Society

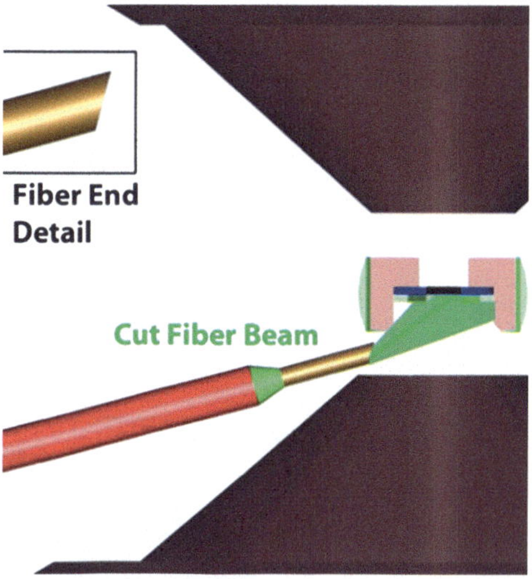

Fig. 23 Cross sectional view of pole piece gap showing fiber orientation relative to upper and lower pole pieces

making it ideal for studying photocatalysts with bandgaps ranging from the infrared to the ultraviolet region. The fiber is clad with aluminum to avoid charging by secondary and backscattered electrons. The alumina cladding was also useful to avoid outgassing in the high vacuum of the TEM.

A more detailed cross sectional illustration of the geometric relationship between the fiber, the pole pieces, and the hot stage furnace is shown in Fig. 23. This illustrates the space constraint problem. Ideally, we would like to bring the fiber towards the sample as close to vertical as possible to avoid shadowing from the hot stage furnace and maximize the light intensity on the sample. In practice, the shallow angle of the objective lens pole piece, the limited opening of the port, and the small space between the upper and lower pole pieces force the fiber to approach at an angle of 15° with respect to the horizontal. The front end of the fiber was cut at a maximum angle of 30° to refract more light onto the middle of the TEM sample. A bellows and micrometer manipulator was constructed to allow the exact position of the fiber to be adjusted to center the light on the optical axis of the microscope. This manipulator structure also provided the high vacuum feedthrough for the fiber. An external fiber was coupled to the fiber inside the TEM at this vacuum feedthrough. This external fiber was used to transport light from a source to the internal fiber.

Any light source can be employed with this arrangement, although it is important to couple the light efficiently to the end of the external fiber (which is also 600 μm in diameter). At present, we use a high brightness white light source manufactured by Energetiq. The emitted intensity varies by only about a factor of 5 over the wavelength range of 200–800 nm. The distribution of emitted intensity is shown in Fig. 24. The high brightness allows a substantial fraction of the emitted intensity to be coupled into the fiber using a parabolic mirror arrangement. With optical bench testing, we were able to map out the distribution of light striking the TEM sample. A photodetector was built into a TEM holder so that the light intensity could be

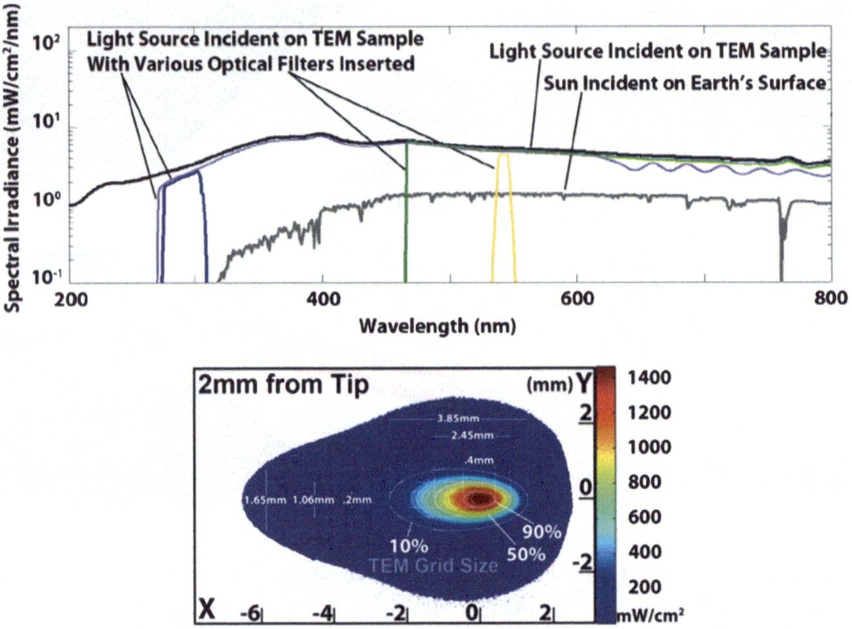

Fig. 24 (*Top*) Maximum intensity of the light incident on the TEM sample as a function of wavelength. (*Bottom*) Light intensity incident on the sample as a function of position. Reprinted with permission from Miller and Crozier [127]. Copyright 2013, Cambridge University Press

measured in situ. The spatial distribution and intensity of light striking the TEM sample is shown as a contour plot in Fig. 24. The intensity "hot spot" region that contains intensity greater than 90% of the maximum is approximately 0.2×0.4 mm in size. The peak intensity is about 1,400 mW/cm^2, which is about a factor of 10 higher than the solar flux. Thus, this system is suitable for investigating changes induced in the catalyst under close to ambient illumination conditions. To achieve higher fluxes on the sample a suitable laser must be used.

5.4 *In Situ Observations of Anatase Nanoparticles*

This system is currently being employed to investigate the evolution of photocatalysts under light and water vapor conditions. Many photocatalysts are composed of a light-harvesting semiconductor coupled to one or more co-catalysts. For reactions in water, oxides such as TiO_2 or Ta_2O_5 can serve as good model systems because of their stability, although with bandgaps in the range 3–4.4 eV they absorb mostly in the ultraviolet region. Figure 25 shows a sequence of images recorded from anatase nanoparticles after in situ exposure to light and water vapor at a pressure of 1 Torr[11]. The initial anatase nanoparticle was well defined and bounded mainly by the low energy {101} and shorter {002} surfaces.

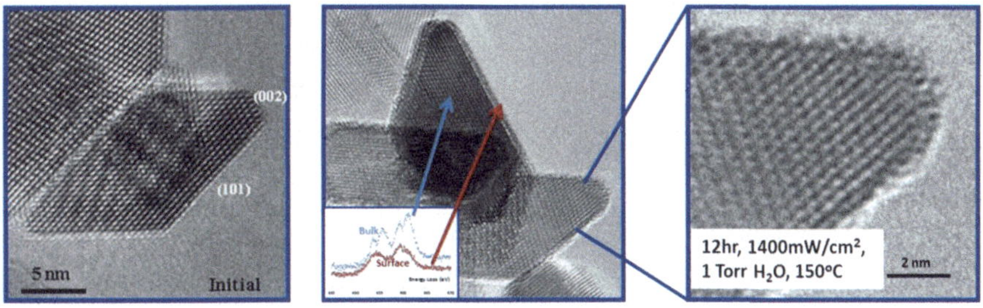

Fig. 25 The change taking place on the surface of anatase nanoparticles in the presence of water vapor and light. (*Left*) Initial anatase particles before illumination. (*Middle*) Anatase particles after 12 h exposure to water and light showing amorphous surface layer. *Inset* is the EELS spectra of the Ti L_{23} edge recorded from the center (*blue*) and surface (*red*) of the particle. (*Right*) Higher magnification image showing disordered layer a few monolayers in thickness. Reproduced (adapted) with permission from Zhang et al. [36]. Copyright 2013, American Chemical Society

The particles appear crystalline on the surface and the surface is smooth and atomically abrupt. There was no obvious evidence for any structural change taking place during this initial observation. Titania is well known to undergo radiolysis under electron beam irradiation in the TEM [128, 129]. To ensure that electron beam effects are negligible, we take advantage of the relative uniformity of the anatase sample to look at the structure of different groups of anatase particles after a prescribed exposure to gas and light. After the initial period of light exposure, an initial set of anatase particles were examined. After a second period of light exposure, a second set of fresh anatase crystals, not previously exposed to the electron beam, were examined. After each observation, the electron beam is turned off. This procedure ensures that electron beam effects are kept small during the in situ processing of the anatase.

Figure 25 also shows that after 12 h exposure to light and H_2O, the first one or two layers at the anatase surface show roughening and amorphization. Over 30 fresh areas which were not previously exposed to the electron beam were checked and all showed some degree of surface amorphization and roughness. No surface change was observed when the anatase crystals were exposed to light in an H_2 atmosphere or in the microscope vacuum, showing that water plays a critical role in the observed phase change on the anatase surface. Spatially resolved EELS was performed on an aberration corrected JEOL ARM 200 F to compare the oxidation state of the Ti in the center and surface of the anatase particle (Fig. 25, middle, inset). The L_{23} edge from the center of the anatase shows the characteristic crystal field splitting in which the L_3 line at 457 eV and L_2 line at 459 eV are split into doublets corresponding to excitations of the $2p_{3/2}$ and $2p_{1/2}$ subshells to unoccupied t_{2g} and e_g orbitals. The Ti L_{23} edge recorded from the surface layer shows pronounced broadening of the L_3 and L_2 lines and the relative peak height is consistent with Ti in the +3 oxidation state [130, 131].

The observed surface amorphization and roughness can be explained by the photogeneration of oxygen vacancies and the resulting dissociation of water at these defect sites. On exposure to UV light or high energy electrons, electron-hole pairs are generated which may diffuse to the crystal surface. The photogenerated electrons tend to reduce Ti^{4+} to Ti^{3+}, whereas trapped holes neutralize surface oxygen leading to emission from the surface, leaving behind oxygen vacancies [132]. High resolution electron energy loss spectroscopy (HREELS) and temperature programmed desorption (TPD) have confirmed that water molecules are dissociatively adsorbed, filling the vacancies and forming hydroxyl groups [133]. STM results and density functional theory (DFT) calculations suggest that one bridging hydroxyl group at the original vacancy site and another at the adjacent bridging oxygen position are formed [134–136]. With continued irradiation, the degrees of hydroxylation of the surface increase and eventually destabilize the crystallinity of the top layer, leading to the formation of a continuous amorphous layer of hydroxylated titania.

Our observations show that this process is self-limiting because the thickness of the disordered layer remains constant under continuous exposure to water and light. This is reasonable because anatase is known to be a stable photocatalyst in an aqueous environment during exposure to light. Continued irradiation of the hydroxylated surface presumably results in the generation of additional oxygen vacancies on the surface. Because the amorphous layer thickness does not significantly increase with time, this implies that most of these vacancies are filled by additional dissociation of water from the gas phase rather than diffusion of oxygen from the interior of the anatase particle. Thus the underlying bulk structure of the anatase is preserved although our TEM images show the TiO_2 layer just below the surface is significantly perturbed by the amorphous top layer.

Acknowledgments This work was supported by the U.S. Department of Energy (DE-SC0004954). We gratefully acknowledge the use of facilities within the LeRoy Eyring Center for Solid State Science at Arizona State University.

6 Advanced Screening and High Throughput Evaluation of Solar Fuel Materials

Joel A. Haber and John M. Gregoire (✉)
Joint Center for Artificial Photosynthesis, California Institute of Technology, Pasadena, CA 9112, USA
gregoire@caltech.edu

6.1 Introduction

In the fields of chemistry and materials science, high throughput experimentation encompasses a suite of techniques typically applied to combinatorial chemistry and materials discovery problems. For well-established technologies such as catalytic synthesis of commodity chemicals, high throughput techniques can be readily applied to optimize materials and processes [137], a common industrial strategy not well documented in the literature. For emerging or future technologies, such as solar fuel generation, judicious application of high throughput methods can accelerate technology development and offer insights into fundamental materials science. As documented in a series of recent review articles [138–142], this mode of high throughput research has been used to explore a wide variety of functional materials. As evident in these reviews and associated literature, applying a high throughput strategy for a particular application requires development of specialized materials synthesis and screening techniques. In the field of solar fuels, the recent introduction of high throughput experimentation, most notably by the McFarland and Parkinson groups [143–148], has been followed by rapid advancement in screening techniques.

As detailed in other sections of this book, solar fuels technology involves the integration of several functional materials. Here we discuss high throughput methods for evaluating solid state materials, in particular photoabsorbers, heterogeneous electrocatalysts, and photocatalysts. Most high throughput solar fuels research has focused on improving materials for anodic OER, although we note that many of the instruments are equally applicable to evaluation of the cathode HER. High throughput discovery of solar fuels materials faces the challenges common to all high throughput materials discovery: (1) instruments and measurements that closely resemble the functional device under operational conditions provide the most relevant information but are exceedingly challenging to make truly high throughput and (2) the highest throughput techniques, in particular parallel measurement schemes, typically provide semi-quantitative or qualitative classification information, requiring additional experimentation to demonstrate material performance credibly. These challenges are met through a twofold approach using creative engineering to accelerate vastly measurements and scientific insight to design screens for specific fundamental performance metrics carefully.

6.2 High Throughput Synthesis

High throughput evaluation of inorganic material libraries requires high throughput syntheses. For a survey of high throughput synthesis techniques, we refer the reader to excellent review articles [138–142] and note here the chemical deposition techniques most utilized for solar fuels applications. Photocatalyst libraries have been synthesized using combinatorial cathodic electrodeposition [144] and thermal calcination of mixtures of metal salts, which can be rapidly deposited using inkjet printing of precursor solutions [146]. This latter technique has also been successfully employed for deposition of electrocatalysts [149] and a related technique involving photochemical decomposition of mixtures of metal complexes has yielded highly active amorphous catalysts [150]. These techniques have generally been used to produce thin film composition samples with lateral dimensions of 1–

10 mm. The associated screening tools must accommodate evaluation of material performance on samples of this size.

6.3 High Throughput Screening

Optical absorption measurements are important for the evaluation of photoabsorbers and photocatalysts. High throughput optical characterization of materials has been employed for mapping composition-dependent optical properties, including the band gap of semiconductors [151] and various performance metrics for transparent conducting oxides [152, 153]. Here we focus on the relatively advanced evaluation techniques involving electrochemical and photoelectrochemical measurements. For high throughput screening of photocatalysts, Jaramillo et al. [145] deployed an automated version of a standard photoelectrochemical cell, as described in Fig. 26.

By using an O-ring seal to isolate a select composition sample, a standard three-electrode photoelectrochemical cell was established with the back contact of the sample serving as the working electrode, enabling a suite of standard techniques including spectral photon-to-current conversion efficiency and characterization of the illuminated open circuit potential. Automation of standard methods provides a direct relationship between the high throughput data and that of conventional research methods, but creating a bulk electrochemical cell for each composition sample requires extensive solution handling and inherently limits measurement throughput.

To accelerate measurement throughput by eliminating the preparation of individual electrochemical cells, a gasket-free design was developed by Jin and collaborators [154] based on a scanning drop cell architecture from the field of electrochemical microscopy [155]. The high throughput scanning drop cell (SDC) establishes sample-specific electrochemical contact to the working electrode by continuously flowing electrolyte solution over a given sample. As shown in Fig. 27, the solution contact is localized by strategically-placed suction tubes that wick away the flowing solution, generating an electrolyte drop that can be rastered across a planar electrode at speeds exceeding 5 mm s^{-1}. This novel drop-restricting scheme circumvents the need for a gasket or similar contact-based solution confinement mechanism, permitting on-demand cell movement without vertical motion. By minimizing the data collection "down time," -this instrument maximizes the throughput for serial three-electrode measurements. Incorporation of a fiber-optic into the cell – perpendicular to the planar working electrode with the illumination at the center of the drop – enables collection of the full range of photoelectrochemical measurements.

Important distinctions remain between these automated serial measurements and traditional electrochemical techniques, including the lack of sample rotation to control mass transport and the lack of membrane-separated anolyte and catholyte to avoid, for example, products of the counter electrode reaction from reaching the

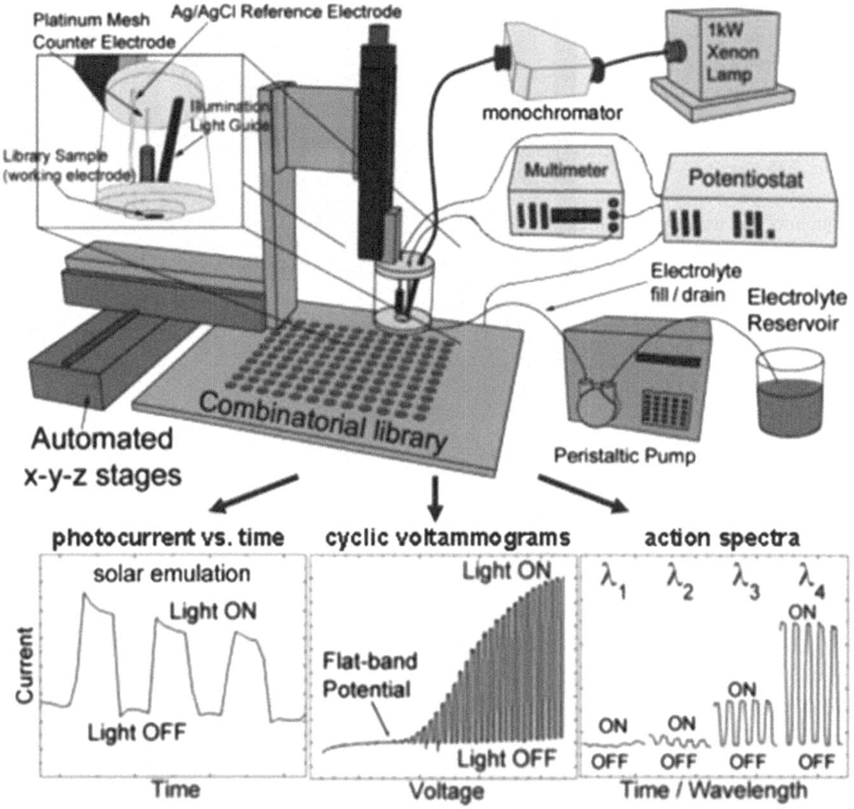

Fig. 26 Serial photoelectrochemical measurement in which a standard O-ring sealed electrochemical cell is established for each sample. Reproduced with permission from Muster et al. [145]. Copyright 2005, American Chemical Society

working electrode. Mass transport issues can be particularly important for high current density or long-duration experiments where electrochemical measurements are compromised by depletion of reactants or saturation of products. The SDC mitigates these effects by maintaining continuous flow over the working electrode to replace the drop volume every ~0.1 s. Addition of the capability to pump differentially fresh electrolyte over the counter electrode sweeps the products of the counter electrode reaction away from the drop and the working electrode. The serial photoelectrochemistry instruments can also be used to evaluate heterogeneous electrocatalysts, particularly for the HER and OER. Evaluation of catalysts generally involves higher current densities, making the SDC flow cell advantageous compared to cell designs with static solution. The excellent performance of this high throughput electrochemical cell can be used to measure not only performance but also fundamental electrochemical properties such as reversible redox potentials of the catalyst film and Tafel parameters [156].

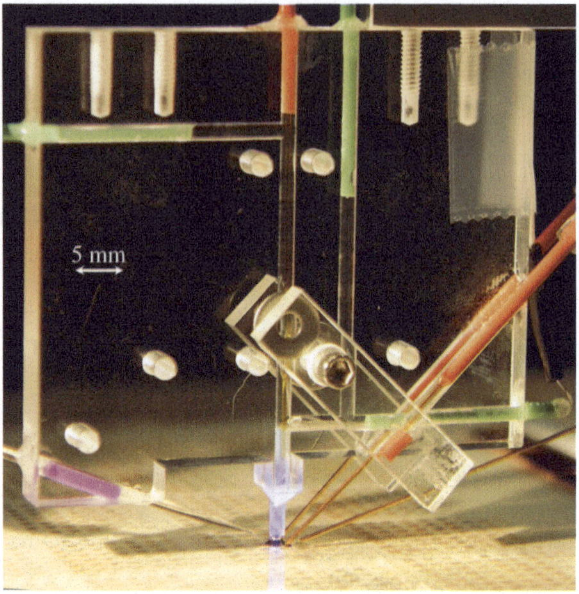

Fig. 27 The high throughput scanning drop cell establishes a three-electrode electrochemical cell for each 1 mm^2 sample in a material library. The cell includes an integrated fiber optic, providing high throughput electrochemical and photoelectrochemical measurements. Reproduced with permission from Gregoire et al. [154]. Copyright 2013, AIP Publishing LLC

Although the SDC architecture maximizes the throughput of serial measurements, several parallel and quasi-parallel measurement techniques have been developed with various compromises in throughput, data quality, and experiment versatility. For screening photocatalysts, the approach taken by the Parkinson group includes parallel sample contact by immersing the entire material library on a common working electrode in electrolyte [146]. Sample-resolved photocurrent measurements are obtained by rastering a light beam matching the sample size. Because the light source can be rastered quickly, the measurement throughput is effectively limited by the response time of the materials. Measurement sensitivity can be enhanced by modulating the illumination and employing lock-in amplification, which is particularly important because of the relatively large dark currents arising from the large unilluminated areas of the working electrode. Modifications to this screening strategy include electrical isolation of the samples on small area working electrodes coupled with multiplexing electronics so that only the active working electrode is chosen to coincide with the illuminated sample, as demonstrated by Katz et al. [157]. Conceivably, the experiment could be further parallelized by simultaneously illuminating all samples and using individual ammeters for each sample. Although the parallel and quasi-parallel measurement schemes provide rapid identification of hit materials, the shared solution contact introduces measurement limitations and opportunities for artifacts such as cross-contamination of the samples. The working electrode cannot be effectively polarized far from open circuit with libraries containing unstable constituents which could leach and redeposit on other samples.

This ensemble of high throughput photoelectrochemical techniques provides limited ability to measure the illuminated open circuit potential or the flat-band potential of a material. Rigorous measurement of these quantities requires that, at the working electrode surface, the electrolyte exclusively contacts the semiconductor material, but high throughput samples and instruments often allow an electrolyte shunt to the back contact or to neighboring samples [157]. As discussed by Jaramillo et al. [143], the serial measurement techniques noted above can be used to measure photocurrent as a function of potential, with the open circuit photovoltage estimated as the extrapolated potential for zero photocurrent.

The high throughput photoelectrochemical experiments referenced above provide screening of *photocatalytic* activity, particularly for the OER. Although stable, efficient photocatalysts are desirable for solar fuels technology, the photoanode of an integrated water-splitting device typically includes an electrocatalyst coupled to the light absorber to enhance reactivity for the OER. The applicability of these techniques for screening *photoabsorbers* for their light absorption and charge separation performance must be considered. For an *n*-type semiconductor, requiring performance of the OER in the screen means that the charge transfer across the solid–liquid interface may be limited by the ability of the particular semiconductor surface to catalyze the requisite reaction. As a result, the measured photocurrent does not reflect the fundamental semiconductor characteristics of most importance when used in an integrated solar fuels device. Consequently, these screens may not identify the materials capable of producing the most efficient integrated device.

To illustrate this limitation, the impact of poor electrocatalytic properties on the short-circuit photocurrent, one of the most common figures of merit used in these measurements, should be considered. The anode photoabsorber must have a valence band potential positive of the equilibrium OER potential to create an energetic driving force for charge transfer across the interface. To maximize efficiency, this valence band potential should be just large enough to provide an optimized OER catalyst with the requisite overpotential to drive the OER at the desired rate. Figure 28 schematically illustrates the impact of band-edge position and of catalysis on the ultimate efficiency of double junction solar fuels devices. Two 1.6 eV band gap photoanodes with different band edge positions are represented by the blue and red solid rectangles. The blue photoanode may yield relatively high performance when screened without a catalyst because its very positive valence band potential creates a large thermodynamic driving force for OER which overcomes kinetic limitations. In contrast, the red photoanode would by itself produce a lower photocurrent in an OER photocatalyst screen, but when coupled with an effective catalyst, photocatalysis of the OER is attained with high efficiency. Although detailed discussion of integrated devices is beyond the scope of this chapter, we have calculated an approximate efficiency for photoelectrolysis of H_2 using the red and blue photoanodes. For both cases, Fig. 28 shows a paired photocathode (dashed lines) for creating a tandem photoelectrochemical cell in which the band overlap at the semiconductor junction is 0.3 V and the photocathode conduction band lies 0.35 V above the HER equilibrium potential. For these tandem cells, the theoretical solar-to-hydrogen conversion efficiency of the blue tandem is approximately 5% whereas the efficiency of the red tandem is in excess of 20% [158]. Considering the

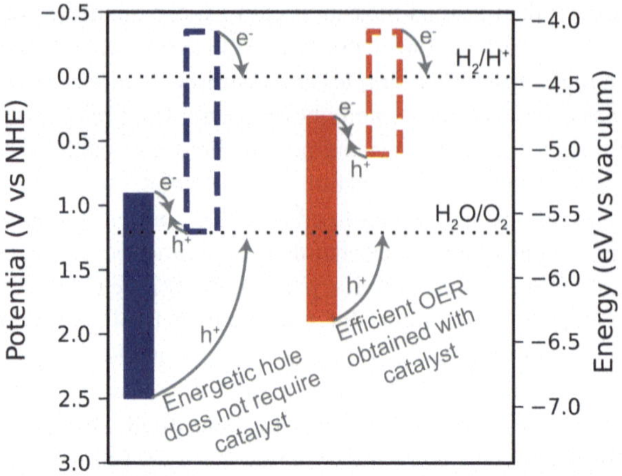

Fig. 28 Illustration of the impact of photoanode valence band position on (1) its performance in a high throughput screen of OER photocatalysis and (2) the efficiency for photoelectrolysis of H_2 in a tandem light absorber device. Both photoanodes (*solid rectangles*) have 1.6 eV gap and the photocathodes (*dashed rectangles*) are matched to provide 0.3 V band overlap at the semiconductor junction and a conduction band 0.35 V negative of the H_2/H^+ potential. The blue scenario has a theoretical efficiency near 5%, and the red tandem has theoretical efficiency in excess of 20%

device-level performance, the high throughput photocatalyst screening techniques would miss potentially high performing photoabsorber materials.

To develop a high throughput screen that avoids selective discovery of sub-optimal photoabsorbers, the kinetic limitation of the interfacial reaction must be removed. One approach is the addition of a sacrificial hole acceptor to the electrolyte, such as Na_2SO_3, which was successfully implemented into a high throughput screen by Ye et al. [159] using a custom SECM instrument. Alternatively, the measurement can benefit from implementing a traditional semiconductor characterization technique in which a well-defined asymmetric junction is introduced to measure the charge separation and voltage generation properties of the semiconductor. One strategy for forming a well-defined junction is to contact the semiconductor with a liquid containing a one-electron reversible redox couple, such as a metallocene system [160]. For experiments employing such semiconductor/liquid contacts, the redox system in the liquid is not photoexcited and serves to transport the charge across the interface with minimal kinetic limitations. Xiang et al. [161] recently developed a high throughput instrument that measures photocurrent generated by a semiconductor with a metallocene junction. The instrument includes a common working electrode contact and parallel electrolyte contact as described above [146], but the measurement requires a carefully tailored cell geometry and small, multiplexed counter electrodes to provide efficient collection of the photogenerated current. By performing a standard semiconductor characterization experiment in a high throughput platform, semiconductors with appropriate photoabsorber properties can be identified.

6.4 High Throughput Evaluation of Electrocatalysts

We note above the efficacy of the SDC instrument for serial screening of electrocatalysts. To attain higher throughput, a parallel scheme is required in which the entire library is used as a common working electrode in an electrochemical cell. The most common strategy for parallel screening is to introduce an imaging technique to identify active catalyst samples. That is, with the entire library held at the same working electrode potential, the measured current indicates the total activity of the library, and a complementary measurement to resolve catalytic activity spatially is required. Several imaging techniques have been employed to image local changes in pH using a dissolved fluorescent molecule [148, 162, 163] or a pH-sensitive membrane [164]. Such techniques provide massive parallelization of catalyst evaluation but suffer from limitations on the electrolyte pH and poor control of the electrochemical environment. Imaging of the local concentration of a reaction product can be employed, alleviating the pH requirements on the electrolyte. Local detection of O_2 generation is an operational principle of the SECM experiment described above [159], which is a scanning probe technique that provide excellent spatial resolution but is not a truly parallel measurement. Recently reported screening approaches include parallel imaging of evolved oxygen by a fluorescence technique [165] or direct imaging of bubbles generated on the catalyst surface [166]. All parallel catalyst screening techniques yield impressive sample throughput, but the disconnect between the high throughput measurement and traditional electrochemistry measurements requires substantial follow-up work to enable the newly discovered electrocatalysts to make an impact on the field.

An important omission from the described set of advanced high throughput instruments for solar fuels materials discovery is one for evaluating catalysts for the CO_2 reduction reaction (CO_2RR). For both photocatalysts and dark electrocatalysts, measurement of catalytic current without product analysis is insufficient to evaluate a material's performance. First, catalysis of the CO_2RR is inherently in competition with the HER. Second, the product distribution from the CO_2RR is of great importance for solar fuels applications. High throughput measurements must be developed for determining the branching ratio between the HER and CO_2RR and between the CO_2RR products at different potentials under different reaction conditions. The ultimate objective is to discover electrocatalyst materials that selectively reduce CO_2 to a single product.

6.5 Concluding Remarks

This brief summary of high throughput measurements highlights the advanced material evaluation experiments being applied to the discovery of solar fuels materials. The discussion emphasizes the need to consider carefully the relationship between the material properties evaluated in high throughput and the ultimate technological application of the materials. A precise description of the ultimate, integrated device greatly facilitates the definition of the desirable materials properties. This in turn enables design of high throughput screens to measure performance metrics which are directly transferable to the target application. Effective high throughput experiments must adhere to the range of operational conditions that are relevant for a deployable technology, which may evolve as the technology matures. For all high throughput materials discovery efforts, the ultimate utility of the high throughput screen hinges upon establishing and reporting the relationship between a material's performance metrics and its operation in the target application.

Acknowledgements Drs. Haber and Gregoire acknowledge support from the Joint Center for Artificial Photosynthesis, a DOE Energy Innovation Hub, supported through the Office of Science of the U.S. Department of Energy (Award No. DE-SC0004993).

7 Screening Photocatalysts Using Scanning Electrochemical Microscopy for Solar Water Splitting

Hyun S. Park
Fuel Cell Research Center, Korea Institute of Science and Technology, 39-1 Hawolgok-dong, Seoul 136-791, Republic of Korea
hspark@kist.re.kr

7.1 Introduction

The development of an efficient and robust photocatalyst with the required factors for the practical generation of solar fuels has been a long-standing challenge of photochemistry and materials science over many decades [1, 167, 168]. In order to convert solar energy to chemical energy efficiently through photolysis of water, photoelectrodes should be (1) chemically stable in highly oxidative and reductive conditions in the aqueous solution, (2) physically resilient to endure high temperature and pressure under strong light irradiation, and (3) characterized with suitable conduction/valence band edge positions to undergo the proton reduction and water oxidation reactions with maximum photovoltages [169]. Many binary, ternary, and quaternary transition metal oxides and chalcogenides have been extensively studied, but no photocatalyst with the optimum stability and activity has been discovered yet [170]. Combinatorial screening coupled with rapid synthesis have been employed as versatile tools for the discovery of many biological, organic, and inorganic materials with specific properties needed for different applications

[171]. The tools have also been utilized to develop promising multi-component photocatalysts for solar energy conversion devices [143, 146, 157, 172, 173]. This chapter mainly deals with the rapid screening of photocatalysts using SECM [174, 175] and also briefly describes in situ electrochemical characterization of photoelectrodes using SECM techniques.

7.2 *Rapid Screening and In Situ Characterizations of Photocatalysts with SECM*

7.2.1 Photocatalyst Array Preparation

The rapid screening of photocatalysts needs fast and facile preparation of numerous photocatalysts with ternary, quaternary, or more compositions on conducting substrates. Different techniques, including ink-jet printing [157, 176], sputtering [172], and electrodeposition [143, 177], have been implemented to create the photocatalyst libraries on a small electrode. The photocatalyst spot arrays fabricated using a piezoelectric dispensing tip is an example of the material collections for the rapid screening with SECM [178]. To create the photocatalyst spot arrays, the piezoelectric tip for dispensing precursor droplets (MJ-AB-01-60, MicroFab) is attached to an XYZ-stage which is controlled by the SECM and personal computer system (Model 1550A Solution Dispenser, CH Instruments, Fig. 29). The substrate for photocatalyst spot arrays should be (1) electrically conductive, (2) chemically and thermally stable for sample preparations, and (3) (photo-)electrochemically inert to record the photocurrent produced only at the photocatalysts without background reactions. Fluorine-doped tin oxide (FTO) coated glass is often employed as the conductive supporting electrode which is photoelectrochemically inert with a large optical band-gap, i.e., approximately 3.6 eV [179]. It is also thermally supportable to form the photocatalysts with heat treatments up to approximately 550°C [146]. Solutions of metal nitrates or metal chlorides dissolved in the aqueous

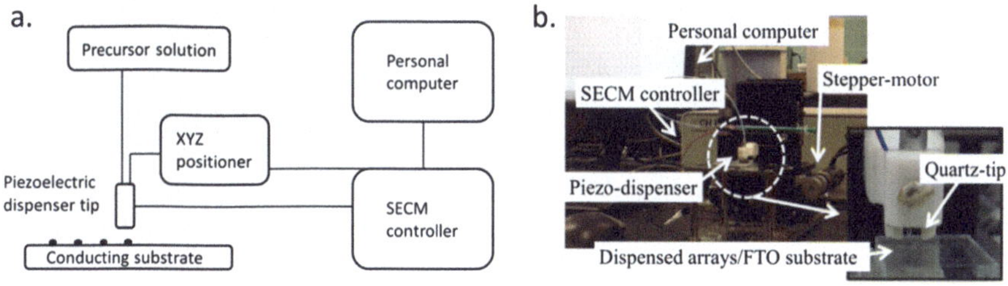

Fig. 29 (**a**) Schematic configurations of the SECM dispenser system to create the photocatalyst arrays. (**b**) Photographic images of the SECM stage equipped with a piezoelectric dispensing tip. The tip is positioned near the FTO glass surface with a distance of a few hundreds of μm between them

or organic solvents are frequently used as precursors to prepare the array electrodes of photocatalysts. The dispensing tip gives solution drops of a few picoliters, and the drop sizes are controlled by varying the potentials and pulse durations applied to the dispensing tip.

The photocatalyst array preparation should be performed carefully so that the efficiency of numerous photocatalysts is accurately measured in the screening process. Each spot should have a uniform size and thickness to ensure identical light absorption conditions over the entire array, because the light absorption efficiency is directly related to the photocatalytic performance. Figure 30a shows a schematic design of the phocatalyst arrays and the photograph in Fig. 30b shows dispensed spot arrays with size and thickness uniformity in accordance with the designed patterns [180]. The spot array electrode in Fig. 30b consists of 55 photocatalyst spots with a uniform thickness of approximately 400 nm and diameter of approximately 250 μm. The distances between spot arrays are large enough, i.e., 800 μm, to record the photoresponse at a single spot electrode under the fiber optic light irradiation without any interference from the photocurrent generated at adjacent spots. The arrays are subsequently annealed in different atmospheric environments to convert the precursor chemicals to oxide, nitride, or chalcogenide semiconductor crystals. For example, quaternary chalcogenide photocatalyst arrays of $Zn_xCd_{1-x}S_ySe_{1-y}$ are prepared in Fig. 30b [180]. The relative atomic compositions in the multi-component materials are controlled by varying the number of dispensed droplets and the concentrations of precursor solutions as programmed in the array patterns.

7.2.2 Rapid Screening of Photocatalyst Arrays

When the array electrodes are produced by the solution dispensing and thermal treatments, the electrodes are placed in the electrolyte solution and an optical

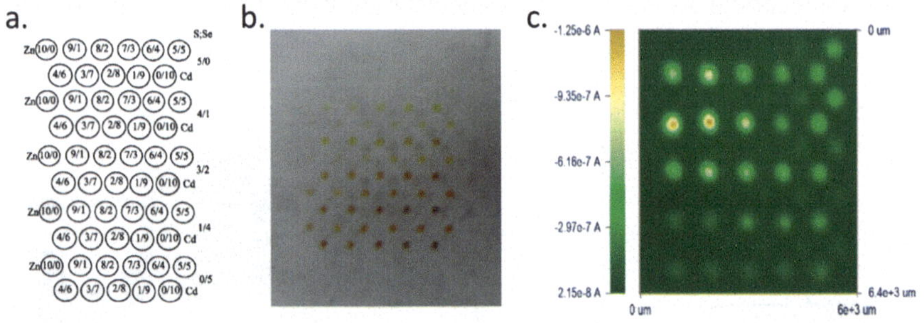

Fig. 30 (**a**) Schematic representation, (**b**) photographic image, and (**c**) SECM photocurrent map of photocatalyst spot arrays of $Zn_xCd_{1-x}S_ySe_{1-y}$. Each spot has different compositions of Zn, Cd, S, and Se as indicated in Fig. 30a. Polysulfide redox electrolytes (S_n^{2-}/S^{2-}) were used to record the photocurrent in Fig. 30c. Reproduced with permission from Liu et al. [180]. Copyright 2010, American Chemical Society

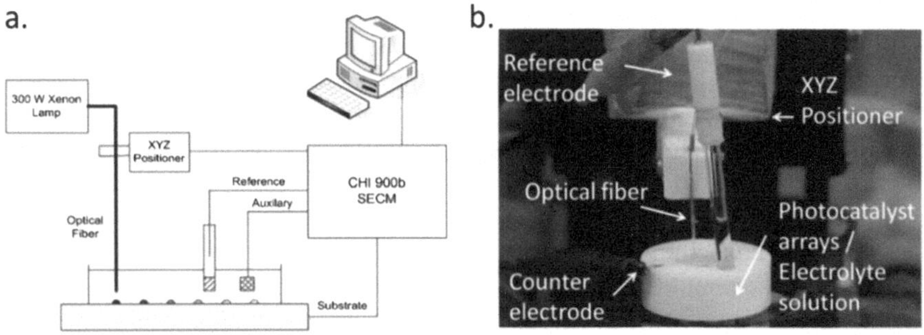

Fig. 31 (**a**) Schematic representation and (**b**) photographic image of SECM equipped with the optical fiber for screening photocatalyst arrays. Reproduced with permission from Lee et al. [175]. Copyright 2008, American Chemical Society

fiber is connected to the XYZ stage of the SECM for scanning the sample arrays (Fig. 31). The optical fiber is also coupled to a light source to irradiate the spot arrays immersed in the electrolyte solution. Scanning parameters, i.e., the optical fiber diameter, distances between the optical fiber and array electrodes, scan rates, and applied potentials, should be determined in consideration of the single spot size and the whole array dimensions [178]. For example, the optical fiber with a diameter of 400 μm was positioned 100 μm above the electrode surface with a scan rate of 500 μm s^{-1} to obtain the SECM image in Fig. 30c [180]. The array electrode serves as the working electrode in a typical three-electrode configuration with reference and counter electrodes in a single compartment cell (Fig. 31b). The photocurrent at individual spots is measured with applied potentials as the optical fiber scans over the arrays. The SECM image obtained from the photocatalyst screening provides quantitative relationships between the photocurrent and material compositions of photocatalysts. For example, $Zn_{0.3}Cd_{0.7}S_{0.8}Se_{0.2}$ was found as the "hot spot" with the highest photoactivity among the prepared materials in Fig. 30c. When promising compositions of the photocatalyst are identified from the screening process, bulk electrodes of the optimum compositions are fabricated to elucidate their physicochemical properties further. By using the combinatorial screening methods, metal dopants or additives to $BiVO_4$ [181], Fe_2O_3 [146, 175], and ZnO [143], oxygen evolution electrocatalysts for $BiVO_4$ [182], and photosensitizers for TiO_2 [183] have been successfully found and studied for solar water splitting.

7.2.3 Other In Situ Characterization with SECM

In addition to the screening of photocatalysts for enhanced performance, other valuable electrochemical characteristics of photocatalysts, including Faradaic and quantum efficiencies [184], minority carrier diffusion length [185], or reaction kinetics and surface coverage of water oxidation intermediates at photoelectrodes [186], have been evaluated using SECM characterization methods combined with various probe techniques. For example, in situ characterization of silicon photocathodes with the combined SECM and scanning photocurrent spectroscopy (SPCM) techniques are shown in Fig. 32. In Fig. 32, an ultramicroelectrode (UME) and a focused laser beam are simultaneously employed to collect the hydrogen molecules generated at Pt electrocatalysts deposited on p-doped Si electrodes (tip collection of SECM) and spatial variation of the external quantum efficiency of the photocathode is recorded under the focused laser beam irradiation (local excitation of SPCM) [185]. By using similar combinations of SECM/SPCM, the Faradaic efficiency of water oxidation reactions at unstable photoanodes, e.g., Ta_3N_5, was obtained by the tip collection/substrate generation mode of SECM [184]. Furthermore, the surface coverage and reaction kinetics of water oxidation intermediates, i.e., hydroxyl radicals, were quantified at $BiVO_4$ and TiO_2 electrodes with the surface interrogation mode of SECM [186]. These

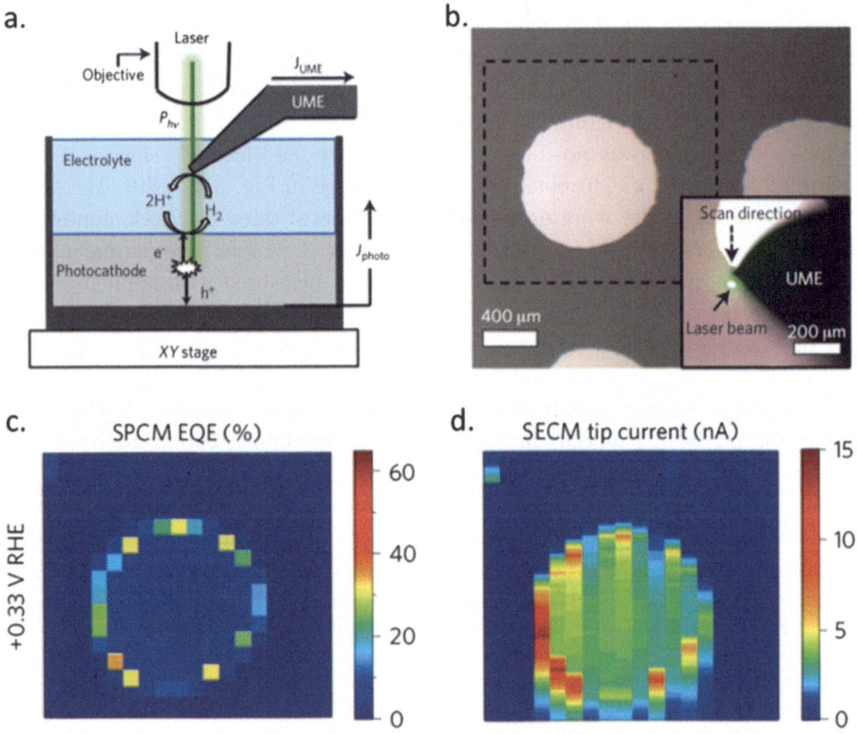

Fig. 32 (**a**) Schematic configuration of SECM combined with SPCM. (**b**) Photograph of Pt/p-doped Si photocathode arrays with an ultramicroelectrode and a focused laser beam positioned for the SECM–SPCM measurements (*inset*). (**c**, **d**) SPCM quantum efficiency and SECM tip-collection images obtained from the SECM–SPCM measurements. Reproduced with permission from Esposito [185]. Copyright 2013, Nature Publishing Group

types of characterization are useful to obtain analytical insights and develop strategies for how the photocatalytic efficiency can be improved further. The development of hydrogen and oxygen evolution (dark-)electrocatalysts are also important applications of the rapid screening techniques for efficient solar water splitting [173, 187].

8 In Situ Characterization of the Optical Properties of Electrocatalysts

Adam S. Batchellor, Lena Trotochaud, and Shannon W. Boettcher (✉)
Department of Chemistry and Biochemistry, University of Oregon, Eugene, OR 97403, USA
swb@uoregon.edu

8.1 Introduction

The development of efficient photoelectrochemical (PEC) water-splitting cells necessitates the use of electrocatalysts at the anode and cathode surfaces to help reduce the overpotentials associated with the oxygen and hydrogen evolution half-reactions. To avoid using spatially separate light-absorbing (photovoltaic) and water-splitting (electrolyzer) devices to drive the evolution of O_2 and H_2, catalysts can be deposited directly onto the light absorbing semiconductor electrodes (or buried photovoltaic junction). One major challenge that arises from physically coupling the catalysts and photoelectrodes in this way is that catalysts parasitically absorb light when operated under the electrochemical conditions required for water splitting [188]. This parasitic absorption by the catalyst reduces the number of photons that can reach the electrode surface, and thereby limits the number of electron-hole pairs that can be created and used to drive the redox reactions.

The efficiency of the individual components (e.g., catalysts, semiconductor electrodes) that would complete a water-splitting cell has been investigated [189, 190]. Recently, an increase in both modeling and experimental work has taken place to see how these components fare in composite systems [182, 191–207]. Although high-throughput screening has been used to predict which catalysts may contain ideal optical properties (low absorption) [208], little work has directly investigated the photocatalytic properties of the current state-of-the-art catalysts. In situ ultraviolet/visible (UV–Vis) spectroscopy is a valuable method for quantitatively evaluating the optical properties of catalysts under working conditions. The light-absorbing effects can then be incorporated into "opto-electrochemical" models of composite water-splitting cells.

8.2 Electrochromism in Catalyst Materials

Electrochromism is the process through which a material changes color in conjunction with a redox reaction. As many catalysts are (semi)transparent upon deposition, and become darker upon oxidation, a general description of their electrochromic behavior is

$$\underset{\text{bleached}}{\mathrm{R}} \rightarrow \underset{\text{colored}}{\mathrm{O}} + ne^{-} \tag{14}$$

This behavior can be valuable for applications including "smart" windows capable of improving the energy efficiency of buildings [209, 210]. However, the same behavior can negatively affect the performance of composite catalyst/semiconductor water-splitting photoelectrodes where the incident photons must first pass through the (colored) catalyst prior to being productively absorbed in the semiconductor.

Seike and Nagai used in situ spectroscopy to investigate the electrochromic properties of 3d transition metal oxides (TMOs) (e.g., NiO_x, CoO_x, TiO_x, MnO_x, CrO_x) [211]. These materials are important in the context of solar fuels because many of the same 3d TMOs have drawn attention as water oxidation catalysts because of their relatively large abundance and low cost compared to precious metal catalysts. In the colored, high-optical-absorption state, electrochromic TMOs such as NiO_x would parasitically absorb photons en route to the semiconductor surface, thus diminishing PEC performance. Seike and Nagai found that NiO_x has a high coloration efficiency compared to the other 3d TMOs studied. Practically, this means that very little charge needs to be injected (via oxidation of the film) to cause a large change in the optical density of NiO_x. The source of this coloration process is not well understood, despite considerable efforts to understand the electronic properties of the material. Possible sources have been postulated to include crystal defects [212], the presence of high valency Ni (to include Ni^{4+}) [213], metal oxygen charge transfer [214], and phase changes arising from ion insertion [215].

NiO_x, which converts to $Ni(OH)_2$ following electrochemical conditioning in basic media [216], is one of the most promising OER catalysts when small concentrations (~10%) of Fe are added [217]. Under conditions relevant to the OER in base, the reaction is

$$\underset{\text{transparent (light green)}}{\mathrm{Ni(OH)_2}} + \mathrm{OH^-} \rightarrow \underset{\text{dark colored}}{\mathrm{NiOOH}} + \mathrm{H_2O} + e^{-} \tag{15}$$

To maintain charge neutrality, the transfer of electrons between the catalyst and the electrode must be accompanied by the transfer of ions into or out of the film. Under OER conditions, every Ni site exposed to electrolyte, whether at the surface of the $Ni(OH)_2$ or as part of a disordered (i.e., only locally ordered) ion-permeable (oxy) hydroxide phase, is oxidized and thus contributes to optical absorption.

8.3 In Situ UV–Vis Absorption and Reflection Spectroscopy

From the above discussion of electrochromism, it is apparent that the optical properties of OER catalysts must be measured in situ to understand and account for their effects on photoabsorption in the underlying semiconductor elements in a photoelectrochemical cell. A spectrophotometer can measure the amount of light that passes through the catalyst, which is the optical transmittance. Example in situ experimental setups employed by Trotochaud et al. and Corrigan and Knight are shown in Fig. 33 [188, 213].

In these cells, the electrochemical potential of the catalyst can be controlled and poised at/near the potential of OER onset while the optical spectra are recorded. The catalysts are typically supported on a transparent conducting oxide electrode. In Fig. 33a the counter electrode (CE) was held above the path of the incident light and in Fig. 33b the CE had a section removed to permit the incident light to pass through. We note that in these setups there can be a significant (diffuse) scattered component to transmission and reflection spectra (especially for thicker films), and therefore these measurements are ideally made in an integrating sphere. In the work by Trotochaud et al., a second cell was designed for use with an integrating sphere to collect reflectance spectra and very thin films were used for the in situ transmission measurements where diffuse scattering was negligible.

The transmittance of a $NiOOH/Ni(OH)_2$ catalyst over a range of potentials is shown in Fig. 34. The transmittance was nearly independent of potential from 0 to 0.45 V vs Hg/HgO, where the catalyst remains in its bleached state and Ni is formally in the 2+ oxidation state. Then, as the Ni in the film was oxidized to 3+/4+ over the potential range from 0.45 to 0.55 V vs Hg/HgO, the catalyst became colored and the transmittance decreased significantly. Once colored (i.e., at

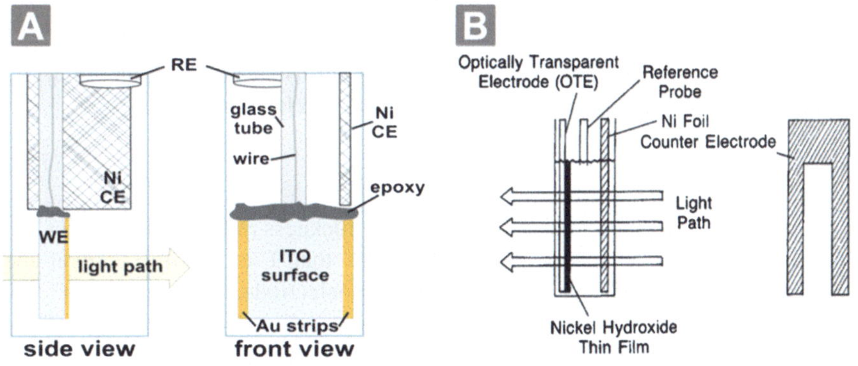

Fig. 33 In situ electrochemical cells employed by (**a**) Trotochaud et al. and (**b**) Corrigan and Knight to collect optical spectroscopy data on catalysts films. In (**a**), thin-film electrocatalysts were deposited by spin-coating onto the ITO substrate prior to electrode fabrication. (**a**) is reproduced with permission from Trotochaud et al. [188]. Copyright 2013, American Chemical Society. (**b**) is reproduced with permission from Corrigan et al. [213]. Copyright 1989, The Electrochemical Society

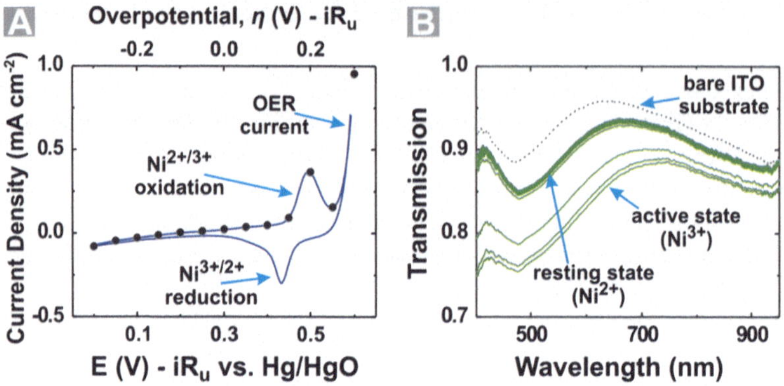

Fig. 34 (**a**) Cyclic voltammetry (CV) and (**b**) transmission spectra of NiO_x film measured at potentials ranging from 0 to 0.6 V vs Hg/HgO in 0.05 V increments. Each *line* in (**b**) was taken at a potential corresponding to the *dots* on the CV, with the *top transmission line* corresponding to the lowest potential. Reproduced with permission from Trotochaud et al. [188]. Copyright 2013, American Chemical Society

potentials higher than the Ni oxidation wave), the transmittance showed little variation as the potential was further increased.

All incident photons on a surface must be absorbed, reflected or transmitted:

$$1 = A + R + T \tag{16}$$

where A, R, and T are the wavelength-dependent absorption, reflection, and transmission processes, respectively. The fraction of light passing through a material I with respect to the incident (non-reflected) light I_o follows Beer's law as given by

$$I = I_o e^{-\alpha t} \tag{17}$$

where α is the absorption coefficient, which is an intrinsic property of the solid, and t is the thickness. The transmittance of a material is related to α by

$$T = (I/I_o) = e^{-\alpha t}. \tag{18}$$

If α is known for a given material, the optimal thickness t of catalyst film for use with a photoelectrode can be determined such that T is maximized over the wavelength region of interest for the photoelectrode, maintaining a low overpotential for the reaction of interest. However, directly determining α for a catalyst in situ can be challenging. This is because, for any particular measurement, the measured value of T is that for the entire electrochemical/optical cell, not simply the transmittance of the catalyst layer T_c. The optical effects of the cell walls (typically glass or quartz) as well as the underlying transparent conductive oxide (TCO) substrate need to be properly accounted for such that the intrinsic wavelength-dependent absorption by the catalyst film $a_c(\lambda)$ can be extracted. This is important if the optical data collected from catalyst films on TCO electrodes are to be used to predict optical effects of the catalyst on different semiconductor photoelectrodes. To accomplish this, an optical model is needed that accounts for the reflection (r) and absorption (a) probabilities at each of the various interfaces the light encounters during the experimental measurement. These events are depicted in Fig. 35. Approximations to such a model (for instance, ignoring various higher order reflection terms) to make its solution analytically tractable are discussed in [188].

From $a_c(\lambda)$, the transmission probability at a given wavelength once the photon has reached the catalyst, $T_c(\lambda)$, can then be found by

$$T_c(\lambda) = 1 - a_c(\lambda) \tag{19}$$

and the absorption coefficient of the catalyst $\alpha(\lambda)$ can be calculated from

$$\alpha(\lambda) = -\ln T_c(\lambda)/t \tag{20}$$

where t is the thickness of the film. Figure 36 shows the extracted absorption coefficients of the ~2 nm thick transition metal oxide water oxidation catalysts studied by Trotochaud et al., as well as the calculated transmission probabilities for a mixed Ni/Fe oxyhydroxide film as a function of film thickness. The transmission losses are significant even for films only a few nanometers thick.

For catalysts, this thickness dependence of the transmission probability poses an interesting problem, as two competing effects are pitted against one another. Increasing the catalyst loading lowers the overpotential required to pass a given oxygen-evolution current, reducing voltage losses from kinetics. However, a thicker catalyst layer also parasitically absorbs more light, thus decreasing the total photocurrent generated by the photoanode. Furthermore, the high coloration efficiency of Ni discussed previously manifests itself here, as the values of $\alpha(\lambda)$ measured in situ are seen to be greater for the Ni-based catalysts than those of the other metal oxides studied.

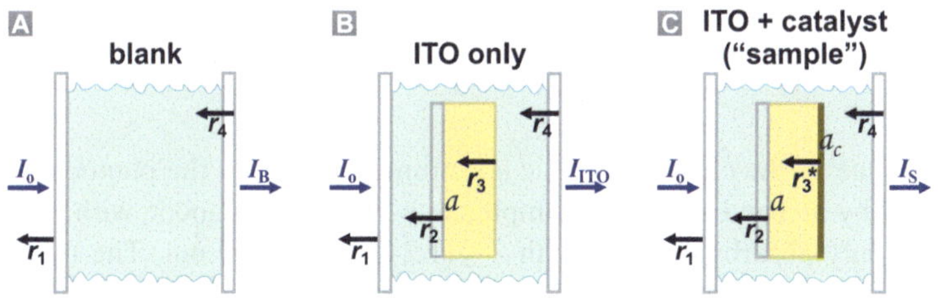

Fig. 35 Cross-sectional view showing the light path through an in situ liquid cell. The symbols a and r signify absorption and reflection probabilities, respectively, with the *subscripts* indicating the location of the optical event as described below. The spatial location of the *arrows* and *symbols* along the vertical scale of the diagrams is not significant. Reproduced with permission from Trotochaud et al. [188]. Copyright 2013, American Chemical Society

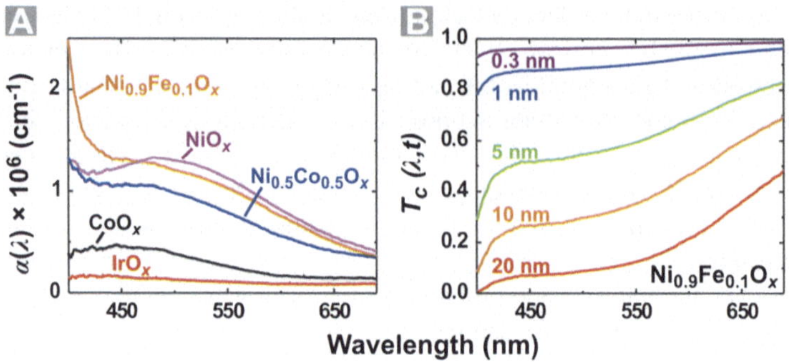

Fig. 36 (**a**) Effective absorption coefficient $\alpha(\lambda)$ calculated for the active catalyst films. (**b**) Transmission probabilities $T_c(\lambda)$ calculated for $Ni_{0.9}Fe_{0.1}O_x$ films of varying thickness. Reproduced with permission from Trotochaud et al. [188]. Copyright 2013, American Chemical Society

8.4 *Identification and Optimization of Catalyst Films for Solar Water-Splitting Photoanodes*

The ultimate measure of an electrocatalyst's utility for photoelectrochemical water splitting is how it performs in a prototype device. Defining reasonable figures of merit allow one to determine which catalysts offer the best combination of high turnover frequency (mol product/mol active site/s) at low overpotential and low parasitic optical absorption, without the need to test each particular catalyst individually in a complete solar water-splitting device as a function of catalyst loading. Trotochaud et al. and Gregoire et al. described figures of merit based on experimentally measured data [188, 208].

The approach described by Trotochaud et al. is summarized here, and later compared to that of Gregoire et al. The amount of light at a given wavelength λ reaching the semiconductor surface is a function of the incoming light flux, $F(\lambda)$, multiplied by the corresponding catalyst transmission probability, $T_c(\lambda, t)$. The optical efficiency, φ_{opt}, is defined as

$$\varphi_{opt}(t) = \int F(\lambda) \cdot T_c(\lambda, t) / \int F(\lambda) \tag{21}$$

over the range of the electromagnetic spectrum available to the photoelectrode as determined by its band gap. For example, an n-Fe_2O_3 photoanode with a band gap of 2.2 eV only absorbs photons with $\lambda \leq (1{,}240/2.2) \approx 560$ nm. The optical efficiency is a function of the catalyst film thickness (i.e., catalyst loading) produced by the dependence of T_c on t, as shown in Fig. 37.

For thin films <1 nm, the $\varphi_{opt}(t)$ of all catalysts approaches unity, as there is too little material to absorb a significant fraction of incoming photons. For films >50–100 nm thick, $\varphi_{opt}(t)$ tends towards zero – even relatively transparent films absorb

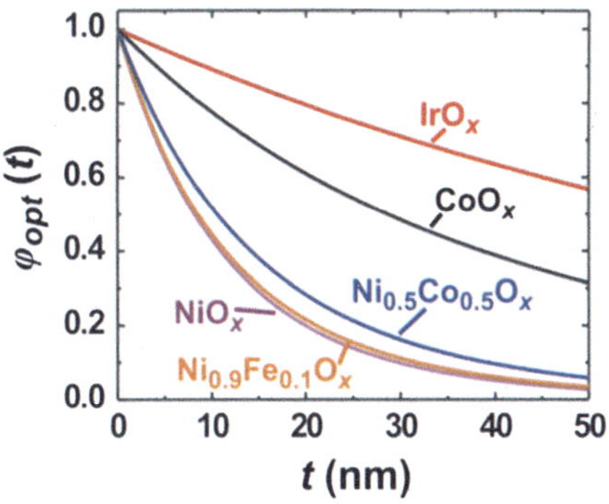

Fig. 37 Optical efficiency $\varphi_{\text{opt}}(t)$ of active catalyst films as a function of thickness t, assuming an optical band gap of 1.8 eV for the absorbing semiconductor and AM 1.5 solar flux. Reproduced with permission from Trotochaud et al. [188]. Copyright 2013, American Chemical Society

the majority of photons as their thickness becomes sufficiently large. It is interesting to note that because of the high absorption coefficient of Ni-based catalyst films, their corresponding $\varphi_{\text{opt}}(t)$ values are much lower than those of the Ni-free films.

Combining the optical properties with the independently measured catalytic properties (we assume the activity, i.e., catalytic current density at a given overpotential scales linearly with film thickness) [216] allows us to define the optocatalytic efficiency of the films, $\Phi_{\text{O–C}}$:

$$\Phi_{\text{O-C}} = P_{\text{max}}/P_{\text{o,max}} \tag{22}$$

where $P_{\text{o,max}}$ is the maximum power achievable in the electrode with a hypothetical "perfect" catalyst that has no parasitic optical absorption and infinitely fast kinetics for O_2 evolution, and P_{max} is the power achievable with the addition of the real catalyst (including optical and kinetic voltage losses). Trotochaud et al. calculated power curves and then determined $\Phi_{\text{O–C}}$ values for some of the most promising catalysts as seen in Fig. 38.

All catalyst films studied reached their maximum $\Phi_{\text{O–C}}$ at less than 10 nm. The highest performing system is predicted to be the Ni/Fe catalyst with a $\Phi_{\text{O–C}}$ value of 0.64. This $\Phi_{\text{O–C}}$ is achieved with a film thickness of less than 0.5 nm. In fact, all Ni-based catalysts had their max $\Phi_{\text{O–C}}$ values at less than 1 nm thickness, with a steep drop-off in efficiency with increasing thickness. This is because of their high absorption coefficient as mentioned previously. IrO_x, on the other hand, is predicted to reach its maximum $\Phi_{\text{O–C}}$ at 8.8 nm because of its lower absorption coefficient. In the case of IrO_x, the advantage of more-transparent catalysts in the trade-off between lower overpotential and higher parasitic absorption is exemplified; an incremental increase in the thickness of IrO_x up to 10 nm decreases the voltage loss more substantially than it increases the parasitic light absorption.

In work by Gregoire et al., a related optocatalytic FOM is proposed [208]. In order to determine the optocatalytic efficiencies of over 5,000 catalysts via high-throughput experimentation, only a snapshot of the catalysts' properties at 0.45 V overpotential was input into the model. The thickness-dependence was also not investigated (which would have been challenging in such a high-throughput

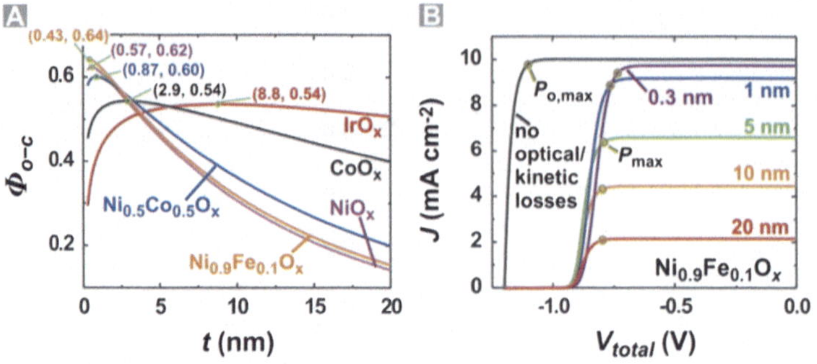

Fig. 38 (**a**) Photoelectrode current voltage curves for $Ni_{0.9}Fe_{0.1}O_x$ and (**b**) optocatalytic efficiency $\Phi_{O\text{-}C}$ as a function of thickness t for different catalysts, assuming an optical band gap 1.8 eV for the absorbing semiconductor and AM 1.5 solar flux. Reproduced with permission from Trotochaud et al. [188]. Copyright 2013, American Chemical Society

screening approach) and thus no ideal parameters were determined for any one particular catalyst. The optical properties were also not screened in situ and therefore any electrochromic behavior was not accounted for. Despite the different goals and approaches, the two works come to some similar conclusions, namely that a Ni/Fe-based film would be a good catalyst for a composite PEC device.

8.5 *Tailoring of Catalyst Optical Properties*

When the relevant properties of a catalyst (e.g., a_c) have been ascertained, it may be possible to tune the optical properties of the catalyst by adjusting composition and not affecting the catalyst behavior. In an effort to develop "optically passive" counter electrodes for electrochromic applications, Azens et al. have shown that, by varying the Ce concentration in Ti and W films, they were able to create a transparent film without the typically observed electrochromic response [218]. In a related study, Monk et al. show that, by mixing various TMOs in amorphous multication oxides, the wavelength of the maximum in the electrochromic absorption band can be tailored [219]. Extending these approaches to water oxidation catalysts, it could be possible to adjust the optical properties of a film without a significant change in catalytic performance. Given the large range of compositions including Ni, Fe, Ce, Ti, Co, etc., now found to have relatively high catalytic activity [156, 190, 208, 220], there are likely mixed oxide compositions that could be found which also have low optical absorption.

For example, Gregoire et al. found two regions of interest in the quaternary system studied [208]. The first is the high catalytic performance of the predominately Ni/Fe system and the other is the high optocatalytic performance with a film composition of $Fe_{0.23}Co_{0.13}Ni_{0.07}Ti_{0.57}O_x$. This particular system is interesting because of the (relatively) poor catalytic performance of films containing significant concentrations of Ti. However, the extremely high transmission properties of films containing large concentrations of Ti are capable of overcoming its catalytic limitations with respect to the combined figure of merit. It is reasonable to assume that a system with a large Ti content would likely never have been explored by researchers searching solely for a highly catalytically active catalyst. This identification opens up a new system to be explored and optimized and exemplifies the value of high throughput investigations.

8.6 New Horizons for In Situ Optical Spectroscopy for Solar Fuels Material Characterization

Beyond determining the optical properties of catalyst films in the context of solar fuels, there are other directions that can be taken to exploit the benefits of in situ optical spectroscopy. Because of the electronic properties of certain elements, analytical techniques are sometimes incapable of acquiring the same data on all materials of interest. Brückner used in situ parallel UV–Vis/EPR/on-line gas chromatography to investigate the oxidation state of a chromium oxide catalyst during the dehydrogenation of propane [221]. Cr^{6+} is not detectable by EPR, but EPR is able to distinguish between Cr^{3+} and Cr^{5+}. Cr^{6+} is identifiable via UV–Vis spectroscopy through its charge-transfer transitions. By measuring the UV–Vis spectra as the temperature was varied, the absorption band was shown to disappear as relevant reaction temperatures were reached, ruling out Cr^{6+} as a possible active site. In a separate work, Brückner used in situ parallel UV–Vis/EPR/Raman spectroscopy to elucidate the oxidation state of V in a V/TiO_2 film [222], showing that several species of V^{4+} are actively engaged in the dehydrogenation of propane.

In composite PEC cells, the catalyst films are not the only material that interfere with the incoming flux of light. Because of the production of H_2 and O_2 at the electrode surfaces, the presence of bubbles also needs to be accounted for. The decrease in efficiency caused by the formation of bubbles comes from both ohmic losses [223] and optical losses because of enhanced scattering [224–228]. As the bubbles are only formed under reaction conditions, the use of in situ optical techniques is paramount to the quantification of these losses.

8.7 Summary

In conclusion, if catalysts are to be utilized in a composite water-splitting cell, both their catalytic and optical properties need to be optimized. In situ optical methods such as UV–Vis spectroscopy provide the basis for combined optocatalytic figures of merit and guide the design of improved materials through tailoring of the catalyst absorption bands and electrochromic properties as well as potentially providing insight into catalyst mechanisms.

Acknowledgements A.S.B. thanks the United State Air Force Academy Faculty Pipeline program for support. L.T. and the synthesis/optical characterization of thin-film electrocatalysts were supported by the National Science Foundation under the Centers for Chemical Innovation Program, grant CHE-1102637. The development of the optocatalytic model was supported by the DOE Basic Energy Sciences grant DE-FG02-12ER16323. S.W.B. thanks the Research Corporation for Science Advancement for a Cottrell Scholar Award.

References

1. Fujishima A, Honda K (1972) Nature 238:37–38
2. Chen Z et al (2010) J Mater Res 25:3–16
3. Chen Z, Dinh H, Miller E (2013) Photoelectrochemical water splitting: standards, experimental methods, and protocols. Springer, New York
4. Kanimura J, Bogdanoff P, Lähnemann J, Hauswald C, Geelhaar L, Fiechter S, Riechert H (2013) J Am Chem Soc 135:10242–10245
5. Hill JC, Choi K-S (2012) J Phys Chem C 116(14):7612–7620
6. Mi Q, Zhanaidarova A, Brunschwig BS, Gray HB, Lewis NS (2012) Energy Environ Sci 5:5694–5700
7. Mi Q, Coridan RH, Brunschwig BS, Gray HB, Lewis NS (2013) Energy Environ Sci 6:2646–2653
8. Somorjai GA, Frei H, Park JY (2009) J Am Chem Soc 131:16589–16605
9. Weinhardt L, Blum M, Fuchs O, Pookpanratana S, George K, Cole B, Marsen B, Gaillard N, Miller E, Ahn K-S, Shet S, Yan Y, Al-Jassim MM, Denlinger JD, Yang W, Bär M, Heske C (2013) J Electron Spectrosc 190:106–112
10. Arrigo R, Havecker M, Schuster ME, Ranjan C, Stotz E, Knop-Gericke A, Schlögl R (2013) Angew Chem Int Ed 52:11660
11. Axnanda S, Crumlin EJ, Mao B, Rani S, Chang R, Karlsson PG, Edwards MOM, Lundqvist M, Moberg R, Ross P, Hussain Z, Liu Z (2015) Sci Rep 5:9788
12. Casalongue HGS, Benck JD, Tsai C, Karlsson RKB, Kaya S, Ng ML, Pettersson LGM, Abild-Pedersen F, Norskov JK, Ogasawara H, Jaramillo TF, Nilsson A (2014) J Phys Chem C 118:29252
13. Bora DK, Braun A, Erat S, Löhnert R, Ariffin AK, Manzke R, Sivula K, Graule T, Grätzel M, Constable E (2011) J Phys Chem C 115(13):5619–5625
14. Kanan MW, Nocera DG (2008) Science 321:1072–1075
15. Kanan MW, Yano J, Surendranath Y, Dinca M, Yachandra VK, Nocera DG (2010) J Am Chem Soc 132(39):13692–13701
16. Yoshida M, Yomogida T, Mineo T, Nitta K, Kato K, Masuda T, Nitani H, Abe H, Takakusagi S, Uruga T, Asakura K, Uosaki K, Kondoh H (2014) J Phys Chem C 118:24302–24309
17. Cox N, Retegan M, Neese F, Pantazis DA, Boussac A, Lubitz W (2014) Science 345:804
18. Hurum DC, Agrios AG, Gray KA, Rajh T, Thurnauer MC (2003) J Phys Chem B 107:4545
19. McAlpin JG, Surendranath Y, Dinca M, Stich TA, Stoian SA, Casey WH, Nocera DG, Britt RD (2010) J Am Chem Soc 132:6882

20. Huang Z, Lin Y, Xiang X, Rodriguez-Cordoba W, McDonald KJ, Hagen KS, Choi KS, Brunschwig BS, Musaev DG, Hill CL, Wang D, Lian T (2012) Energy Environ Sci 5:8923
21. Waegele MM, Chen X, Herlihy DH, Cuk T (2014) J Am Chem Soc 136:10632–10639
22. Cooper JK, Ling Y, Longo C, Li Y, Zhang JZ (2012) J Phys Chem C 116:17360–17368
23. Sivasankar N, Weare WW, Frei H (2011) J Am Chem Soc 133:12976
24. Zhang M, de Respinis M, Frei H (2014) Nat Chem 6:362
25. Andrews E, Ren MM, Wang F, Zhang FZY, Sprunger P, Kurtz R, Flake J (2013) J Electrochem Soc 160:H841
26. Klahr B, Gimenez S, Fabregat-Santiago F, Hamann T, Bisquert J (2012) J Am Chem Soc 134 (9):4294–4302
27. Tan SJ, Feng H, Ji YF, Wang Y, Zhao J, Zhao AD, Wang B, Luo Y, Yang JL, Hou JG (2012) J Am Chem Soc 134:9978
28. Lee J, Sorescu DC, Deng XY, Jordan KD (2013) J Phys Chem Lett 4:53
29. Tan SJ, Zhao Y, Zhao J, Wang Z, Ma CX, Zhao AD, Wang B, Luo Y, Yang JL, Hou JG (2011) Phys Rev B 84:155418
30. Liu C, Hwang YJ, Jeong HE, Yang P (2011) Nano Lett 11:3755
31. Baker RTK, Harris PS (1972) J Phys E 5(8):793–797
32. Parkinson GM (1989) Catal Lett 2:303–307
33. Yuk JM, Park J, Ercius P, Kim K, Hellebusch DJ, Crommie MF, Lee JY, Zettl A, Alivisatos AP (2012) Science 336:61–64
34. Novoselov KS, Geim AK, Morozov SV, Jiang D, Zhang Y, Dubonos SV, Grigorieva IV, Firsov AA (2004) Science 306:666–669
35. Wang C-M (2015) J Mater Res 30(3):326–339
36. Zhang LX, Miller BK, Crozier PA (2013) Nano Lett 13(2):679–684
37. Liu Q, Zhang L, Crozier PA (2015) Appl Catal B Environ 172–173:58–64
38. Zhang L, Liu Q, Aoki T, Crozier PA (2015) J Phys Chem C 119:7207–7214
39. Liu Y, Dillon SJ (2014) Chem Commun 50:1761–1763
40. Erni R, Rossell MD, Kisielowski C, Dahmen U (2009) Phys Rev Lett 102:096101
41. Krivanek OL et al (2014) Nature 514:209–212
42. Crozier PA, Chenna S (2011) Ultramicroscopy 111:177–185
43. Chenna S, Crozier PA (2012) ACS Catal 2(11):2395–2402
44. Miller BK, Crozier PA (2014) Microsc Microanal 20:815–824
45. Woodhouse M, Parkinson BA (2009) Chem Soc Rev 38:197–210
46. Drews J (2000) Science 287:1960–1964
47. Sliozberg K, Stein HS, Khare C, Parkinson BA, Ludwig A, Schuhmann W (2015) ACS Appl Mater Interfaces 7(8):4883–4889
48. Smith JL, Hendrickson WA, Terwilliger TC, Berendzen J (2006) MAD and MIR. In: International tables for crystallography, vol F, pp 299–309 (Chapter 14.2)
49. Krishna Murthy HM (1996) Use of multiple-wavelength anomalous diffraction measurements in ab initio phase determination for macromolecular structures. In: Jones C, Mulloy B, Sanderson MR (eds) Crystallographic methods and protocols, vol 56, Methods in Molecular Biology™. Humana, Totowa, pp 127–151. doi:10.1385/0-89603-259-0:127, Print ISBN 978-0-89603-259-0, Online ISBN 978-1-59259-543-3
50. Koningsberger DC, Prins R (eds) (1988) X-Ray absorption: principles, applications, techniques of EXAFS, SEXAFS and XANES. Wiley, New York. ISBN 978-0471875475
51. Gu W, Jacquamet L, Patil DS, Wang H-X, Evans DJ, Smith MC, Millar M, Koch S, Eichhorn DM, Latimer M, Cramer SP (2003) J Inorg Biochem 93:41–51
52. Gu W (2003) Study of active sites in Ni enzymes using X-ray absorption spectroscopy. PhD thesis, UC Davis
53. Richter J, Braun A, Harvey AS, Holtappels P, Graule T, Gauckler LJ (2008) Physica B 403 (1):87–94
54. Braun A, Wang H, Bergmann U, Tucker MC, Gu W, Cramer SP, Cairns EJ (2003) J Power Sources 112(1):231–235

55. Bergmann U, Grush MM, Horne CR, DeMarois P, Penner-Hahn JE, Yocum CF, Wright DW, Dube CE, Armstrong WH, Christou G, Eppley HJ, Cramer SP (1998) J Phys Chem B 102 (42):8350–8352
56. Wang H, Ralston CY, Patil DS, Jones RM, Gu W, Verhagen M, Adams M, Ge P, Riordan C, Marganian CA, Mascharak P, Kovacs J, Miller CG, Collins TJ, Brooker S, Croucher PD, Wang K, Stiefel EI, Cramer SP (2000) J Am Chem Soc 122:10544–10552
57. Weigand W, Schollhammer P (2014) Bioinspired catalysis: metal-sulfur complexes. Wiley-VCH, Weinheim, 440 pages. ISBN 978-3-527-33308-0
58. de Groot FMF, Kotani A (2008) Core level spectroscopy of solids. CRC, New York
59. Wang H, Ge P, Riordan CG, Brooker S, Woomer CG, Collins T, Melendres CA, Graudejus O, Bartlett N, Cramer SP (1998) J Phys Chem B 102(42):8343–8346
60. Braun A, Bayraktar D, Harvey AS, Beckel D, Purton JA, Holtappels P, Gauckler LJ, Graule T (2009) Appl Phys Lett 94:202102
61. Braun A, Erat S, Bayraktar D, Harvey A, Graule T (2012) Chem Mater 24(8):1529–1535
62. Jiang P, Chen J-L, Borondics F, Glans P-A, West MW, Chang C-L, Salmeron M, Guo J (2010) Electrochem Commun 12(6):820–822
63. Myneni S, Luo Y, Näslund LÅ, Cavalleri M, Ojamäe L, Ogasawara H, Pelmenschikov A, Wernet P, Väterlein P, Heske C, Hussain Z, Pettersson LGM, Nilsson A (2002) J Phys Condens Matter 14:L213–L219
64. Hetényi B, De Angelis F, Giannozzi P, Car R (2004) J Chem Phys 120(18):8632–8637
65. Braun A, Sivula K, Bora DK, Zhu J, Zhang L, Grätzel M, Guo J, Constable EC (2012) Direct observation of two electron holes in hematite during photo-electrochemical water splitting. J Phys Chem C 116(23):16870–16875
66. Bora DK, Hu Y, Thiess S, Erat S, Feng X, Mukherjee S, Fortunato G, Gaillard N, Toth R, Gajda-Schrantz K, Drube W, Grätzel M, Guo J, Zhu J, Constable EC, Sarma DD, Wang H, Braun A (2013) J Electron Spectrosc Relat Phenom 190A:93–105
67. Sivula K (2013) J Phys Chem Lett 4(10):624–1633
68. Cavalleri M, Ogasawara H, Pettersson LGM, Nilsson A (2002) Chem Phys Lett 364:363–370
69. Lanyi JK (1997) J Biol Chem 272:31209–31212
70. Balasubramanian S, Wang P, Schaller RD, Rajh T, Rozhkova EA (2013) Nano Lett 13 (7):3365–3371
71. Pieper J, Buchsteiner A, Dencher NA, Lechner RE, Hauß T (2009) Photochem Photobiol 85:590–597
72. Patzelt H, Simon B, terLaak A, Kessler B, Kuhne R, Schmieder P, Oesterhaelt D, Oschkinat H (2002) Proc Natl Acad Sci U S A 99:9765–9770
73. Dencher NA, Dresselhaus D, Zaccai G, Büldt G (1989) Proc Natl Acad Sci U S A 86:7876–7879
74. Horn C, Steinem C (2005) Biophys J 89:1046–1054
75. Allam NK, Yen C-W, Near RD, El-Sayed MA (2011) Energy Environ Sci 4:2909–2914
76. Chen Q, Holdsworth S, Embs J, Pomjakushin V, Frick B, Braun A (2012) High Press Res 32 (4):471–481
77. Chen Q, Banyte J, Zhang X, Embs JP, Braun A (2013) Solid State Ionics 252:2–6
78. Fabbri E, Pergolesi D, Traversa E (2010) Sci Technol Adv Mater 11:044301
79. Chen Q, El Gabaly F, Aksoy Akgul F, Liu Z, Mun BS, Yamaguchi S, Braun A (2013) Chem Mater 25(23):4690–4696
80. Braun A, Embs J-P, Remhof A (2014) 2013 ESS science symposium: neutrons for future energy strategies. Neutron News (Taylor & Francis) 25(1):6–7
81. Pieper J (1804) Biochim Biophys Acta 2010:83–88
82. Raman CV, Krishnan KS (1928) Nature 121:501
83. Fleischmann M, Hendra PJ, Mcquilla AJ (1974) Chem Phys Lett 26:163
84. Albrecht AC (1961) J Chem Phys 34:1476
85. Albrecht AC, Hutley MC (1971) J Chem Phys 55:4438
86. Pettinger B, Ren B, Picardi G, Schuster R, Ertl G (2004) Phys Rev Lett 92:096101

87. Moskovits M (2005) J Raman Spectrosc 36:485
88. Pettinger B, Domke KF, Zhang D, Schuster R, Ertl G (2007) Phys Rev B 76:113409
89. Zhang R, Zhang Y, Dong ZC, Jiang S, Zhang C, Chen LG, Zhang L, Liao Y, Aizpurua J, Luo Y, Yang JL, Hou JG (2013) Nature 498:82
90. Murgida DH, Hildebrandt P (2004) Acc Chem Res 37:854
91. Chatterjee S, Sengupta K, Samanta S, Das PK, Dey A (2013) Inorg Chem 52:9897
92. Schlögl R (2010) ChemSusChem 3:209
93. Yeo BS, Bell AT (2011) J Am Chem Soc 133:5587
94. Yeo BS, Bell AT (2012) J Phys Chem C 116:8394
95. Ranjan C, Zoran P, Schloegl R (2014) E-MRS 2014 spring meeting, Lille, France
96. Schmitt KG, Gewirth AA (2014) J Phys Chem C 118:17567
97. Klink S, Höche D, La Mantia F, Schuhmann W (2013) J Power Sources 240:273–280
98. Klink S, Madej E, Ventosa E, Lindner A, Schuhmann W, La Mantia F (2012) Electrochem Commun 22:120–123
99. Agarwal P, Orazem ME, Garcia-Rubio LH (1995) J Electrochem Soc 142(12):4159–4168
100. Agarwal P, Orazem ME, Garcia-Rubio LH (1992) J Electrochem Soc 139(7):1917–1927
101. La Mantia F, Vetter J, Novák P (2008) Electrochim Acta 53(12):4109–4121
102. Schiller CA, Richter F, Gulzow E, Wagner N (2001) Phys Chem Chem Phys 3(3):374–378
103. Newman JS, Tobias CW (1962) J Electrochem Soc 109(12):1183–1191
104. de Levie R (1963) Electrochim Acta 8(10):751–780
105. Darby R (1966) J Electrochem Soc 113(4):392–396
106. De Vidts P, White RE (1997) J Electrochem Soc 144(4):1343–1353
107. Fabregat-Santiago F, Garcia-Belmonte G, Mora-Sero I, Bisquert J (2011) Phys Chem Chem Phys 13(20):9083–9118
108. Dewald JF (1960) J Phys Chem Solid 14:155–161
109. Abram RA, Doherty PJ (1982) Philos Mag B 45(2):167–176
110. Archibald IW, Abram RA (1983) Philos Mag B 48(2):111–125
111. Archibald IW, Abram RA (1986) Philos Mag B 54(5):421–438
112. Cohen JD, Lang DV (1982) Phys Rev B 25(8):5321–5350
113. Di Quarto F, La Mantia F, Santamaria M (2005) Electrochim Acta 50(25–26):5090–5102
114. La Mantia F, Habazaki H, Santamaria M, Di Quarto F (2010) Russ J Electrochem 46 (11):1306–1322
115. La Mantia F, Stojadinović J, Santamaria M, Di Quarto F (2010) ChemPhysChem 13 (12):2910–2918
116. Huang VMW, Vivier V, Orazem ME, Pebere N, Tribollet B (2007) J Electrochem Soc 154 (2):C81–C88
117. Hirschorn B, Orazem ME, Tribollet B, Vivier V, Frateur I, Musiani M (2010) Electrochim Acta 55(21):6218–6227
118. Ertl G, Knözinger H, Weitkamp J (1997) Handbook of heterogeneous catalysis. VCH, Weinheim
119. Somorjai GA, Li Y (2010) Introduction to surface chemistry and catalysis. Wiley-VCH, New York
120. Crozier PA, Hansen TW (2015) MRS Bull 40:38–45
121. Zhang L, Liu Q, Aoki T, Crozier PA (2015) J Phys Chem C. doi:10.1021/jp512907g
122. Liu Q, Zhang L, Crozier PA (2015) Appl Catal Environ 172:58–64
123. Yoshida K, Yamasaki J, Tanaka N (2004) Appl Phys Lett 84:2542–2544
124. Yoshida K, Nozaki T, Hirayama T, Tanaka N (2007) J Electron Microsc (Tokyo) 56:177–180
125. Cavalca F, Laursen AB, Kardynal BE, Dunin-Borkowski RE, Dahl S, Wagner JB, Hansen TW (2012) Nanotechnology 23:6
126. Cavalca F, Laursen AB, Wagner JB, Damsgaard CD, Chorkendorff I, Hansen TW (2013) ChemCatChem 5:2667–2672
127. Miller BK, Crozier PA (2013) Microsc Microanal 19:461–469
128. Crozier PA, McCartney MR, Smith DJ (1990) Surf Sci 237:232–240

129. McCartney MR, Crozier PA, Weiss JK, Smith DJ (1991) Vacuum 42:301–308
130. Brydson R, Sauer H, Engel W, Thomas JM, Zeitler E, Kosugi N, Kuroda H (1989) J Phys Condens Matter 1:797–812
131. Lazar S, Botton GA, Wu M-Y, Tichelaar FD, Zandbergen HW (2003) Ultramicroscopy 96:535–546
132. Sakai N, Fujishima A, Watanabe T, Hashimoto K (2003) J Phys Chem B 107:1028–1035
133. Henderson MA (1996) Langmuir 12:5093–5098
134. Wendt S, Matthiesen J, Schaub R, Vestergaard EK, Laegsgaard E, Besenbacher F, Hammer B (2006) Phys Rev Lett 96:4
135. Wendt S, Schaub R, Matthiesen J, Vestergaard EK, Wahlstrom E, Rasmussen MD, Thostrup P, Molina LM, Laegsgaard E, Stensgaard I, Hammer B, Besenbacher F (2005) Surf Sci 598:226–245
136. Bikondoa O, Pang CL, Ithnin R, Muryn CA, Onishi H, Thornton G (2006) Nat Mater 5:189–192
137. Bohmer N, Roussiere T, Kuba M, Schunk SA (2012) Comb Chem High Throughput Screen 15:123
138. Green ML, Takeuchi I, Hattrick-Simpers JR (2013) J Appl Phys 113:231101
139. Barber ZH, Blamire MG (2008) Mater Sci Technol 24:757
140. Koinuma H, Takeuchi I (2004) Nat Mater 3:429
141. Muster TH, Trinchi A, Markley TA, Lau D, Martin P, Bradbury A, Bendavid A, Dligatch S (2011) Electrochim Acta 56:9679
142. Sun S, Ding J, Bao J, Luo Z, Gao C (2011) Comb Chem High Throughput Screen 14:160
143. Jaramillo TF, Baeck S-H, Kleiman-Shwarsctein A, McFarland EW (2004) Macromol Rapid Commun 25:297–301
144. Baeck SH, Jaramillo TF, Brändli C, McFarland EW (2002) J Comb Chem 4:563
145. Jaramillo TF, Baeck S-H, Kleiman-Shwarsctein A, Choi K-S, Stucky GD, McFarland EW (2005) J Comb Chem 7:264
146. Woodhouse M, Herman GS, Parkinson BA (2005) Chem Mater 17:4318–4324
147. Woodhouse M, Parkinson BA (2008) Chem Mater 20:2495
148. Seley D, Ayers K, Parkinson BA (2013) ACS Comb Sci 15:82
149. Haber JA, Cai Y, Jung S, Xiang C, Mitrovic S, Jin J, Bell AT, Gregoire JM (2014) Energy Environ Sci 7:682
150. Smith RDL, Prévot MS, Fagan RD, Zhang Z, Sedach PA, Siu MKJ, Trudel S, Berlinguette CP (2013) Science 340:60
151. Gregoire JM, Kirby SD, Turk ME, van Dover RB (2009) Thin Solid Films 517:1607
152. Perkins JD, Teplin CW, van Hest MF, Alleman JL, Li X, Dabney MS, Keyes BM, Gedvilas LM, Ginley DS, Lin Y, Lu Y (2004) Appl Surf Sci 223:124
153. Sigdel AK, Ndione PF, Perkins JD, Gennett T, van Hest MF, Shaheen SE, Ginley DS, Berry JJ (2012) J Appl Phys 111:093718
154. Gregoire JM, Xiang C, Liu X, Marcin M, Jin J (2013) Rev Sci Instrum 84:024102
155. Hassel AW, Lohrengel MM (1997) Electrochim Acta 42:3327
156. Haber JA, Xiang C, Guevarra D, Jung S, Jin J, Gregoire JM (2014) ChemElectroChem 1:524–528
157. Katz JE, Gingrich TR, Santori EA, Lewis NS (2009) Energy Environ Sci 2:103–112
158. Hu S, Xiang C, Haussener S, Berger AD, Lewis NS (2013) Energy Environ Sci 6:2984
159. Ye H, Lee L, Jang JS, Bard AJ (2010) J Phys Chem C 114:13322
160. Lewis NS (1984) J Electrochem Soc 131:2496
161. Xiang C, Haber J, Marcin M, Mitrovic S, Jin J, Gregoire JM (2014) ACS Comb Sci 16:120
162. Chen G, Bare SR, Mallouk TE (2002) J Electrochem Soc 149:A1092
163. Dokoutchaev AG, Abdelrazzaq F, Thompson ME, Willson J, Chang C, Bocarsly A (2002) Chem Mater 14:3343
164. Nakayama A, Suzuki E, Ohmori T (2002) Appl Surf Sci 189:260
165. Gerken JB, Chen JYC, Massé RC, Powell AB, Stahl SS (2012) Angew Chem 51:6676

166. Xiang C, Suram SK, Haber JA, Guevarra DW, Jin J, Gregoire JM (2014) ACS Comb Sci 16:47
167. Nozik AJ (1975) Nature 257:383–386
168. Parkinson B (1984) Annu Rev Phys Chem 17:431–437
169. Bard AJ, Fox MA (1995) Annu Rev Phys Chem 28:141–145
170. McKone JR, Lewis NS, Gray HB (2013) Chem Mater 26:407–414
171. Xiang X-D, Sun X, Briceño G, Lou Y, Wang K-A, Chang H, Wallace-Freedman WG, Chen S-W, Schultz PG (1995) Science 268:1738–1740
172. Stepanovich A, Sliozberg K, Schuhmann W, Ludwig A (2012) Int J Hydrogen Energy 37:11618–11624
173. Seyler M, Stoewe K, Maier WF (2007) Appl Catal B Environ 76:146–157
174. Bard AJ, Fan FRF, Kwak J, Lev O (1989) Anal Chem 61:132–138
175. Lee J, Ye H, Pan S, Bard AJ (2008) Anal Chem 80:7445–7450
176. Reddington E, Sapienza A, Gurau B, Viswanathan R, Sarangapani S, Smotkin ES, Mallouk TE (1998) Science 280:1735–1737
177. Baeck S-H, Jaramillo TF, Jeong DH, McFarland EW (2004) Chem Commun 390–391
178. Bard A, Lee HC, Leonard K, Park HS, Wang S (2013) Rapid screening methods in the discovery and investigation of new photocatalyst compositions. In: Lewerenz HJ, Peter L (eds) Photoelectrochemical water splitting: materials processes and architectures. The Royal Society of Chemistry, London, pp 132–153
179. Batzill M, Diebold U (2005) Prog Surf Sci 79:47–154
180. Liu G, Liu C, Bard AJ (2010) J Phys Chem C 114:20997–21002
181. Park HS, Kweon KE, Ye H, Paek E, Hwang GS, Bard AJ (2011) J Phys Chem C 115:17870–17879
182. Ye H, Park HS, Bard AJ (2011) J Phys Chem C 115(25):12464–12470
183. Zhang F, Roznyatovskiy V, Fan F-RF, Lynch V, Sessler JL, Bard AJ (2011) J Phys Chem C 115:2592–2599
184. Cong Y, Park HS, Wang S, Dang HX, Fan F-RF, Mullins CB, Bard AJ (2012) J Phys Chem C 116:14541–14550
185. Esposito DV, Levin I, Moffat TP, Talin AA (2013) Nat Mater 12:562–568
186. Park HS, Leonard KC, Bard AJ (2013) J Phys Chem C 117:12093–12102
187. Fosdick SE, Berglund SP, Mullins CB, Crooks RM (2014) ACS Catal 4:1332–1339
188. Trotochaud L, Mills TJ, Boettcher SW (2013) J Phys Chem Lett 4(6):931–935
189. Walter MC et al (2010) Chem Rev 110(11):6446–6473
190. McCrory CC et al (2013) J Am Chem Soc 135(45):16977–16987
191. Sun J, Zhong DK, Gamelin DR (2010) Energy Environ Sci 3(9):1252–1261
192. Klahr B et al (2012) J Am Chem Soc 134(40):16693–16700
193. Seabold JA, Choi K-S (2011) Chem Mater 23(5):1105–1112
194. Liu R et al (2011) Angew Chem Int Ed 50(2):499–502
195. Kay A, Cesar I, Grätzel M (2006) J Am Chem Soc 128(49):15714–15721
196. Tilley SD et al (2010) Angew Chem Int Ed 49(36):6405–6408
197. Abdi FF, van de Krol R (2012) J Phys Chem C 116(17):9398–9404
198. Gamelin DR (2012) Nat Chem 4(12):965–967
199. Zhong DK, Gamelin DR (2010) J Am Chem Soc 132(12):4202–4207
200. McDonald KJ, Choi K-S (2011) Chem Mater 23(7):1686–1693
201. Barroso M et al (2011) J Am Chem Soc 133(38):14868–14871
202. Steinmiller EMP, Choi K-S (2009) Proc Natl Acad Sci U S A 106(49):20633–20636
203. Surendranath Y, Bediako DK, Nocera DG (2012) Proc Natl Acad Sci U S A 109:15617–15621
204. Pijpers JJH et al (2011) Proc Natl Acad Sci U S A 108(25):10056–10061
205. Reece SY et al (2011) Science 334(6056):645–648
206. Young ER et al (2011) Energy Environ Sci 4(6):2058–2061
207. Zhong DK, Choi S, Gamelin DR (2011) J Am Chem Soc 133(45):18370–18377

208. Gregoire JM et al (2013) J Electrochem Soc 160(4):F337–F342
209. Malara F et al (2014) ACS Appl Mater Interfaces 6(12):9290–9297
210. Dyer AL et al (2014) Adv Mater 26(28):4895–4900
211. Seike T, Nagai J (1991) Solar Energy Mater 22:107–117
212. Hugot-Le Goff A, Cordoba de Torresi S (1990) Electrochromic materials. In: Proceedings of the electrochemical society, Pennington, NJ
213. Corrigan DA, Knight SL (1989) J Electrochem Soc 136(3):613–619
214. McIntyre JDE, Kolb DM (1970) Specular reflection spectroscopy of electrode surface films. Symp Faraday Soc 4:99–113
215. Decker F et al (1992) Electrochim Acta 37(6):1033–1038
216. Trotochaud L et al (2012) J Am Chem Soc 134(41):17253–17261
217. Trotochaud L et al (2014) J Am Chem Soc 136(18):6744–6753
218. Azens A et al (1998) Sol Energy Mater Sol Cells 54(1–4):85–91
219. Monk PMS et al (2001) Electrochim Acta 46(13–14):2091–2096
220. Smith RDL et al (2013) J Am Chem Soc 135(31):11580–11586
221. Brückner A (2001) Chemic Commun 2122–2123
222. Brückner A (2005) Chem Commun 1761–1763
223. Zeng K, Zhang DK (2010) Prog Energy Combust Sci 36(3):307–326
224. Leenheer AJ, Atwater HA (2010) J Electrochem Soc 157(9):B1290–B1294
225. Sides PJ, Tobias CW (1982) J Electrochem Soc 129(12):2715–2720
226. Yu HT, Shen JQ, Wei YH (2008) Particuology 6(5):340–346
227. Balzer RJ, Vogt H (2003) J Electrochem Soc 150(1):E11–E16
228. Eigeldinger J, Vogt H (2000) Electrochim Acta 45(27):4449–4456

Index

The manufacturer's authorised representative in the EU is Springer Nature Customer Service Centre GmbH, Europaplatz 3, 69115 Heidelberg, Germany. If you have any concerns regarding our products, please contact ProductSafety@springernature.com

Printed and bound by CPI Group (UK) Ltd, Croydon, CR0 4YY
30/09/2026
02225652-0011